# Viscoelasticity and Collective Cell Migration

# VISCOELASTICITY AND COLLECTIVE CELL MIGRATION

## An Interdisciplinary Perspective Across Levels of Organization


Edited by

**IVANA PAJIC-LIJAKOVIC**

*Laboratoire Interdisciplinaire de Physique, CNRS, Université Grenoble Alpes, Grenoble, France*

**ELIAS H BARRIGA**

*Mechanisms of Morphogenesis Lab, Gulbenkian Institute of Science (IGC), Oeiras, Portugal*



ELSEVIER

ACADEMIC PRESS

An imprint of Elsevier

Academic Press is an imprint of Elsevier
125 London Wall, London EC2Y 5AS, United Kingdom
525 B Street, Suite 1650, San Diego, CA 92101, United States
50 Hampshire Street, 5th Floor, Cambridge, MA 02139, United States
The Boulevard, Langford Lane, Kidlington, Oxford OX5 1GB, United Kingdom



**British Library Cataloguing-in-Publication Data**
A catalogue record for this book is available from the British Library

**Library of Congress Cataloging-in-Publication Data**
A catalog record for this book is available from the Library of Congress

ISBN: 978-0-12-820310-1

For Information on all Academic Press publications
visit our website at https://www.elsevier.com/books-and-journals

*Publisher:* Mara Conner
*Acquisitions Editor:* Fiona Geraghty
*Editorial Project Manager:* Charlotte Rowley
*Production Project Manager:* Sojan P. Pazhayattil
*Cover Designer:* Victoria Pearson

Typeset by MPS Limited, Chennai, India

# Contents

# List of Contributors

**Teckla Akinyi**
Laboratory of Biorheology and Medical Ultrasonics, University of Montreal Hospital Research Center, Montréal, QC, Canada

**Martial Balland**
Laboratoire Interdisciplinaire de Physique, Centre national de la recherche scientifique, Université Grenoble Alpes, Grenoble, France

**Elias H. Barriga**
Mechanisms of Morphogenesis Lab, Gulbenkian Institute of Science (IGC), Oeiras, Portugal

**Timo Betz**
Third Institute of Physics, University of Göttingen, Göttingen, Germany; Institute of Cell Biology, ZMBE, University of Münster, Münster, Germany

**Manish Bhatt**
Laboratory of Biorheology and Medical Ultrasonics, University of Montreal Hospital Research Center, Montréal, QC, Canada

**Cristian Borja**
Barcelona University (UB), Spain

**Thomas Boudou**
Laboratoire Interdisciplinaire de Physique, Centre national de la recherche scientifique, Université Grenoble Alpes, Grenoble, France

**Françoise Brochard-Wyart**
Laboratoire Physico-Chimie Curie, Institut Curie, PSL Research University, Sorbonne Université, CNRS, Paris, France

**Giovanni Cappello**
Laboratoire Interdisciplinaire de Physique, Centre national de la recherche scientifique, Université Grenoble Alpes, Grenoble, France

**Stefan Catheline**
LabTAU, INSERM unit 1032, Université de Lyon, Lyon, France

**Andrew G. Clark**
Institut Curie, PSL Research University, CNRS, UMR 144 - Cell Biology and Cancer, Paris, France

**Guy Cloutier**
Laboratory of Biorheology and Medical Ultrasonics, University of Montreal Hospital Research Center, Montréal, QC, Canada; Department of Radiology, Radio-oncology and Nuclear Medicine, and Institute of Biomedical Engineering, University of Montreal, Montréal, QC, Canada

**Jaime A. Espina**
Mechanisms of Morphogenesis Lab, Gulbenkian Institute of Science (IGC), Oeiras, Portugal

**David Gonzalez-Rodriguez**
Université de Lorraine, LCP-A2MC, F-57000 Metz, France

**Pol Grasland-Mongrain**
ENS de Lyon, Université Claude Bernard, CNRS, Laboratoire de Physique, Lyon, France

**Karine Guevorkian**
Laboratoire Physico-Chimie Curie, Institut Curie, PSL Research University, Sorbonne Université, CNRS, Paris, France

**Milan Milivojevic**
Faculty of Technology and Metallurgy, Belgrade University, Belgrade, Serbia

**Elena Moral**
Polytechnic University of Catalonia (UPC), Barcelona, Spain. Currently at Technical University of Munich (TUM), Germany

**Jose J. Muñoz**
Laboratory of Numerical Analysis (LaCàN). Polytechnic University of Catalonia (UPC), Barcelona, Spain

**Ivana Pajic-Lijakovic**
Faculty of Technology and Metallurgy, Belgrade University, Belgrade, Serbia

**Vanni Petrolli**
Laboratoire Interdisciplinaire de Physique, Centre national de la recherche scientifique, Université Grenoble Alpes, Grenoble, France

**Joana E. Saraiva**
Mechanisms of Morphogenesis Lab, Gulbenkian Institute of Science (IGC), Oeiras, Portugal

CHAPTER ONE

# The basics of collective cell migration: unity makes strength

**Joana E. Saraiva and Elias H. Barriga**
Mechanisms of Morphogenesis Lab, Gulbenkian Institute of Science (IGC), Oeiras, Portugal

## 1.1 Introduction

### 1.1.1 Collective cell migration

The collective migration of cells as a cohesive and synchronized group is a hallmark of several tissue remodeling events that underlie embryogenesis, tissue repair, and cancer metastasis [1]. While individually migrating cells move fast and in a rather less directional manner, the migration of cellular clusters is slower but highly persistent in terms of directionality and in turn, extremely efficient. To achieve this, cells within the cluster require to move in the same direction and at similar speeds. Hence, for the group to move as a synchronized unit, cells need to fluently transmit the information that they survey from their environment [1–3]. This level of cooperative migration does not only involve finely tuned mechanical coupling between neighboring cells, but it also requires fluent molecular communication among cells and the coordination of their cytoskeletal activity [4]. Hence, collective cell migration (CCM) can be defined as a highly synchronized and cooperative migratory mode [1,3]. Due to its wide biological relevance, CCM is currently under intense research and several advances have been made to understand the mechanisms that help cells to "stay together" and efficiently communicate while collectively migrating. Thus in this introductory chapter we discuss basic and rather canonical concepts about the cellular structures that cells use to connect and communicate while collectively migrating. Furthermore, we briefly comment on the role that molecular components of these cellular structures play during CCM in different in vitro and in vivo systems.

*Viscoelasticity and Collective Cell Migration.*
DOI: https://doi.org/10.1016/B978-0-12-820310-1.00001-X

## 1.2 Experimental models to study collective cell migration

### 1.2.1 Studying collective cell migration in vitro

Diverse in vitro and in vivo experimental systems are available to study the mechanisms governing CCM. 2D in vitro models include the classic scratch wound assay, in which epithelial cells at the edge of the wound become the leader cells and more internal cells are designated as followers [2,5]. A common feature of leader cells is to extend lamellipodia in the direction of the free space generated by the wound. While this role of leader cells during CCM is clearer, the contribution of follower cells seems more controversial, that is, some studies suggest that they are passively dragged by leaders [6,7], and others suggest that follower cells actually contribute by generating traction forces [8] via cryptic lamellipodia [9]. Wound assays have been used to study cell polarity [2], traction force generation across the cluster [8], and cell–cell adhesion during the migration of confluent monolayers [1,10] (extensively reviewed in Stamm et al., 2016) [11]. For instance, using this assay it was demonstrated that the microtubule-organizing center and the Golgi apparatus of collectively migrating cells reposition themselves at the leading edge during cell polarization [12]. Collective invasion assays have also been used to study 2D sheets of cancer cells. In this case, the authors showed that follower cells move along a track generated by leader cells that extend lamellipodia, as observed in the first row of migrating cells moving into the wound space [10]. Similarly, cells can follow gradients of stiffness as a cluster by transmitting force from the substrate across cell–cell junctions [13,14]. These works represent the first demonstrations of collective durotaxis—the migration of cell collectives along a stiffness gradient by crawling from soft to stiff substrates.

Not only 2D environments have been used to study CCM. Indeed, in vitro 3D experiments have nicely reproduced what was already known in 2D, showing that cells in the leading edge perform traction forces, that align 3D matrices to facilitate the migration of rear cells [15]. Remarkably, 3D environments have not only reproduced what was known in 2D but have also revealed striking results about novel migratory and invasive strategies of cellular clusters in physiology and disease [16,17] (extensively reviewed in Yamada and Sixt 2019) [18]. For instance, cancer explants have been shown to collectively invade 3D extracellular matrices [19], providing an important tool to further study the mechanisms of cancer invasion and metastasis.

Thus depending on the biological context in which a scientific question is formulated, CCM can be studied in 2D or 3D in vitro or *ex vivo* environments. Consequently, 2D or 3D substrates are powerful tools which nicely complement in vivo studies, by providing a controlled platform where the response of collectively migrating cells to a particular cue can be easily assessed in a simplified context (further discussed in Box 1).

### 1.2.2 Models to study collective cell migration in vivo

Due to its relevance and potential as a therapeutic target, the mechanisms of CCM are extensively studied in vivo by using diverse models, including epithelial collective migration of *Drosophila* border cells [31], lateral line primordia migration in zebrafish [32,33], branching and sprouting morphogenesis of *Drosophila* trachea [34,35], rodent retina and cornea cells [36], cranial neural crest cells [37,38] and mesendoderm of *Xenopus* and zebrafish [39,40], among other systems. Although the strategies used by these cell types to successfully reach their target tissues are diverse and context-dependent, the core mechanisms required to maintain CCM are well conserved among these animal systems [1,41,42]. Here we discuss the collective migration of *Drosophila* border cells, zebrafish lateral line primordia, and *Xenopus* cranial neural crest cells, as their migratory strategies represent most of the behaviors displayed by other models of CCM (Fig. 1.1A and B). We also provide details about collective cancer invasion (Fig. 1.1C).

*Drosophila* border cells delaminate from the follicular epithelium of the fly's ovary and collectively migrate toward the oocyte, squeezing between the giant nurse cells that surround them (Fig. 1.1A). Therefore border cells need to resist high deformation levels imposed by their neighboring nurse cells, developing mechanisms to properly maintain their shape, even while exchanging positions [31]. Recent findings show that these border cells rely both on chemical signals, such as gradients of EGF/PVF1 [43], and mechanical cues, such as pulling forces exerted by the front cells [44], to efficiently trigger a directional collective motion. Hence, these cells are a great system to study epithelial CCM in complex topologies and under strong mechanical stress.

Zebrafish posterior lateral line primordia (pLLP) migrates as a collective from anterior regions to the caudal end of the embryo (Fig. 1.1B). At

**Box 1 Selecting the right assay to study collective cell migration.**

The use of 3D assays has not only brought novel information, but it also opened some controversy in the cell migration field; as when comparing migration of the same cell type in 2D versus 3D environments does not always yield matching results [16,18]. This has generated a certain degree of skepticism about the extent in which 2D results may reproduce migration of cells in native contexts. Nonetheless is important to clarify that the use of 2D systems is still a powerful tool that nicely reproduces the environment of cells migrating in 2D-like native contexts. Hence, one should be just careful when choosing the in vitro assay that will best reproduce the context in which their questions are framed [20]. For instance, if one aims to study specific wound healing of rat cornea [21,22] or pig skin [23] under a controlled environment, using a 3D gel may not be the most suitable platform. In this case, using a 2D wound healing assay would result more informative. Worth mentioning, there are very novel 3D wound healing assays that can be also used to study CCM during wound healing of tissues that possess a more complex geometry [24].

Finally, one may wonder why to use these in vitro assays if CCM could be studied in native conditions? that is, using the models shown in Fig. 1.1. The answer is relatively simple and despite the fact that in vivo systems can be highly informative, their inherent complexity may preclude of dissecting cell behavior in response to a specific cue. Hence the need for a simplified system that provides a controlled environment and where the migratory response of cells to a given cue can be studied. In this context, a system that would offer the chance to do both approaches, studying CCM in vivo and in simplified *ex vivo* environments, can result in a powerful tool. That is, *Xenopus* cranial neural crest cells [25,26], rat corneal cells [27], tumor cells [28], or *Drosophila* follicle cells [29] stand as a good platform to study collective migration by comparing in vivo and *ex vivo* results obtained from the same model [30]. Finding more systems where CCM could be studied by contrasting in vitro with in vivo results by using the same cell type could deeply impact the advance of the cell migration field and cell biology in general.

the front of the pLLP, cells remain undifferentiated and highly migratory while at the back, cells remain migratory but with a precise periodicity they lag behind to differentiate in clusters known as neuromasts, which are the sensory structures of the lateral line [32,45]. This front to back polarity is explained by an asymmetric expression of the receptors Cxcr4b, at the leading edge of the migrating pLLP, and Cxcr7b, at the rear edge

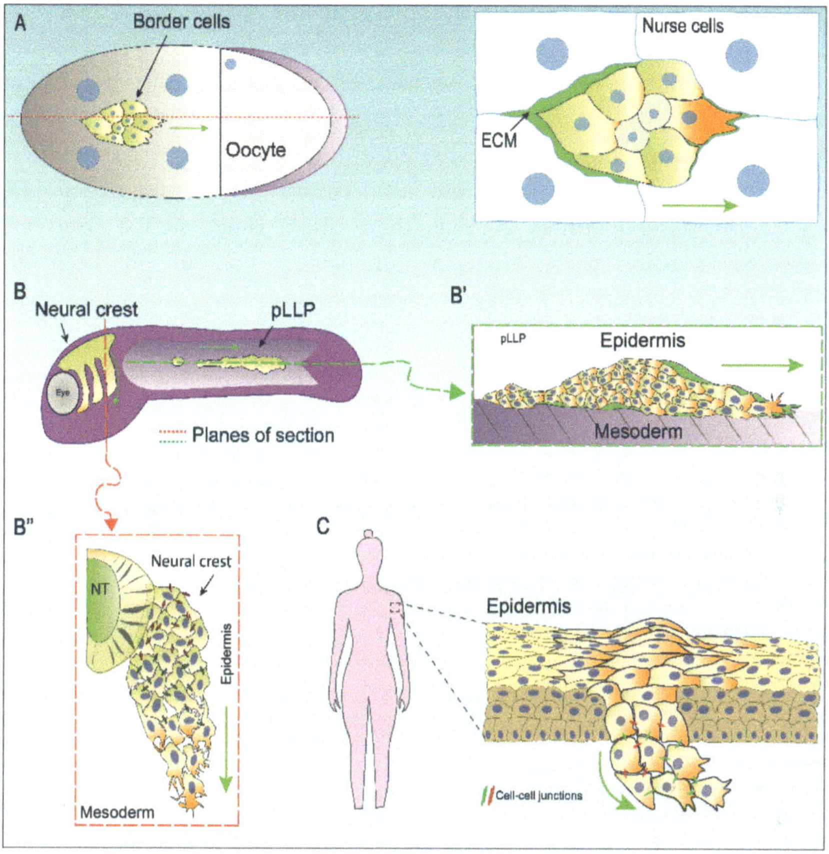


**Figure 1.1** Models to study collective cell migration in vivo. (A) Schematics of *Drosophila* border cell migration. Border cells delaminate from the anterior pole of the egg chamber toward the oocyte, squeezing between the nurse cells. (B) The drawing represents a hypothetical vertebrate embryo in which cranial neural crest cells migrate as a collective from dorsal toward ventral territories. (B′) Diagram representing a sagittal section across the embryo shows how the pLLP migrates in a highly confined space between the epidermis and the underlying mesoderm. (B″) Diagram representing transversal section showing how the neural crest delaminates from the neural plate and migrates dorsoventrally by forming streams and tunneling itself in between the epidermis and the underlying head mesoderm. (C) Schematics exemplifying how cancer cells loose apicobasal polarity, acquire a migratory state and collectively invade. Green arrows show the direction of migration in each case.

of the cell cluster [46]. Thereby, Cxcr7b expression sequesters Cxcl12a at the back of the cluster, generating a chemokine gradient across the pLLP [47], which guides a directional tissue migration [48]. Thus these cells are a great system to study how chemokine gradients control the direction of CCM in vivo.

*Xenopus* cranial neural crest cells are induced at the border of the neural plate, from where they collectively migrate to form distinct tissues. While migrating, these cells experience high degrees of confinement between the epidermis and the underneath head mesoderm [49] (Fig. 1.1B). Thus in order to reach their target tissues, neural crest cells need to modify their migratory behavior via an epithelial-to-mesenchymal transition (EMT), a transition from an epithelial adhesive phenotype to a migratory mesenchymal state, while they also adapt to a challenging environment [42]. These cells are an excellent system to study the molecular and biophysical aspects of CCM and offer the opportunity to contrast both in vivo and in vitro results in a single cell type [25,26,50].

Furthermore, cancer cells have been also observed to perform CCM [51–54]. These cells display a range of CCM and invasion strategies, that is, some tumor cells display multicellular streams, migrating in a "train-like" motion, with loose adherens junctions (AJs) and exhibiting a higher persistence and lower speed than single cells. This strategy has been observed within human breast cancer cells grafted into adult mice, revealing a higher invasive phenotype [55]. Cancer cells can also perform collective invasion, in which cells migrate as a coordinated collective, displaying a high degree of communication, directionality, and speed of migration [1,56] (Fig. 1.1C). Direct evidence for this migratory mode has been provided by injecting 3D spheroids into the deep dermis of mice, followed by imaging with intravital microscopy [57]. Remarkably, these advances are possible due to improvements in culture methods that enable the generation of cancer spheroids [58,59] and imaging tools such as intravital microscopy [60,61], which allow now to image cancer cell behavior as well as immune cell migration in vivo [62,63]. Such improvements are bringing our understanding of cancer metastasis to new levels—these advances, in combination with novel therapies are considerably improving the way we approach this disease and its prognosis.

During the last decade, further improvements have been made in our comprehension of CCM in physiology and disease. Using the systems mentioned above, and eventually new models, is key to decipher the whole range of mechanisms inducing, maintaining, and propagating CCM in different tissue and organismal contexts [41,64]. Furthermore, cancer and several other diseases have been mostly approached from molecular perspectives, hence the urgency of combining novel imaging and molecular methods with the advances from the biomechanics field in order to tackle diseases with more integrative biomedical approaches. Advances in biomechanical mechanisms of CCM in various contexts are further discussed in the following chapters of this book.

## 1.3 Using cell–cell junctions to stay as a group and communicate

### 1.3.1 Cell–cell junctions

While migrating as a collective, cells remain connected via cell–cell junctions, structures that not only ensure that cells adhere to each other to "stay together," but they also mediate organization of cell polarity and participate in the mechanical coupling required for cells to sense, integrate, and transmit environmental cues across the cluster [42,64]. Although cell–cell junctions provide connections of cellular membranes, they can also directly or indirectly connect the cytoskeleton of cells within the cluster, further ensuring coordination of the collective [64] (Fig. 1.2A). Cell–cell junction dynamics of migratory clusters can vary depending on the context in which they have been studied [65]. For example, while epithelial cell monolayers display less dynamic junctions [66], other cell types such as border cells display more dynamic junctions [67]. Dynamic junctions are thought to allow cells within the cluster to exchange their positions while migrating, allowing the cluster to behave with higher fluidity rates [68,69]. Thus fine-tuning cell–cell junctions can modulate the ability of collectively migrating cells to deform and move as a collective across challenging native microenvironments (extensively reviewed in Barriga and Mayor [42] and discussed along the different chapters of this book).

A variety of cell–cell junction types have been described, and for each one of these types, an assortment of molecules has been shown to mediate the proper formation and dynamics of cell–cell junctions [70]. Each one of these molecules undergoes context-dependent changes and modulate an array of signaling cascades, whose fine regulation is essential for CCM to emerge and proceed [4,66]. Here we summarize the most canonical cell–cell adhesion systems and functions that enable cells to stay together during different types of CCM.

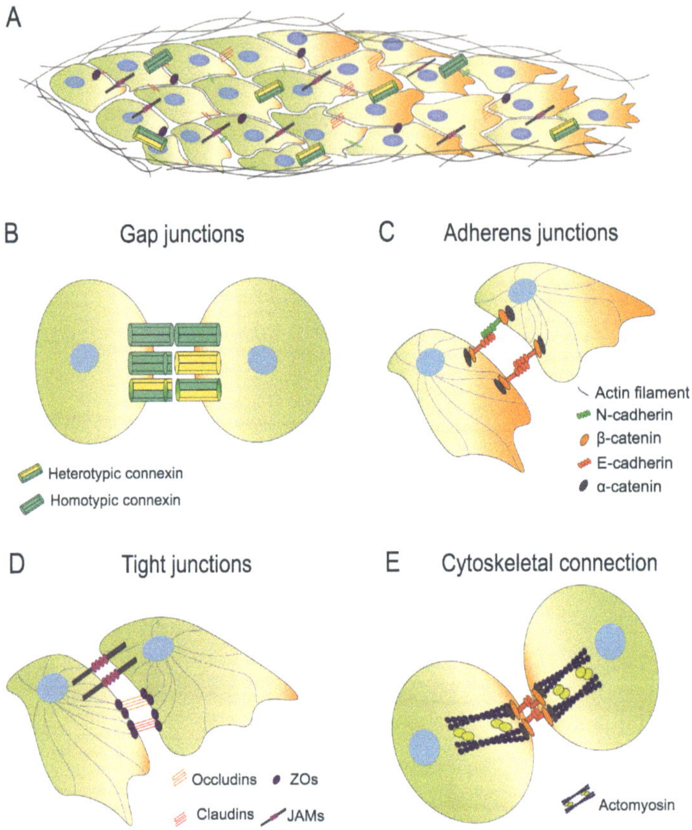


**Figure 1.2** Canonical components of cell–cell junctions. (A) Schematic representing a cluster of migrating cells. (B) Connexins assemble into channels called connexons, which then align into the extracellular space to form gap junction channels. Depending on whether the same type or different types of connexins interact with each other, they can form a homotypic or a heterotypic interaction, respectively. (C) Adherens junctions are mainly composed by transmembrane proteins, such as E-cadherin, and adapter proteins, such as β-catenin and α-catenin, which link to the underlying actin cytoskeleton. Cadherins selectively interact with each other to form homotypic or heterotypic channels. (D) Tight junctions are mainly composed by transmembrane proteins, such as claudin, occludin and JAMs, and by adapter proteins, such as ZOs, which connect to the actin cytoskeleton. (E) Cell–cell junctions link actin filaments between neighboring cells, here exemplified by adherens junctions connected to the actomyosin cytoskeleton.

## 1.3.2 Cell–cell gap junctions

Gap junctions (GJs) are intercellular membrane channels mostly composed of connexins, which are pore-forming proteins that transfer small molecules and ions between neighboring cells [71] (Fig. 1.2B). To form these pores, six connexins oligomerize giving rise to channels called "connexons" (or hemichannels), which then align in the extracellular space to form GJ channels [72]. Depending on the content and spatial arrangement of connexin subunits, these channels can be homotypic or heterotypic. Homotypic channels are formed from identical connexons, while heterotypic channels contain different hemichannels [72] (Fig. 1.2B).

The extracellular domains of connexins form a tight connection between adjacent cells, allowing an efficient cell adhesion [73]. In addition to their role in cell adhesion, connexins also mediate intercellular signaling, controlling the formation and permeability of channels, which allows a coordinated cell activity [74]. Consequently, it has been proposed that the channel activity of GJs could be relevant for cell coupling and coordination during glioma CCM, as these channels allow an efficient and fluent molecular communication among cells within the cluster [75]. Indeed, many cancer cells demonstrate gap junctional communication, which suggests cell–cell coupling, communication and with that multicellular organization [76]. Interestingly, depending on the tissue and stage of disease, the role of connexins in cancer may be either favoring or inhibitory [77]. Moreover, connexins have been also shown to both physically and functionally interact with other cell–cell adhesion proteins. For instance, Cx43 interacts with N-cadherin and other proteins of the AJ complex [78] at regions of cell–cell contact [79]. Moreover, Cx43 carboxy-tail has been shown to regulate the transcription of *n-cadherin* to control cell polarity and CCM [80]. GJs' proteins can also interact with the cytoskeleton as it has been shown that connexins can bind to microfilaments and microtubules [74].

Besides their role in cancer, GJs are also involved in CCM during development and tissue repair across different animal systems [81,82]. In vitro studies demonstrated that Gap27 inhibition enhances cell migration and wound healing. This was observed in 2D and 3D organotypic mouse skin models [83], human juvenile keratinocytes and fibroblasts [84], and human 3D *ex vivo* organotypic wound healing models [85]. Conversely, other in vivo studies reported that inhibiting Cx43 dramatically inhibits CCM [80,86]. Therefore GJs play diverse roles in morphogenetic events, including cell proliferation, adhesion, extracellular matrix deposition, and very importantly in CCM [82].

### 1.3.3 Cell–cell adherens junctions

AJs are composed of a complex of proteins that require to be dynamically regulated to allow mechanical coupling and cell signaling in a variety of biological processes [87,88]. One of the main molecular components of AJs are cadherins, which are calcium-dependent transmembrane proteins. These proteins contain an extracellular domain that is normally associated with cell–cell adhesion, and an intracellular domain that associates with several adapter and signaling proteins [42,89] (Fig. 1.2C). Furthermore, AJs can also help cellular clusters to synchronize their cytoskeletal activities. For instance, it has been shown that cadherins connect to the actomyosin cytoskeleton by interacting with the adapters α-, β-, and γ-catenin. Similarly, Nectin, another member of AJs [90], connects to the cytoskeleton by interacting with the adapter protein Afadin [91]. Although the formation and fine-tuning of an AJ is a complex and context-dependent process, we can simplify their formation in three main steps—first, cells contact each other and mechanically couple by engaging the extracellular domains of their cadherins [92,93]. This first step is known as "initiation" and the strength of this initial engagement will be defined by the type of cadherin expressed by interacting cells [94]. Hence, cadherin content will determine the formation of a heterotypical or homotypical AJ, and with that the strength and dynamic of cell–cell adhesion of interacting cells [95]. Second, additional cadherin molecules engage to laterally expand the contact area [92,96]. Third, AJs engage to the cytoskeleton to further coordinate cellular activities—as it has been shown that downregulating actin bundle turnover at the contact sites is essential for efficient CCM in *Drosophila* epithelial cells [70,97,98].

In vitro and in vivo studies have shown that both epithelial and mesenchymal cells can collectively migrate upon modulation of their protein content at the transcriptional [99,100], or posttranslational level [101,102], regulating the turnover of AJs' proteins. A hallmark of cell delamination is a cadherin switch in the AJs, that is, a reduction in the amount of type-I (E-cadherin) and, in some cases increase in type-II cadherins (N-cadherin) [103]. Other AJs' proteins such as members of the immunoglobulin family, or neural cell adhesion molecule, can be also involved in these switches [71,72]. For example, in order to start migrating, pancreatic tumor cells require decreasing their levels of the AJs' protein E-cadherin. Upon reduction of E-cadherin, cells increase the levels of CAM proteins, correlating with tumor invasion [104]. Another example is neural crest cells, where a reduction of E-cadherin has been shown to precede migration in *Xenopus* [26], and a switch from E- to N-cadherin has been shown to occur in chick embryos [105]. Across species, transcription factors such as Zeb, Snail, and Twist families are canonical regulators of these cadherins or other AJs' proteins switch [106]. This has been for a while a canonical view of how cells transit from an epithelial to a mesenchymal phenotype in order to migrate as single cells. Nonetheless, cells also migrate as collectives and the migration of some cell types actually relies on an increase of E-cadherin levels [107,108]. Thus the concept of EMT is currently under extensive discussion, as it has been recently and thoroughly reviewed in Yang et al. [109]. Regardless these almost semantic issues, the spacing, duration, and strength of AJs mediating cell–cell contacts is tightly regulated in a context-dependent manner [110]. Hence, positioning AJs as key players in diverse processes that require cells to collectively migrate, such as in branching morphogenesis of the mammary ducts [111] and the trachea [112], in epidermal repair [113], sprouting of blood vessels [114], and in several cancer types [115–117].

### 1.3.4 Cell–cell tight junctions

Tight junctions (TJs) are adhesion complexes that form a semipermeable barrier to regulate liquid, ion, and nutrient flow, and they are commonly observed in polarized epithelial and endothelial cells [118]. TJs are composed of transmembrane proteins (claudin, JAM-A, etc.) (Fig. 1.2D), which contribute to junction stability by being part of adhesion complexes, and to cell polarity by limiting intermixing of apical and basolateral transmembrane components [118–120]. These proteins are linked to scaffolding proteins, such as zonula occludens, which are in turn connected to actin and microtubules through several linkers [121] (Fig. 1.2D). This multifunctional complex enables the control of signaling pathways regulating actin organization [122,123], cell proliferation [124], cell polarity [125,126], gene expression [127], and cell migration [128–130].

Epithelial cell adhesion is maintained via two intercellular complexes—TJs and AJs [91]. At the onset of CCM, cells at the leading edge often display a mesenchyme-like state deprived of TJs, whereas cells at the rear end tend to form more packed assemblies, with TJs [131]. Indeed, claudins and occludins are downregulated by Snail, a known transcriptional mediator of cadherin switch, prior to neural crest cells migration [132,133]. For example, the downregulation of Claudin-1 promotes migration of chick cranial neural crest cells, while its overexpression blocks cell migration [134]. Similarly, in zebrafish TJ proteins are specifically expressed in epithelial neuromast cells, not in the migrating primordium [131]. Therefore TJs are known to be relevant in diverse contexts where CCM occurs. For instance, disruption of TJs is thought to reduce cell adhesion and promote cancer invasion, which has been observed among several primary human cancers and model systems [135]. Furthermore, in vivo and *ex vivo* studies have shown that TJs are also implicated in wound healing [136]. Hence, the multifunctional complex that constitutes TJs enables both cytoskeleton anchorage and the formation of intercellular junctional complexes (Fig. 1.2D and E), which are essential to regulate the diverse processes above mentioned [137].

More broadly, cell–cell adhesion is a key component of directional collective migration, enabling cells to stay together and allowing communication across the cluster, which in turn is translated into a front-rear polarity that directs CCM. However, cell–cell junctions can be a double-edged sword for CCM, since "too much" adhesion may preclude the cluster to migrate in complex geometries and "too little" may induce cell delamination and a loss of the collectiveness. Therefore CCM is thought to emerge from the finely balanced degree of cell–cell strength and AJs dynamics [42]. Exposing how this fine-tuning of junctional complexes is achieved and the role of the viscoelasticity of cells and tissues in this process is one of the scopes of this book.

## 1.4 Future perspectives

A large body of evidence about the molecular nature of CCM has been accumulated, but little is known about how the mechanical state of cellular clusters is determined by their interaction with the mechanical cues of their surrounding tissues [138]. This is relevant as collectively migrating cells not only experience molecular cues or undergo molecular modifications, but they also endure the inevitable mechanical interaction with their environment, which in turn triggers mechanical responses in the cell and with that modify the cellular and molecular behavior of the migratory cluster [25,138,139]. Thus integrating mechanical and molecular frameworks is relevant to fully understand how CCM is induced, sustained, and propagated in different contexts [41,64]. Further knowledge on mechanical and molecular crosstalk will require a better determination of force transmission at cell–cell contacts and its modulation by mechanical properties of the extracellular matrix and cell–matrix adhesion.

Recent advances have been generated about the mechanical and molecular interplay of cellular and subcellular mechano-sensing at cell–cell and cell–matrix contacts [140–142]. These new concepts are discussed in detail in the following chapters of this book, as they are crucial to understand the general principles underlying morphogenetic events orchestrated by collectively migrating cells in diverse relevant contexts, such as embryogenesis, wound healing, regeneration, and metastasis.

## Acknowledgments

The authors thank Cristian Marchant for providing templates for Figure 1.1 and Ailin Buzzi for comments on the manuscript.

## References

[1] P. Friedl, D. Gilmour, Collective cell migration in morphogenesis, regeneration and cancer, Nat. Rev. Mol. Cell Biol. 10 (7) (2009) 445–457.
[2] R. Mayor, S. Etienne-Manneville, The front and rear of collective cell migration, Nat. Rev. Mol. Cell Biol. 17 (2) (2016) 97–109.
[3] P. Rorth, Fellow travellers: emergent properties of collective cell migration, EMBO Rep. 13 (11) (2012) 984–991.
[4] O. Ilina, P. Friedl, Mechanisms of collective cell migration at a glance, J. Cell Sci. 122 (Pt 18) (2009) 3203–3208.
[5] M. Poujade, et al., Collective migration of an epithelial monolayer in response to a model wound, Proc. Natl. Acad. Sci. U. S. A. 104 (41) (2007) 15988–15993.
[6] M. Reffay, et al., Interplay of RhoA and mechanical forces in collective cell migration driven by leader cells, Nat. Cell Biol. 16 (3) (2014) 217–223.
[7] N.S. Gov, Collective cell migration patterns: follow the leader, Proc. Natl. Acad. Sci. U. S. A. 104 (41) (2007) 15970–15971.
[8] X. Trepat, et al., Physical forces during collective cell migration, Nat. Phys. 5 (6) (2009) 426–430.
[9] R. Farooqui, G. Fenteany, Multiple rows of cells behind an epithelial wound edge extend cryptic lamellipodia to collectively drive cell-sheet movement, J. Cell Sci. 118 (Pt 1) (2005) 51–63.
[10] F. Biname, et al., What makes cells move: requirements and obstacles for spontaneous cell motility, Mol. Biosyst. 6 (4) (2010) 648–661.
[11] A. Stamm, et al., In vitro wound healing assays—state of the art, BioNanoMaterials 17 (2016) 1–2.
[12] J. Magdalena, T.H. Millard, L.M. Machesky, Microtubule involvement in NIH 3T3 golgi and MTOC polarity establishment, J. Cell Sci. 116 (Pt 4) (2003) 743–756.
[13] R. Sunyer, et al., Collective cell durotaxis emerges from long-range intercellular force transmission, Science 353 (6304) (2016) 1157–1161.
[14] M.R. Ng, et al., Substrate stiffness regulates cadherin-dependent collective migration through myosin-II contractility, J. Cell Biol. 199 (3) (2012) 545–563.
[15] N. Gjorevski, et al., Dynamic tensile forces drive collective cell migration through three-dimensional extracellular matrices, Sci. Rep. 5 (2015) 11458.
[16] R.J. Petrie, K.M. Yamada, Multiple mechanisms of 3D migration: the origins of plasticity, Curr. Opin. Cell Biol. 42 (2016) 7–12.
[17] T.A. Collins, B.M. Yeoman, P. Katira, To lead or to herd: optimal strategies for 3D collective migration of cell clusters, Biomech. Model. Mechanobiol. (2020).

[18] K.M. Yamada, M. Sixt, Mechanisms of 3D cell migration, Nat. Rev. Mol. Cell Biol. 20 (12) (2019) 738–752.
[19] P. Friedl, et al., Migration of coordinated cell clusters in mesenchymal and epithelial cancer explants in vitro, Cancer Res. 55 (20) (1995) 4557–4560.
[20] K. Duval, et al., Modeling physiological events in 2D vs. 3D cell culture, Physiology (Bethesda) 32 (4) (2017) 266–277.
[21] H.L. Chandler, et al., MG53 promotes corneal wound healing and mitigates fibrotic remodeling in rodents, Commun. Biol. 2 (2019) 71.
[22] M.A. Stepp, et al., Wounding the cornea to learn how it heals, Exp. Eye Res. 121 (2014) 178–193.
[23] P. Patil, et al., Porcine ischemic wound-healing model for preclinical testing of degradable biomaterials, Tissue Eng. Part C Methods 23 (11) (2017) 754–762.
[24] Z. Chen, et al., A novel three-dimensional wound healing model, J. Dev. Biol. 2 (4) (2014) 198–209.
[25] E.H. Barriga, et al., Tissue stiffening coordinates morphogenesis by triggering collective cell migration in vivo, Nature 554 (7693) (2018) 523–527.
[26] E.H. Barriga, et al., The hypoxia factor Hif-1alpha controls neural crest chemotaxis and epithelial to mesenchymal transition, J. Cell Biol. 201 (5) (2013) 759–776.
[27] M. Zhao, et al., Electrical signals control wound healing through phosphatidylinositol-3-OH kinase-gamma and PTEN, Nature 442 (7101) (2006) 457–460.
[28] K. Hotary, et al., A cancer cell metalloprotease triad regulates the basement membrane transmigration program, Genes Dev. 20 (19) (2006) 2673–2686.
[29] D.Y. Chen, J. Crest, D. Bilder, A cell migration tracking tool supports coupling of tissue rotation to elongation, Cell Rep. 21 (3) (2017) 559–569.
[30] P. Friedl, et al., New dimensions in cell migration, Nat. Rev. Mol. Cell Biol. 13 (11) (2012) 743–747.
[31] M. Prasad, et al., Border cell migration: a model system for live imaging and genetic analysis of collective cell movement, Methods Mol. Biol. 1328 (2015) 89–97.
[32] D. Dalle Nogare, A.B. Chitnis, A framework for understanding morphogenesis and migration of the zebrafish posterior Lateral Line primordium, Mech. Dev. 148 (2017) 69–78.
[33] A.B. Chitnis, D.D. Nogare, M. Matsuda, Building the posterior lateral line system in zebrafish, Dev. Neurobiol. 72 (3) (2012) 234–255.
[34] C. Cela, M. Llimargas, Egfr is essential for maintaining epithelial integrity during tracheal remodelling in Drosophila, Development 133 (16) (2006) 3115–3125.
[35] M. Affolter, E. Caussinus, Tracheal branching morphogenesis in Drosophila: new insights into cell behaviour and organ architecture, Development 135 (12) (2008) 2055–2064.
[36] M. Hellstrom, et al., Dll4 signalling through Notch1 regulates formation of tip cells during angiogenesis, Nature 445 (7129) (2007) 776–780.
[37] D. Alfandari, H. Cousin, M. Marsden, Mechanism of Xenopus cranial neural crest cell migration, Cell Adh. Migr. 4 (4) (2010) 553–560.
[38] E.H. Barriga, R. Mayor, Embryonic cell-cell adhesion: a key player in collective neural crest migration, Curr. Top. Dev. Biol. 112 (2015) 301–323.
[39] M. Tada, C.P. Heisenberg, Convergent extension: using collective cell migration and cell intercalation to shape embryos, Development 139 (21) (2012) 3897–3904.
[40] J.G. Dumortier, et al., Collective mesendoderm migration relies on an intrinsic directionality signal transmitted through cell contacts, Proc. Natl. Acad. Sci. U. S. A. 109 (42) (2012) 16945–16950.
[41] C. Norden, V. Lecaudey, Collective cell migration: general themes and new paradigms, Curr. Opin. Genet. Dev. 57 (2019) 54–60.

[42] E.H. Barriga, R. Mayor, Adjustable viscoelasticity allows for efficient collective cell migration, Semin. Cell Dev. Biol. 93 (2019) 55–68.
[43] P. Duchek, P. Rorth, Guidance of cell migration by EGF receptor signaling during Drosophila oogenesis, Science 291 (5501) (2001) 131–133.
[44] D. Cai, et al., Mechanical feedback through E-cadherin promotes direction sensing during collective cell migration, Cell 157 (5) (2014) 1146–1159.
[45] A. Nechiporuk, D.W. Raible, FGF-dependent mechanosensory organ patterning in zebrafish, Science 320 (5884) (2008) 1774–1777.
[46] C. Dambly-Chaudiere, N. Cubedo, A. Ghysen, Control of cell migration in the development of the posterior lateral line: antagonistic interactions between the chemokine receptors CXCR4 and CXCR7/RDC1, BMC Dev. Biol. 7 (2007) 23.
[47] B. Boldajipour, et al., Control of chemokine-guided cell migration by ligand sequestration, Cell 132 (3) (2008) 463–473.
[48] E. Dona, et al., Directional tissue migration through a self-generated chemokine gradient, Nature 503 (7475) (2013) 285–289.
[49] A. Szabo, et al., In vivo confinement promotes collective migration of neural crest cells, J. Cell Biol. 213 (5) (2016) 543–555.
[50] A. Shellard, R. Mayor, Integrating chemical and mechanical signals in neural crest cell migration, Curr. Opin. Genet. Dev. 57 (2019) 16–24.
[51] P. Friedl, S. Alexander, Cancer invasion and the microenvironment: plasticity and reciprocity, Cell 147 (5) (2011) 992–1009.
[52] A.G. Clark, D.M. Vignjevic, Modes of cancer cell invasion and the role of the microenvironment, Curr. Opin. Cell Biol. 36 (2015) 13–22.
[53] M. Lintz, A. Munoz, C.A. Reinhart-King, The mechanics of single cell and collective migration of tumor cells, J. Biomech. Eng. 139 (2) (2017) 0210051–0210059.
[54] X. Wang, et al., Collective invasion of cancer: perspectives from pathology and development, Pathol. Int. 66 (4) (2016) 183–192.
[55] A. Patsialou, et al., Intravital multiphoton imaging reveals multicellular streaming as a crucial component of in vivo cell migration in human breast tumors, Intravital 2 (2) (2013) e25294.
[56] P. Friedl, et al., Classifying collective cancer cell invasion, Nat. Cell Biol. 14 (8) (2012) 777–783.
[57] S. Alexander, et al., Dynamic imaging of cancer growth and invasion: a modified skin-fold chamber model, Histochem. Cell Biol. 130 (6) (2008) 1147–1154.
[58] S. Raghavan, et al., Comparative analysis of tumor spheroid generation techniques for differential in vitro drug toxicity, Oncotarget 7 (13) (2016) 16948–16961.
[59] A. Nagelkerke, et al., Generation of multicellular tumor spheroids of breast cancer cells: how to go three-dimensional, Anal. Biochem. 437 (1) (2013) 17–19.
[60] M.J. Pittet, R. Weissleder, Intravital imaging, Cell 147 (5) (2011) 983–991.
[61] R. Weigert, et al., Intravital microscopy: a novel tool to study cell biology in living animals, Histochem. Cell Biol. 133 (5) (2010) 481–491.
[62] C. Scheele, C. Maynard, J. van Rheenen, Intravital insights into heterogeneity, metastasis, and therapy responses, Trends Cancer 2 (4) (2016) 205–216.
[63] V. Naumenko, C. Jenne, D.J. Mahoney, Intravital microscopy for imaging the tumor microenvironment in live mice, Methods Mol. Biol. 1458 (2016) 217–230.
[64] A. Haeger, et al., Collective cell migration: guidance principles and hierarchies, Trends Cell Biol. 25 (9) (2015) 556–566.
[65] C.J. Weijer, Collective cell migration in development, J. Cell Sci. 122 (Pt 18) (2009) 3215–3223.
[66] P. Friedl, R. Mayor, Tuning collective cell migration by cell-cell junction regulation, Cold Spring Harb. Perspect. Biol. 9 (4) (2017).

[67] D.J. Montell, Morphogenetic cell movements: diversity from modular mechanical properties, Science 322 (5907) (2008) 1502–1505.
[68] A. Cliffe, et al., Quantitative 3D analysis of complex single border cell behaviors in coordinated collective cell migration, Nat. Commun. 8 (2017) 14905.
[69] S. Kuriyama, et al., In vivo collective cell migration requires an LPAR2-dependent increase in tissue fluidity, J. Cell Biol. 206 (1) (2014) 113–127.
[70] M. Cavey, T. Lecuit, Molecular bases of cell-cell junctions stability and dynamics, Cold Spring Harb. Perspect. Biol. 1 (5) (2009) a002998.
[71] T.S. Lawrence, W.H. Beers, N.B. Gilula, Transmission of hormonal stimulation by cell-to-cell communication, Nature 272 (5653) (1978) 501–506.
[72] G. Mese, G. Richard, T.W. White, Gap junctions: basic structure and function, J. Invest. Dermatol. 127 (11) (2007) 2516–2524.
[73] M.L. Cotrina, J.H. Lin, M. Nedergaard, Adhesive properties of connexin hemichannels, Glia 56 (16) (2008) 1791–1798.
[74] C.J. Wei, X. Xu, C.W. Lo, Connexins and cell signaling in development and disease, Annu. Rev. Cell Dev. Biol. 20 (2004) 811–838.
[75] Q. Aftab, W.C. Sin, C.C. Naus, Reduction in gap junction intercellular communication promotes glioma migration, Oncotarget 6 (13) (2015) 11447–11464.
[76] J. Czyz, The stage-specific function of gap junctions during tumourigenesis, Cell Mol. Biol. Lett. 13 (1) (2008) 92–102.
[77] C.C. Naus, D.W. Laird, Implications and challenges of connexin connections to cancer, Nat. Rev. Cancer 10 (6) (2010) 435–441.
[78] X. Xu, et al., Modulation of mouse neural crest cell motility by N-cadherin and connexin 43 gap junctions, J. Cell Biol. 154 (1) (2001) 217–230.
[79] C.J. Wei, et al., Connexin43 associated with an N-cadherin-containing multiprotein complex is required for gap junction formation in NIH3T3 cells, J. Biol. Chem. 280 (20) (2005) 19925–19936.
[80] M. Kotini, et al., Gap junction protein Connexin-43 is a direct transcriptional regulator of N-cadherin in vivo, Nat. Commun. 9 (1) (2018) 3846.
[81] M. Kotini, R. Mayor, Connexins in migration during development and cancer, Dev. Biol. 401 (1) (2015) 143–151.
[82] C. Lorraine, C.S. Wright, P.E. Martin, Connexin43 plays diverse roles in co-ordinating cell migration and wound closure events, Biochem. Soc. Trans. 43 (3) (2015) 482–488.
[83] E.E. Kandyba, M.B. Hodgins, P.E. Martin, A murine living skin equivalent amenable to live-cell imaging: analysis of the roles of connexins in the epidermis, J. Invest. Dermatol. 128 (4) (2008) 1039–1049.
[84] C.S. Wright, et al., Connexin mimetic peptides improve cell migration rates of human epidermal keratinocytes and dermal fibroblasts in vitro, Wound Repair. Regen. 17 (2) (2009) 240–249.
[85] S. Pollok, et al., Connexin 43 mimetic peptide Gap27 reveals potential differences in the role of Cx43 in wound repair between diabetic and non-diabetic cells, J. Cell Mol. Med. 15 (4) (2011) 861–873.
[86] H.P. Ehrlich, T. Diez, Role for gap junctional intercellular communications in wound repair, Wound Repair. Regen. 11 (6) (2003) 481–489.
[87] B. Ladoux, et al., The mechanotransduction machinery at work at adherens junctions, Integr. Biol. (Camb.) 7 (10) (2015) 1109–1119.
[88] G. Charras, A.S. Yap, Tensile forces and mechanotransduction at cell-cell junctions, Curr. Biol. 28 (8) (2018) R445–R457.
[89] D.E. Leckband, J. de Rooij, Cadherin adhesion and mechanotransduction, Annu. Rev. Cell Dev. Biol. 30 (2014) 291–315.

[90] I. Indra, et al., The adherens junction: a mosaic of cadherin and nectin clusters bundled by actin filaments, J. Invest. Dermatol. 133 (11) (2013) 2546–2554.
[91] C.M. Niessen, Tight junctions/adherens junctions: basic structure and function, J. Invest. Dermatol. 127 (11) (2007) 2525–2532.
[92] A.R. Harris, A. Daeden, G.T. Charras, Formation of adherens junctions leads to the emergence of a tissue-level tension in epithelial monolayers, J. Cell Sci. 127 (Pt 11) (2014) 2507–2517.
[93] S. Yamada, et al., Deconstructing the cadherin-catenin-actin complex, Cell 123 (5) (2005) 889–901.
[94] A. Nose, A. Nagafuchi, M. Takeichi, Expressed recombinant cadherins mediate cell sorting in model systems, Cell 54 (7) (1988) 993–1001.
[95] M.S. Steinberg, M. Takeichi, Experimental specification of cell sorting, tissue spreading, and specific spatial patterning by quantitative differences in cadherin expression, Proc. Natl. Acad. Sci. U. S. A. 91 (1) (1994) 206–209.
[96] A. Vaezi, et al., Actin cable dynamics and Rho/Rock orchestrate a polarized cytoskeletal architecture in the early steps of assembling a stratified epithelium, Dev. Cell 3 (3) (2002) 367–381.
[97] F. Pilot, et al., Spatial control of actin organization at adherens junctions by a synaptotagmin-like protein Btsz, Nature 442 (7102) (2006) 580–584.
[98] N.L. Zandy, M. Playford, A.M. Pendergast, Abl tyrosine kinases regulate cell-cell adhesion through Rho GTPases, Proc. Natl. Acad. Sci. U. S. A. 104 (45) (2007) 17686–17691.
[99] A.F. Paulson, et al., Regulation of cadherin expression in nervous system development, Cell Adh. Migr. 8 (1) (2014) 19–28.
[100] H. Peinado, F. Portillo, A. Cano, Transcriptional regulation of cadherins during development and carcinogenesis, Int. J. Dev. Biol. 48 (5-6) (2004) 365–375.
[101] C.L. Fairchild, L.S. Gammill, Tetraspanin18 is a FoxD3-responsive antagonist of cranial neural crest epithelial-to-mesenchymal transition that maintains cadherin-6B protein, J. Cell Sci. 126 (Pt 6) (2013) 1464–1476.
[102] B.M. Gumbiner, Regulation of cadherin adhesive activity, J. Cell Biol. 148 (3) (2000) 399–404.
[103] P. Ruiz, U. Gunthert, The cellular basis of metastasis, World J. Urol. 14 (3) (1996) 141–150.
[104] U. Cavallaro, G. Christofori, Cell adhesion and signalling by cadherins and Ig-CAMs in cancer, Nat. Rev. Cancer 4 (2) (2004) 118–132.
[105] C.D. Rogers, A. Saxena, M.E. Bronner, Sip1 mediates an E-cadherin-to-N-cadherin switch during cranial neural crest EMT, J. Cell Biol. 203 (5) (2013) 835–847.
[106] S. Ansieau, G. Collin, L. Hill, EMT or EMT-promoting transcription factors, where to focus the light? Front. Oncol. 4 (2014) 353.
[107] E. Theveneau, R. Mayor, Cadherins in collective cell migration of mesenchymal cells, Curr. Opin. Cell Biol. 24 (5) (2012) 677–684.
[108] Y. Arboleda-Estudillo, et al., Movement directionality in collective migration of germ layer progenitors, Curr. Biol. 20 (2) (2010) 161–169.
[109] J. Yang, et al., Guidelines and definitions for research on epithelial-mesenchymal transition, Nat. Rev. Mol. Cell Biol. 21 (6) (2020) 341–352.
[110] E. Theveneau, R. Mayor, Collective cell migration of epithelial and mesenchymal cells, Cell Mol. Life Sci. 70 (19) (2013) 3481–3492.
[111] A.J. Ewald, et al., Collective epithelial migration and cell rearrangements drive mammary branching morphogenesis, Dev. Cell 14 (4) (2008) 570–581.
[112] M. Shindo, et al., Dual function of Src in the maintenance of adherens junctions during tracheal epithelial morphogenesis, Development 135 (7) (2008) 1355–1364.

[113] P. Young, et al., E-cadherin controls adherens junctions in the epidermis and the renewal of hair follicles, EMBO J. 22 (21) (2003) 5723–5733.
[114] F. Neto, et al., YAP and TAZ regulate adherens junction dynamics and endothelial cell distribution during vascular development, Elife (2018) 7.
[115] W. Weichert, et al., ALCAM/CD166 is overexpressed in colorectal carcinoma and correlates with shortened patient survival, J. Clin. Pathol. 57 (11) (2004) 1160–1164.
[116] N. Gavert, et al., L1-CAM in cancerous tissues, Expert Opin. Biol. Ther. 8 (11) (2008) 1749–1757.
[117] L.C.L.T. van Kempen, et al., Activated leukocyte cell adhesion molecule/CD166, a marker of tumor progression in primary malignant melanoma of the skin, Am. J. Pathol. 156 (3) (2000) 769–774.
[118] C. Zihni, et al., Tight junctions: from simple barriers to multifunctional molecular gates, Nat. Rev. Mol. Cell Biol. 17 (9) (2016) 564–580.
[119] S. Garrido-Urbani, P.F. Bradfield, B.A. Imhof, Tight junction dynamics: the role of junctional adhesion molecules (JAMs), Cell Tissue Res. 355 (3) (2014) 701–715.
[120] M.A. Schluter, B. Margolis, Apicobasal polarity in the kidney, Exp. Cell Res. 318 (9) (2012) 1033–1039.
[121] C.M. Van Itallie, J.M. Anderson, Architecture of tight junctions and principles of molecular composition, Semin. Cell Dev. Biol. 36 (2014) 157–165.
[122] K. Matter, M.S. Balda, Signalling to and from tight junctions, Nat. Rev. Mol. Cell Biol. 4 (3) (2003) 225–236.
[123] A.A. Bhat, et al., Tight junction proteins and signaling pathways in cancer and inflammation: a functional crosstalk, Front. Physiol. 9 (2018) 1942.
[124] M. Diaz-Coranguez, X. Liu, D.A. Antonetti, Tight junctions in cell proliferation, Int. J. Mol. Sci. 20 (23) (2019).
[125] K. Ebnet, et al., The cell polarity protein ASIP/PAR-3 directly associates with junctional adhesion molecule (JAM), EMBO J. 20 (14) (2001) 3738–3748.
[126] M. Itoh, et al., Junctional adhesion molecule (JAM) binds to PAR-3: a possible mechanism for the recruitment of PAR-3 to tight junctions, J. Cell Biol. 154 (3) (2001) 491–497.
[127] L. Gonzalez-Mariscal, et al., Tight junctions and the regulation of gene expression, Semin. Cell Dev. Biol. 36 (2014) 213–223.
[128] E.A. Severson, et al., Junctional adhesion molecule A interacts with Afadin and PDZ-GEF2 to activate Rap1A, regulate beta1 integrin levels, and enhance cell migration, Mol. Biol. Cell 20 (7) (2009) 1916–1925.
[129] O. Tornavaca, et al., ZO-1 controls endothelial adherens junctions, cell-cell tension, angiogenesis, and barrier formation, J. Cell Biol. 208 (6) (2015) 821–838.
[130] L. Huo, et al., Cdc42-dependent formation of the ZO-1/MRCKbeta complex at the leading edge controls cell migration, EMBO J. 30 (4) (2011) 665–678.
[131] V. Lecaudey, et al., Dynamic Fgf signaling couples morphogenesis and migration in the zebrafish lateral line primordium, Development 135 (16) (2008) 2695–2705.
[132] J. Ikenouchi, et al., Regulation of tight junctions during the epithelium-mesenchyme transition: direct repression of the gene expression of claudins/occludin by Snail, J. Cell Sci. 116 (Pt 10) (2003) 1959–1967.
[133] O.M. Martinez-Estrada, et al., The transcription factors Slug and Snail act as repressors of Claudin-1 expression in epithelial cells, Biochem. J. 394 (Pt 2) (2006) 449–457.
[134] K.J. Fishwick, et al., The tight junction protein claudin-1 influences cranial neural crest cell emigration, Mech. Dev. 129 (9-12) (2012) 275–283.
[135] T.A. Martin, W.G. Jiang, Loss of tight junction barrier function and its role in cancer metastasis, Biochim. Biophys. Acta 1788 (4) (2009) 872–891.

[136] T. Volksdorf, et al., Tight junction proteins Claudin-1 and Occludin are important for cutaneous wound healing, Am. J. Pathol. 187 (6) (2017) 1301–1312.
[137] E.E. Schneeberger, R.D. Lynch, The tight junction: a multifunctional complex, Am. J. Physiol. Cell Physiol 286 (6) (2004) C1213–C1228.
[138] P. Roca-Cusachs, R. Sunyer, X. Trepat, Mechanical guidance of cell migration: lessons from chemotaxis, Curr. Opin. Cell Biol. 25 (5) (2013) 543–549.
[139] B. Ladoux, R.M. Mege, X. Trepat, Front-rear polarization by mechanical cues: from single cells to tissues, Trends Cell Biol. 26 (6) (2016) 420–433.
[140] F. Martino, et al., Cellular mechanotransduction: from tension to function, Front. Physiol. 9 (2018) 824.
[141] V.P. Hytonen, B. Wehrle-Haller, Mechanosensing in cell-matrix adhesions—converting tension into chemical signals, Exp. Cell Res. 343 (1) (2016) 35–41.
[142] A.D. Doyle, K.M. Yamada, Mechanosensing via cell-matrix adhesions in 3D microenvironments, Exp. Cell Res. 343 (1) (2016) 60–66.

CHAPTER TWO

# The basic concept of viscoelasticity

**Ivana Pajic-Lijakovic**
Faculty of Technology and Metallurgy, Belgrade University, Belgrade, Serbia

## 2.1 Introduction

Viscoelasticity of contractile cells has been considered within three time regimes [1]: (1) short-time regime (subcellular level) for milliseconds to seconds time scale [2–4]), (2) middle-time regime (cellular level) for several tens of seconds to several minutes time scale [5–7], and (3) long-time regime (supracellular level) for several tens of minutes to several hours [5,6,8,9]. Every time-regime has its own space scale, and consequently, the different structural entities that contribute to viscoelasticity. Cumulative effects of structural changes at a subcellular level influence viscoelasticity at a supracellular level. At the same time established configurations of migrating cells, which significantly influence viscoelasticity at a supracellular level, have the feedback impact to the cohesiveness inhomogeneity and various mechanical and biochemical processes at a subcellular level. Milliseconds scale corresponds to protein interchain interactions. Seconds scale corresponds to collective phenomena caused primarily by protein intrachain interactions. The scale of several tens of seconds corresponds to shape relaxation of passive (resting) cells [6]. The shape relaxation of active cells corresponds to minutes time scale [7]. Minutes scale corresponds to single-cell migration time [10], cell signaling [11], cell polarization [12], and cadherin turnover time [13]. Several tens of minutes time scale corresponds to cell persistence time [14]. Collective cell migration has been studied at a time scale of hours [15]. The special interest here is middle and long-time viscoelastic regimes, which will be elaborated through various examples.

Viscoelastic behavior reflects the combined viscous and elastic responses which are intermediate between liquid-like and solid-like behavior [1,15,16]. Deeper insight into rheological response of

*Viscoelasticity and Collective Cell Migration.*
DOI: https://doi.org/10.1016/B978-0-12-820310-1.00006-9

multicellular systems under various loading conditions in the context of stress and strain relaxation phenomena is the prerequisite for understanding various biological processes such as embryo morphogenesis, wound healing, regeneration, and also in pathological conditions, such as cancer [10,15,17–21].

The framework for further interpretation of the problem includes (1) discussion of the basic concept of viscoelasticity [22] and (2) consideration of rheological responses of various multicellular systems under various loading conditions in the context of the proposed concept. Various types of constitutive models which have been applied in the literature for describing viscoelasticity of multicellular systems under various experimental conditions such as (1) linear models proposed for viscoelastic liquids, that is, the Maxwell model and the Jeffreys model, (2) linear models proposed for viscoelastic solids, that is, the Kelvin–Voigt model, the Zener model, and the Four-parameter model, and (3) nonlinear viscoelastic model for jamming state were discussed. Jamming state corresponds satisfies the following conditions: (1) migration of the system constituents is significantly damped such that the diffusion coefficient tends to zero and (2) relaxation time tends to infinity [23]. The corresponding constitutive model was formulated by Pajic-Lijakovic and Milivojevic [19]. Viscoelastic models are analog models formulated by series, parallel, and series–parallel combinations of rheological elements such as (1) spring, (2) dashpot, and (3) spring-pot. Spring element accounts for reversible, elastic behavior while dashpot accounts for irreversible viscous behavior. Spring-pot is between them. This element accounts for the anomalous nature of energy dissipation which can be modeled by applying fractional derivatives [24]. Every model is discussed in the time domain and frequency domain. Time domain accounts for stress and strain change versus time while frequency domain obtained after Fourier transform accounts for interrelation between storage and loss moduli versus angular velocity. The storage modulus represents the measure of storage energy while the loss modulus represents the measure of energy dissipation during system structural changes. The general result of this theoretical consideration is presented in the context of the main characteristics of viscoelastic liquids and viscoelastic solids. This interpretation can be useful for researchers to recognize corresponding rheological behavior of multicellular systems in their experiments.

The theoretical concept of viscoelasticity in the form of various constitutive models is applied to multicellular systems such as cell aggregates or tumor spheroids under various stress conditions. I am interested here in the rheological response of multicellular systems at a time scale from a few minutes to a few hours. The basic concept of viscoelasticity in the context of linear models for viscoelastic liquids and viscoelastic solids and jamming state was discussed in the context of rheological response of multicellular systems caused under various stress conditions proposed in the literature. Stress has been applied to cell aggregates or tumor spheroids in various ways, that is, under (1) uniaxial compression between parallel plates [5,8], (2) hyperosmotic stress [25–27], (3) compression caused by hydrogel matrix rheological response induced by cell expansion [28,29], and (4) micropipette aspiration [30].

## 2.2 Linear viscoelasticity: constitutive models

I consider and compare various types of viscoelastic behaviors in order to clarify the main characteristics of viscoelastic liquid and viscoelastic solid in the context of system ability to relax. The relaxation ability represents a consequence of reversible, elastic and irreversible, viscous nature of the system structural changes under various loading conditions. Relaxation phenomena of viscoelastic systems can be analyzed by applying two types of experiments in time domain and frequency domain such as (1) strain relaxation under constant stress conditions (creep experiments); stress relaxation under constant strain conditions, and (2) change of storage and loss moduli under low oscillator strain conditions. Both strategies lead to determine the rheological parameters and relaxation times. Linear viscoelastic models can be formulated by series or parallel combining the rheological elements: (1) spring and (2) dashpot (Fig. 2.1).

Spring element accounts for the reversible, elastic nature of a system structural change and is expressed as $\tilde{\boldsymbol{\sigma}}_S(t) = G_s\tilde{\boldsymbol{\varepsilon}}_S(t)$ (where $\tilde{\boldsymbol{\sigma}}_S(t)$ is the stress at spring, $\tilde{\boldsymbol{\varepsilon}}_S(t)$ is the strain at spring, and $G_s$ is the elastic shear

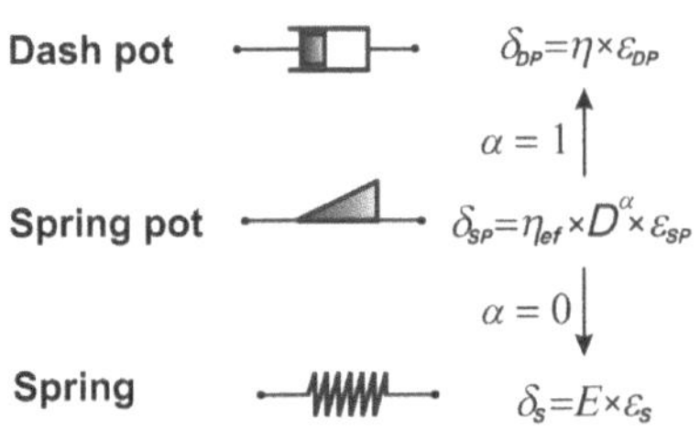


**Figure 2.1** The rheological elements spring, spring-pot, and dashpot.

modulus). Dashpot element accounts for the irreversible, viscous nature of a system structural change and is expressed as $\tilde{\sigma}_{DP}(t) = \eta\dot{\tilde{\varepsilon}}_{DP}(t)$ (where $\tilde{\sigma}_{DP}(t)$ is the stress at a dashpot, $\tilde{\varepsilon}_{DP}(t)$ is the strain at a dashpot, $\eta$ is the viscosity, and $\dot{\tilde{\varepsilon}}_{DP}(t) = \frac{d\tilde{\varepsilon}_{DP}(t)}{dt}$ is the strain rate).

### 2.2.1 Viscoelastic liquid behavior

I discussed the main characteristics of the Newtonian liquid and compared it with characteristics of viscoelastic liquids. The Newtonian liquid is presented by dashpot element (Fig. 2.2).

Linear constitutive models for describing rheological behavior of viscoelastic liquids, considered here, are the Maxwell model and the Jeffreys model. The similarities and differences of these models will be discussed in order to extract the general characteristics of viscoelastic liquids. The Newtonian liquid is described by dashpot element as:

$$\tilde{\sigma}_{NL}(t) = \tilde{\sigma}_{DP}(t) = \eta\dot{\tilde{\varepsilon}}_{NL}(t), \tag{2.1}$$

where $\tilde{\sigma}_{NL}(t)$ is the stress, $\tilde{\varepsilon}_{NL}(t)$ is the strain, $\dot{\tilde{\varepsilon}}_{NL}(t) = \frac{d\tilde{\varepsilon}_{NL}(t)}{dt}$ is strain rate, and $\eta$ is the viscosity of the Newtonian liquid (Fig. 2.2). Strain change under constant stress condition $\tilde{\sigma}_0$ for the initial condition $\tilde{\varepsilon}_{NL}(0) = 0$ is equal to $\tilde{\varepsilon}_{NL}(t) = \frac{\tilde{\sigma}_0}{\eta}t$, while the strain rate is constant and equal to $\dot{\varepsilon}_{NL}(t) = \frac{\tilde{\sigma}_0}{\eta}$. Strain increases during the time period $\Delta t$ from $\tilde{\varepsilon}_{NL}(0) = 0$ to $\tilde{\varepsilon}_{NL}(\Delta t)$ without the ability to relax. If the system undergoes free relaxation at $t = \Delta t$ such that $\tilde{\sigma}_{NL} = 0$ and $\tilde{\varepsilon}_{NL}(\Delta t) = \tilde{\varepsilon}_0$, the strain stays constant for $t > \Delta t$ and equal to $\tilde{\varepsilon}_{NL}(t) = \tilde{\varepsilon}_0$. Stress cannot relax under constant strain rate condition.

The Maxwell model equation is the simplest linear model suitable for describing viscoelastic liquid behavior. This model represents a series combination of the rheological elements spring and dashpot (Fig. 2.2). The series combination indicates that $\tilde{\sigma}_{MM}(t) = \tilde{\sigma}_S(t) + \tilde{\sigma}_{DP}(t)$ and $\tilde{\varepsilon}_{MM}(t) = \tilde{\varepsilon}_S(t) = \tilde{\varepsilon}_{DP}(t)$

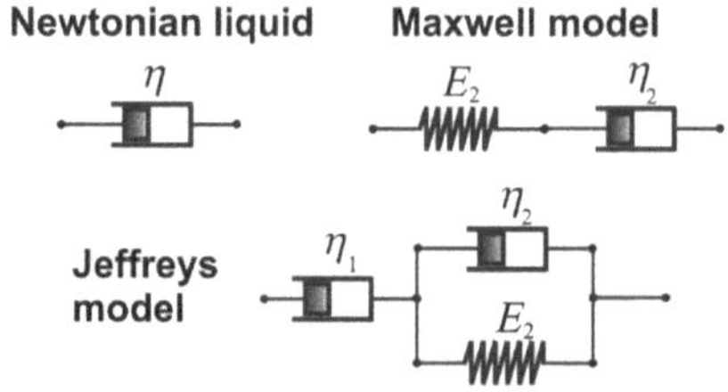


**Figure 2.2** Schematic representations of the Newtonian liquid model [Eq. (2.1)], the Maxwell model [Eq. (2.2)], and the Jeffreys model [Eq. (2.5)].

(where $\tilde{\boldsymbol{\sigma}}_{\boldsymbol{MM}}(t)$ is the stress and $\tilde{\varepsilon}_{\boldsymbol{MM}}(t)$ is the strain for the Maxwell model). Consequently, the formulated Maxwell model is expressed as:

$$\tilde{\boldsymbol{\sigma}}_{\boldsymbol{MM}}(t) + \tau_{R\,MM}\dot{\tilde{\boldsymbol{\sigma}}}_{\boldsymbol{MM}}(t) = \eta\dot{\tilde{\varepsilon}}_{\boldsymbol{MM}}(t) \tag{2.2}$$

where $\tau_{R\,MM}$ is the stress relaxation time equal to $\tau_{R\,MM} = \frac{\eta_2}{E_2}$, $E_2$ is the elastic modulus, and $\eta_2$ is the viscosity (Fig. 2.2). Strain change under constant stress condition $\tilde{\sigma}_0$ for the initial condition $\tilde{\varepsilon}_{\boldsymbol{MM}}(0) = 0$ shows similar behavior as the one previously described for Newtonian liquid. Corresponding strain increase is equal to $\tilde{\varepsilon}_{\boldsymbol{MM}}(t) = \frac{\tilde{\sigma}_0}{\eta_2}t$. Strain increases during the time period $\Delta t$ from $\tilde{\varepsilon}_{\boldsymbol{MM}}(0) = 0$ to $\tilde{\varepsilon}_{\boldsymbol{MM}}(\Delta t)$ without the ability to relax. If the system undergoes free relaxation at $t = \Delta t$ such that $\tilde{\boldsymbol{\sigma}}_{\boldsymbol{MM}} = 0$ and $\tilde{\varepsilon}_{\boldsymbol{MM}}(\Delta t) = \tilde{\varepsilon}_0$, the strain stays constant for $t > \Delta t$ and equal to $\tilde{\varepsilon}_{\boldsymbol{MM}}(t) = \tilde{\varepsilon}_0$, similarly as for the Newtonian liquid. Consequently, a strain cannot relax under constant stress condition due to sample fluidity. It is the main characteristic of both viscoelastic liquid and viscous, the Newtonian liquid.

On the contrary to the Newtonian liquid, viscoelastic liquids show the ability of stress relaxation. Stress relaxation under constant strain rate $\dot{\tilde{\varepsilon}}_0$ for the Maxwell model can be expressed starting from the initial condition $\tilde{\boldsymbol{\sigma}}_{\boldsymbol{MM}}(t = 0) = \tilde{\boldsymbol{\sigma}}_0$ as $\tilde{\boldsymbol{\sigma}}_{\boldsymbol{MM}}(t) = \tilde{\boldsymbol{\sigma}}_0 e^{-\frac{t}{\tau_{RMM}}} + \tilde{\boldsymbol{\sigma}}_{\boldsymbol{R}}(1 - e^{-\frac{t}{\tau_{RMM}}})$ (where $\tilde{\boldsymbol{\sigma}}_{\boldsymbol{R}}$ is the residual stress equal to $\tilde{\boldsymbol{\sigma}}_{\boldsymbol{R}} = \eta_2\dot{\tilde{\varepsilon}}_0$).

Eq. (2.2) can be transformed from the time domain into the frequency domain using the Fourier integral transform. Transforming equation is expressed in the form $F[\sigma(t)] = G*(\omega)F[\varepsilon(t)]$ (where $F[\Delta]$ is the Fourier transform, $\omega$ is the angular velocity, and $G^*(\omega)$ is the complex modulus). The complex modulus is equal to:

$$G^*(\omega) = G'(\omega) + iG''(\omega), \tag{2.3}$$

where $G'(\omega)$ is the storage modulus, $G''(\omega)$ is the loss modulus, and $i = \sqrt{-1}$ is the imaginary unit. The storage modulus $G'(\omega)$ quantifies elastic behavior while the loss modulus $G''(\omega)$ quantifies viscous behavior of the examined system. A higher value of the storage modulus indicates more intensive solid-like behavior while a higher value of the loss modulus indicates more intensive liquid-like behavior. Storage and loss moduli for the Maxwell model can be expressed from Eqs. (2.2) and (2.3):

$$\begin{aligned} G'_{MM}(\omega) &= \frac{\eta_2\tau_{RMM}\omega^2}{1 + \tau_{RMM}^2\omega^2} \\ G''_{MM}(\omega) &= \frac{\eta_2\omega}{1 + \tau_{RMM}^2\omega^2}, \end{aligned} \tag{2.4}$$

where $G'_{MM}(\omega)$ is the storage modulus and $G''_{MM}(\omega)$ is the loss modulus for the Maxwell model.

On the contrary to viscoelastic liquid behavior described by the Maxwell model, the Newtonian liquid shows only viscous behavior quantified by loss modulus $G''_{NL}(\omega) = \eta\omega$, while $G'_{NL}(\omega) = 0$. Viscoelastic liquids described by the Maxwell model show: (1) liquid-like behavior for lower angular velocities $G'_{MM}(\omega)/G''_{MM}(\omega) < 1$ and (2) solid-like behavior $G'_{MM}(\omega)/G''_{MM}(\omega) > 1$ for larger angular velocities. The crossover point corresponds to the condition $G'(\omega^*) = G''(\omega^*)$ (where $\omega^* = \frac{2\pi}{\tau^*_{RMM}}$). The crossover time is equal to the relaxation time for stress, that is, $\tau^*_{R\,MM} = \tau_{R\,MM}$.

The Jeffreys model for viscoelastic liquid is more complex than the Maxwell model. This model is formulated based on a series combination of dashpot and the Kelvin–Voigt element (Fig. 2.2). It is a three-parameter model expressed as:

$$\tilde{\boldsymbol{\sigma}}_{\boldsymbol{JM}}(t) + \tau_{RJM}\dot{\tilde{\boldsymbol{\sigma}}}_{\boldsymbol{JM}}(t) = \eta_1 \dot{\tilde{\boldsymbol{\varepsilon}}}_{\boldsymbol{JM}}(t) + \eta_1 t_R \ddot{\tilde{\boldsymbol{\varepsilon}}}_{\boldsymbol{JM}}(t), \tag{2.5}$$

where $\dot{\tilde{\boldsymbol{\sigma}}}_{\boldsymbol{JM}}(t) = \frac{d\tilde{\boldsymbol{\sigma}}_{\boldsymbol{JM}}(t)}{dt}$, $t_R$ is equal to $t_R = \frac{\eta_2}{E_2}$ while $E_2$, $\eta_1$, and $\eta_2$ are parameters from Fig. 2.2, $\tau_{R\,JM} = \frac{\eta_1}{E_2}(1 + \frac{\eta_2}{\eta_1})$, $\tilde{\boldsymbol{\varepsilon}}_{\boldsymbol{JM}}(t)$ is the strain, $\dot{\tilde{\boldsymbol{\varepsilon}}}_{\boldsymbol{JM}}(t) = \frac{d\tilde{\boldsymbol{\varepsilon}}_{\boldsymbol{JM}}(t)}{dt}$ is the strain rate, and $\ddot{\tilde{\boldsymbol{\varepsilon}}}_{\boldsymbol{JM}}(t) = \frac{d^2\tilde{\boldsymbol{\varepsilon}}_{\boldsymbol{JM}}(t)}{dt^2}$. In contrary to the Maxwell model, this model has an additional second term on the right-hand side. Stress relaxation ability under constant strain rate $\dot{\tilde{\boldsymbol{\varepsilon}}}_0$ is the same as the one described by the Maxwell model. Strain rate can relax under constant stress condition $\tilde{\boldsymbol{\sigma}}_0$ for the initial condition $\tilde{\boldsymbol{\varepsilon}}_{\boldsymbol{JM}}(t=0) = 0$ such that: $\dot{\tilde{\boldsymbol{\varepsilon}}}_{\boldsymbol{JM}}(t) = \frac{\tilde{\boldsymbol{\sigma}}_0}{\eta_1}(1 - e^{-\frac{t}{t_R}})$. The relaxed (equilibrium) value of the strain rate is equal to $\frac{\dot{\tilde{\boldsymbol{\varepsilon}}}_\infty = \tilde{\boldsymbol{\sigma}}_0}{\eta}$. Strain rate free relaxation for the condition $\tilde{\boldsymbol{\sigma}}_{\boldsymbol{JM}} = 0$ and initial strain rate $\dot{\tilde{\boldsymbol{\varepsilon}}}_0$ is equal to $\dot{\tilde{\boldsymbol{\varepsilon}}}_{\boldsymbol{JM}}(t) = \dot{\tilde{\boldsymbol{\varepsilon}}}_0 e^{-\frac{t}{t_R}}$. Strain rate can relax while the strain cannot relax under constant stress conditions. This is the main characteristic of a liquid. Storage and loss moduli can be expressed from Eq. (2.5) and simplified as:

$$\begin{aligned} G'_{JM}(\omega) &= G'_{MM}(\omega) - \Delta G'_{JM}(\omega) \\ G''_{JM}(\omega) &= G''_{MM}(\omega) + \Delta G''_{JM}(\omega), \end{aligned} \tag{2.6}$$

where $G'_{JM}(\omega)$ is the storage modulus and $G''_{JM}(\omega)$ is the loss modulus for the Jeffreys model, $\Delta G'_{JM}(\omega)$ is the contribution to storage modulus equal to $\Delta G'_{JM}(\omega) = \frac{\eta_1 t_R \omega^2}{1 + \tau^2_{R\,JM}\omega^2}$, and $\Delta G''_{JM}(\omega)$ is the contribution to loss modulus equal to $\Delta G''_{JM}(\omega) = \frac{\eta_1 \tau_{RJM} t_R \omega^3}{1 + \tau^2_{RJM}\omega^2}$. Moduli $G'_{MM}(\omega)$ and $G''_{MM}(\omega)$ are defined by Eq. (2.4). Storage modulus for the Jeffreys model is lower than that expressed for the Maxwell model and functional dependence $G'_{JM}(\omega)$ versus $\omega$ has the similar shape as the one expressed for the Maxwell model. Loss modulus for the Jeffreys model is larger than that expressed for the Maxwell model. The deviation of $G''_{JM}(\omega)$ from $G''_{MM}(\omega)$ is pronounced for lower angular velocities.

## 2.2.2 Various viscoelastic solid states

I discussed the main characteristics of the Hookean elastic solid and compared that with the characteristics of viscoelastic solids. Linear constitutive models for describing rheological behavior of viscoelastic solid considered here are the Kelvin–Voigt model, the Zener model, and the Four-parameter model. The similarities and differences of these models will be discussed in order to point out the general characteristics of viscoelastic solids. A constitutive model for the Hookean elastic solid represents the spring element (Fig. 2.3):

$$\tilde{\sigma}_{ES}(t) = \tilde{\sigma}_S(t) = E\tilde{\varepsilon}_{ES}(t), \tag{2.7}$$

where $\tilde{\sigma}_{ES}(t)$ is the stress, $\tilde{\varepsilon}_{ES}(t)$ is the strain of the elastic solid, and $E$ is the elastic shear modulus. Strain instantaneously increases under constant stress condition. Stress also instantaneously increases under constant strain condition. The relaxation times for stress and for a strain are equal to zero. Corresponding storage and loss moduli are equal to:

$$\begin{aligned} G'_{ES}(\omega) &= E \\ G''_{ES}(\omega) &= 0, \end{aligned} \tag{2.8}$$

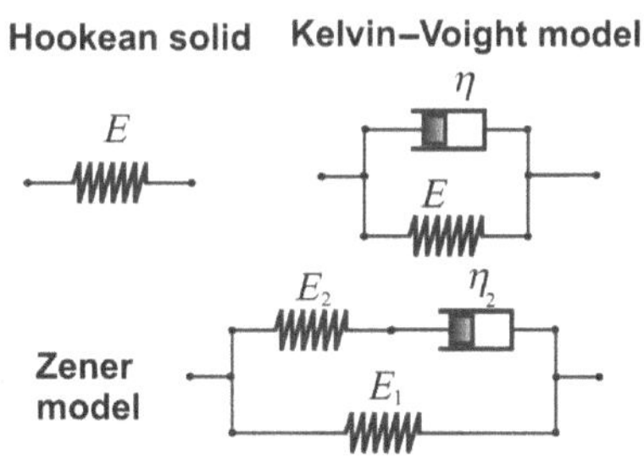


**Figure 2.3** Schematic representations of the Hookean elastic solid model [Eq. (2.7)], the Kelvin–Voigt model [Eq. (2.9)], and the Zener model [Eq. (2.11)].

where $G'_{ES}(\omega)$ is the storage modulus and $G''_{ES}(\omega)$ is the loss modulus for the elastic solid.

The simplest linear model suitable for describing viscoelastic solid behavior is the Kelvin–Voigt model. This model represents a parallel combination of the rheological elements spring and dashpot (Fig. 2.3).

The parallel combination indicates that $\tilde{\varepsilon}_{KVM}(t) = \tilde{\varepsilon}_S(t) + \tilde{\varepsilon}_{DP}(t)$ and $\tilde{\sigma}_{KVM}(t) = \tilde{\sigma}_S(t) = \tilde{\sigma}_{DP}(t)$ (where $\tilde{\sigma}_{KVM}(t)$ is the stress and $\tilde{\varepsilon}_{KVM}(t)$ is the strain for the Kelvin–Voigt model). Consequently, the formulated Kelvin–Voigt model is expressed as:

$$\tilde{\sigma}_{KVM}(t) = E\tilde{\varepsilon}_{KVM}(t) + \eta\dot{\tilde{\varepsilon}}_{KVM}(t), \tag{2.9}$$

where $\tilde{\sigma}_{KVM}(t)$ is the stress, $\tilde{\varepsilon}_{KVM}(t)$ is the strain, and $\dot{\tilde{\varepsilon}}_{KVM}(t) = \frac{d\tilde{\varepsilon}_{KVM}(t)}{dt}$ is strain rate for the Kelvin–Voigt model.

The strain relaxation under constant stress condition $\tilde{\sigma}_0$ (i.e., creep experiments) could be expressed starting from the initial condition $\tilde{\varepsilon}_{KVM}(t=0) = 0$ as: $\tilde{\varepsilon}_{KVM}(t) = \frac{\tilde{\sigma}_0}{E}(1 - e^{-\frac{t}{\tau_{R\,KVM}}})$ (where $\tilde{\varepsilon}_{KVM}(t)$ is the strain and $\tau_{R2}$ is the strain relaxation time equal to $\tau_{R\,KVM} = \frac{\eta}{E}$). Strain increases from $\tilde{\varepsilon}_{KVM}(t=0) = 0$ to $\tilde{\varepsilon}_{KVM}(\Delta t)$. If the system undergoes free relaxation at $t = \Delta t$ such that $\tilde{\sigma}_{KVM} = 0$ and $\tilde{\varepsilon}_{KVM}(\Delta t) = \tilde{\varepsilon}_0$, the strain relaxation is expressed as $\tilde{\varepsilon}_{KVM}(t) = \tilde{\varepsilon}_0 e^{-t/\tau_{R\,KVM}}$. The Kelvin–Voigt model pointed out that stress cannot relax under constant strain conditions. The ability of a strain to relax under constant stress conditions is the main characteristic of viscoelastic solid. Complex modulus $G^*(\omega)$ could be formulated after Fourier transform of Eq. (2.9). Corresponding storage and loss moduli can be expressed after Fourier transform of Eq. (2.9) as:

$$\begin{aligned} G'_{KVM}(\omega) &= G'_{ES}(\omega) = E \\ G''_{KVM}(\omega) &= G''_{NL}(\omega) = \eta\omega, \end{aligned} \tag{2.10}$$

where $G'_{KVM}(\omega)$ is the storage modulus and $G''_{KVM}(\omega)$ is the loss modulus for the Kelvin–Voigt model. Viscoelastic solids described by Kelvin–Voigt model show: (1) solid-like behavior for lower angular velocities $G'_{KVM}(\omega)/G''_{KVM}(\omega) > 1$ and (2) liquid-like behavior $G'_{KVM}(\omega)/G''_{KVM}(\omega) < 1$ for larger angular velocities. The crossover point corresponds to the condition $G'(\omega^*) = G''(\omega^*)$ (where $\omega^* = \frac{2\pi}{\tau^*_{R\,KVM}}$). The crossover time is equal to the relaxation time for strain, that is, $\tau^*_{R\,KVM} = \frac{\eta}{E}$.

The Zener model for viscoelastic solid is more complex than the Kelvin–Voigt model. This model represents the series combination of spring and dashpot combined in parallel with another spring element (Fig. 2.3). It is the simplest model for a viscoelastic solid capable to describe the stress relaxation under constant strain conditions. It is a three-parameter model expressed as:

$$\tilde{\boldsymbol{\sigma}}_{\boldsymbol{ZM}}(t) + \tau_{RZM1}\dot{\tilde{\boldsymbol{\sigma}}}_{\boldsymbol{ZM}}(t) = E_{eff}\tilde{\boldsymbol{\varepsilon}}_{\boldsymbol{ZM}}(t) + \eta_{eff}\dot{\tilde{\boldsymbol{\varepsilon}}}_{\boldsymbol{ZM}}(t), \tag{2.11}$$

where $\tau_{R\,ZM1} = \frac{\eta_2}{E_2}$, $\eta_2$ and $E_2$ are parameters from Fig. 2.3, $E_{eff} \equiv E_1$ (Fig. 2.3), $\eta_{eff} = \eta_2(1 + \frac{E_1}{E_2})$, $\tilde{\boldsymbol{\sigma}}_{\boldsymbol{ZM}}(t)$ is the stress, $\dot{\tilde{\boldsymbol{\sigma}}}_{\boldsymbol{ZM}} = \frac{d\tilde{\boldsymbol{\sigma}}_{\boldsymbol{ZM}}(t)}{dt}$, $\tilde{\boldsymbol{\varepsilon}}_{\boldsymbol{ZM}}(t)$ is the strain, $\frac{\dot{\tilde{\boldsymbol{\varepsilon}}}_{\boldsymbol{ZM}}(t) = d\tilde{\boldsymbol{\varepsilon}}_{\boldsymbol{ZM}}(t)}{dt}$ is the strain rate for the Zener model. The strain relaxation under constant stress condition $\tilde{\boldsymbol{\sigma}}_0$ (i.e., creep experiments) is similar as the one described by the Kelvin–Voigt model and can be expressed starting from the initial condition $\tilde{\boldsymbol{\varepsilon}}_{\boldsymbol{ZM}}(t = 0) = 0$ as: $\tilde{\boldsymbol{\varepsilon}}_{\boldsymbol{ZM}}(t) = \frac{\tilde{\boldsymbol{\sigma}}_0}{G_{seff}}(1 - e^{-\frac{t}{\tau_{RZM2}}})$ (where $\tilde{\boldsymbol{\varepsilon}}_{\boldsymbol{ZM}}(t)$ is the strain and $\tau_{RZM2}$ is the strain relaxation time equal to $\tau_{R\,ZM2} = \frac{\eta_{eff}}{E_{eff}}$). Strain increases from $\tilde{\boldsymbol{\varepsilon}}_{\boldsymbol{ZM}}(t = 0) = 0$ to $\tilde{\boldsymbol{\varepsilon}}_{\boldsymbol{ZM}}(\Delta t)$. If the system undergoes free relaxation at $t = \Delta t$ such that $\tilde{\boldsymbol{\sigma}}_{\boldsymbol{ZM}} = 0$ and $\tilde{\boldsymbol{\varepsilon}}_{\boldsymbol{ZM}}(\Delta t) = \tilde{\boldsymbol{\varepsilon}}_0$, the strain relaxation is expressed as $\tilde{\boldsymbol{\varepsilon}}_{\boldsymbol{ZM}}(t) = \tilde{\boldsymbol{\varepsilon}}_0 e^{-t/\tau_{RZM2}}$. Stress relaxation under constant strain condition $\tilde{\boldsymbol{\varepsilon}}_0$ for the Zener model can be expressed starting from the initial condition $\tilde{\boldsymbol{\sigma}}_{\boldsymbol{ZM}}(t = 0) = \tilde{\boldsymbol{\sigma}}_0$ as $\tilde{\boldsymbol{\sigma}}_{\boldsymbol{ZM}}(t) = \tilde{\boldsymbol{\sigma}}_0 e^{-\frac{t}{\tau_{RZM1}}} + \tilde{\boldsymbol{\sigma}}_{\boldsymbol{R}}(1 - e^{-\frac{t}{\tau_{RZM1}}})$ (where $\tilde{\boldsymbol{\sigma}}_{\boldsymbol{R}}$ is the residual stress equal to $\tilde{\boldsymbol{\sigma}}_{\boldsymbol{R}} = G_s\tilde{\boldsymbol{\varepsilon}}_0$). Consequently, the Zener model introduces one relaxation time for strain and the other for stress.

Complex modulus $G^*(\omega)$ could be formulated after Fourier transform of Eq. (2.11). Corresponding storage and loss moduli could be simplified as:

$$\begin{aligned} G'_{ZM}(\omega) &= G'_{MM}(\omega) + \Delta G'_{ZM}(\omega) \\ G''_{ZM}(\omega) &= G''_{MM}(\omega) - \Delta G''_{ZM}(\omega), \end{aligned} \tag{2.12}$$

where $G'_{ZM}(\omega)$ is the storage modulus and $G''_{ZM}(\omega)$ is the loss modulus for the Zener model, $\Delta G'_{ZM}(\omega)$ is the contribution to storage modulus equal to $\Delta G'_{ZM}(\omega) = \frac{E_{eff}}{1 + \tau^2_{RZM1}\omega^2}$, and $\Delta G''_{ZM}(\omega)$ is the contribution to loss modulus equal to $\Delta G''_{ZM}(\omega) = \frac{E_{eff}\tau_{RZM1}\omega}{1 + \tau^2_{RZM1}\omega^2}$. Storage modulus for the Zener model is larger than the one formulated for the Maxwell model [Eq. (2.4)]. Loss modulus for the Zener model is lower than the one expressed for the Maxwell model. The deviation of $G'_{ZM}(\omega)$ from $G'_{MM}(\omega)$ is pronounced for lower angular velocities, while for higher angular velocities $\Delta G'_{ZM}(\omega) \to 0$. The decrease of the loss modulus

$G''_{ZM}(\omega)$ in comparison with $G''_{MM}(\omega)$ is pronounced for lower angular velocities.

The Four-parameter model for viscoelastic solid is more complex than the Zener and the Kelvin–Voigt models. This model can be formulated by a parallel combination of the Zener element and dashpot element (Fig. 2.4).

It accounts for two relaxation times for strain and one relaxation time for stress. This model is expressed as:

$$\tilde{\boldsymbol{\sigma}}_{\boldsymbol{FPM}}(t) + \tau_{RZM}\dot{\tilde{\boldsymbol{\sigma}}}_{\boldsymbol{FPM}}(t) = E_{eff}\tilde{\boldsymbol{\varepsilon}}_{\boldsymbol{FPM}}(t) + (\eta_1 + \eta_{eff})\dot{\tilde{\boldsymbol{\varepsilon}}}_{\boldsymbol{FPM}}(t) + \eta_1\tau_{RZM}\ddot{\tilde{\boldsymbol{\varepsilon}}}_{\boldsymbol{FPM}}(t), \quad (2.13)$$

where $\eta_{eff}$ and $E_{eff}$ are the effective viscosity and the effective elastic modulus previously defined for the Zener model such that $\eta_{eff} = \eta_{22}(1 + \frac{E_{21}}{E_{22}})$ and $E_{eff} \equiv E_{21}$ (Fig. 2.4), $\eta_1$ is the parameter from Fig. 2.4, $\tilde{\boldsymbol{\sigma}}_{\boldsymbol{FPM}}(t)$ is the stress, $\dot{\tilde{\boldsymbol{\sigma}}}_{\boldsymbol{FPM}} = \frac{d\tilde{\boldsymbol{\sigma}}_{\boldsymbol{FPM}}(t)}{dt}$, $\tilde{\boldsymbol{\varepsilon}}_{\boldsymbol{FPM}}(t)$ is the strain, $\dot{\tilde{\boldsymbol{\varepsilon}}}_{\boldsymbol{FPM}}(t) = \frac{d\tilde{\boldsymbol{\varepsilon}}_{\boldsymbol{FPM}}(t)}{dt}$ is strain rate, $\ddot{\tilde{\boldsymbol{\varepsilon}}}_{\boldsymbol{FPM}}(t) = \frac{d\dot{\tilde{\boldsymbol{\varepsilon}}}_{\boldsymbol{FPM}}(t)}{dt}$, $\tau_{RZM}$ is the relaxation time for stress previously defined for the Zener model. Stress relaxation under constant strain condition $\varepsilon_0$ is similar as the one previously defined for the Zener model and for the initial condition $\tilde{\boldsymbol{\sigma}}_{\boldsymbol{FPM}}(t=0) = \tilde{\boldsymbol{\sigma}}_0$ can be expressed as $\tilde{\boldsymbol{\sigma}}_{\boldsymbol{FPM}}(t) = \tilde{\boldsymbol{\sigma}}_0 e^{-\frac{t}{\tau_{RZM}}} + \tilde{\boldsymbol{\sigma}}_{\boldsymbol{R}}(1 - e^{-\frac{t}{\tau_{RZM}}})$ (where $\tilde{\boldsymbol{\sigma}}_{\boldsymbol{R}}$ is the residual stress equal to $\tilde{\boldsymbol{\sigma}}_{\boldsymbol{R}} = E_{eff}\tilde{\boldsymbol{\varepsilon}}_0$). The strain relaxation under constant stress condition $\tilde{\boldsymbol{\sigma}}_0$ (i.e., creep experiments) can be expressed starting from the initial conditions $\tilde{\boldsymbol{\varepsilon}}_{\boldsymbol{FPM}}(t=0) = 0$ and $\dot{\tilde{\boldsymbol{\varepsilon}}}_{\boldsymbol{FPM}}(t=0) = \dot{\tilde{\boldsymbol{\varepsilon}}}_0$ as: $\tilde{\boldsymbol{\varepsilon}}_{\boldsymbol{FPM}}(t) = C_1 e^{-\frac{t}{\tau^*_{R1}}} + C_2 e^{-\frac{t}{\tau^*_{R2}}} + \frac{\tilde{\boldsymbol{\sigma}}_0}{E_{eff}}$ (where $C_1 = \frac{\dot{\tilde{\boldsymbol{\varepsilon}}}_0 - \frac{\tilde{\boldsymbol{\sigma}}_0}{E_{eff}}\frac{1}{\tau^*_{R1}}}{\frac{1}{\tau^*_{R2}} - \frac{1}{\tau^*_{R1}}}$, $C_2 = \frac{\dot{\tilde{\boldsymbol{\varepsilon}}}_0 - \frac{1}{\tilde{\boldsymbol{\sigma}}_0}E_{eff}\frac{1}{\tau^*_{R2}}}{\frac{1}{\tau^*_{R2}} - \frac{1}{\tau^*_{R1}}}$ are constants, while

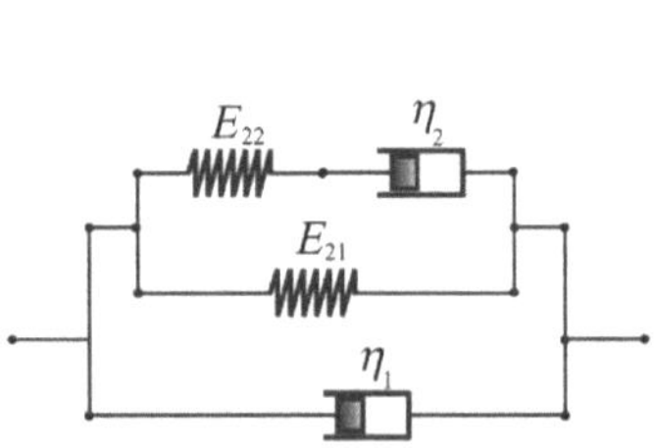


**Figure 2.4** Schematic representations of the Four-parameter model for viscoelastic solid [Eq. (2.13)].

$\tau^*_{R1} = \frac{2\frac{\eta_1}{E_{eff}}\tau_{RZM}}{\frac{\eta_1}{E_{eff}} + \sqrt{\left(\frac{\eta_1}{E_{eff}}\right)^2 - 4\frac{\eta_1}{E_{eff}}\tau_{RZM}}}$, and $\tau^*_{R2} = \frac{2\frac{\eta_1}{E_{eff}}\tau_{RZM}}{\frac{\eta_1}{E_{eff}} - \sqrt{\left(\frac{\eta_1}{E_{eff}}\right)^2 - 4\frac{\eta_1}{E_{eff}}\tau_{RZM}}}$ are the short and long relaxation times for strain.

Complex modulus $G^*(\omega)$ could be formulated after Fourier transform of Eq. (2.13). Corresponding storage and loss moduli could be simplified as:

$$\begin{aligned} G'_{FPM}(\omega) &= G'_{MM}(\omega) + \Delta G'_{ZM}(\omega) - \Delta G'_{JM}(\omega) \\ G''_{FPM}(\omega) &= G''_{MM}(\omega) - \Delta G''_{ZM}(\omega) + \Delta G''_{JM}(\omega), \end{aligned} \tag{2.14}$$

where $G'_{FPM}(\omega)$ is the storage modulus and $G''_{FPM}(\omega)$ is the loss modulus for the four-parameter model. This model shows lower storage modulus and larger loss modulus than the Zener model described by Eq. (2.11). Consequently, the Four-parameter model is suitable for viscoelastic solid with more intensive liquid-like behavior than the Zener model.

## 2.3 Characteristics of the jamming state as the nonlinear viscoelastic solid

Jamming state, similarly as glass state corresponds to a much-damped movement of the system constituents compared with other visco-elastic solid states. This state can be realized under isothermal conditions (the jamming state transition) and by decreasing the temperature (the glass state transition). The jamming state transition has been considered in vibrational flow field for various systems such as: (1) entangled polymer systems and coacervate-based systems [31–32]; (2) granular systems and microgels [33–35]), (3) polymer nanocomposites [36], and (4) multicellular systems [37–40]. The jamming state transition in granular systems is stimulated by increasing the packing density of grans in vibration field, which intensifies grain–grain interconnectivity [35]. The phenomenon is described thermodynamically by introducing the configurational temperature.

Liu et al. [32] considered the rheological response of coacervate-based systems in oscillatory shear field and distinguished four regimes: (1) low-frequency terminal regime described by the Maxwell model, (2) middle-frequency plateau regime described by the Kelvin–Voigt model, (3) higher-frequency transition regime, and (4) high-frequency jamming regime (Fig. 2.5). The terminal regime is characterized by intensive motion of the system constituents and intrachain interactions, which induces significant energy dissipation. In the case, these intrachain interactions have been interpreted as the chain reptation [41]. Reptation is the thermal motion of very long linear, entangled macromolecules in polymer melts or concentrated polymer solutions. Plateau regime corresponds to intensive entropic effects caused by interchain interactions. Transition regime corresponds to reduced mobility of the system constituents and accounts for local conformation changes of the chain parts described by entanglement relaxation time [31]. In the case of polymer nanocomposites, the addition of nanoparticles such as silica or carbon black intensifies systems interconnectivity in low shear oscillator field in the context of chain–particle and particle–particle interactions. The critical concentration of nanoparticles (fillers) induces jamming state transition [36].

The main characteristics of the transition regime are (1) storage modulus is equal to loss modulus, that is, $G'(\omega) = G''(\omega)$ and (2) the scaling exponent for storage and loss moduli versus angular velocity is equal to 1/2 [42,43]. Jamming regime corresponds to totally suppressed migration of the system constituents. The main characteristics of the jamming regime are (1) storage modulus is higher than loss modulus, that is, $\frac{G'(\omega)}{G''(\omega)} = \text{const.} > 1$, (2) the scaling exponent for storage and loss moduli versus angular velocity is higher than zero and lower than 1/2 [19,25]). The jamming state transition induces system stiffening as a consequence of

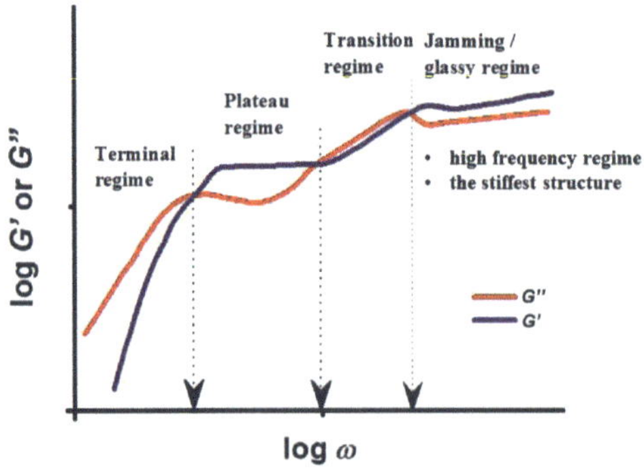


**Figure 2.5** Schematic presentation of storage and loss moduli versus angular velocity for (1) the terminal regime [the Maxwell model, Eqs. (2.2)], (2) the plateau regime [the Kelvin–Voigt model, Eqs. (2.9)], (3) the transition regime [Eq. (2.16) for the damping coefficient equal to $\alpha = 1/2$], and (4) the jamming/glass regime [Eq. (2.17) for the damping coefficient in the range of $0 < \alpha < 1/2$].

the interconnectivity of the constituents (Fig. 2.5) [42, 43]. Characteristics of the transition regime and the jamming/glass regime are: (1) correlation length "symbol from the text" increases rapidly and (2) the relaxation time "symbol from the text" tends to infinity [44]. Correspondingly, elastic modulus is equal to $E = \frac{k_B T}{\xi^3} \to 0$.

The transition regime and the jamming/glass regime (Fig. 2.5) have been described by a spring-pot element [19] (Fig. 2.1). This model accounts for all necessary conditions for the jamming state. Spring-pot element accounts for the anomalous nature of energy dissipation during system structural change expressed by fractional derivatives. The stress—strain relation for spring-pot element is $\tilde{\boldsymbol{\sigma}}_{SP}(t) = \eta_\alpha D^\alpha(\tilde{\varepsilon}_{SP}(t))$ (where $\tilde{\boldsymbol{\sigma}}_{SP}(t)$ is the stress and $\tilde{\varepsilon}_{SP}(t)$ is the strain of the spring-pot, $\eta_\alpha$ is the effective modulus, while $D^\alpha \tilde{\varepsilon}(t) = \frac{d^\alpha \tilde{\varepsilon}(t)}{dt^\alpha}$ is the fractional derivative, and $\alpha$ is the order of fractional derivative, that is, the damping coefficient of system structural changes). Caputo's definition of the fractional derivative of a function $\tilde{\varepsilon}(t)$ was used and it is given as [24]:

$$D^\alpha \tilde{\varepsilon}(t) = \frac{1}{\Gamma(1-\alpha)} \frac{d}{dt} \int_0^t \frac{\tilde{\varepsilon}(t')}{(t-t')^\alpha} dt', \tag{2.15}$$

where $\Gamma(1-\alpha)$ is a gamma function. Viscoelasticity described by the spring-pot element depends on the value of the parameter $\alpha$ as was shown in Fig. 2.6.

If the parameter is $\alpha = 0$, it is obtained that $D^0 \tilde{\varepsilon}(t) = \tilde{\varepsilon}(t)$ while the effective modulus tends to elastic modulus $\eta_\alpha \to E$ (where $E$ is the

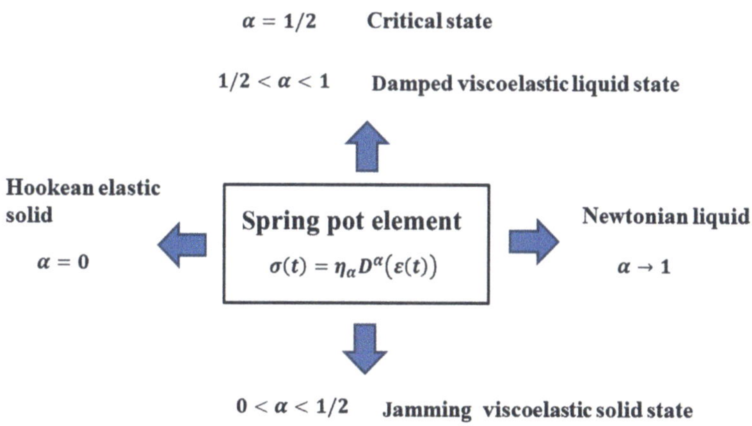


**Figure 2.6** The spring-pot element describes various viscoelastic states depending on the value of the model parameter $\alpha$.

modulus of elasticity). For this condition, the spring-pot element becomes the spring element. Corresponding system structural change is purely reversible, elastic. For $0 < \alpha < 1/2$, storage and loss moduli are $\frac{G'(\omega)}{G''(\omega)} = \text{const.} > 1$. This is the jamming regime, described as the significantly damped, nonlinear viscoelastic solid state. This is the unique viscoelastic solid state which satisfies the condition $E \to 0$ (where $E$ is the elastic modulus). On the contrary, other viscoelastic solid states satisfy the condition $E > 0$ while $E \to 0$ is the characteristic of viscoelastic liquids. For $\alpha = 1/2$, storage and loss moduli are equal $G'(\omega) = G''(\omega)$. This state corresponds to the transition regime. For $1/2 < \alpha < 1$, storage and loss moduli are $\frac{G'(\omega)}{G''(\omega)} = \text{const.} < 1$. This state corresponds to a damped viscoelastic liquid state. When $\alpha \to 1$, the corresponding fractional derivative is equal to $D^1 \tilde{\varepsilon}(t) \frac{= d\tilde{\varepsilon}(t)}{dt}$ and (2) effective modulus tends to viscosity $\eta_\alpha \to \eta$. For this condition, the spring-pot element becomes the dashpot element.

Consequently, the transition regime (Fig. 2.5) is expressed by the constitutive model:

$$\tilde{\boldsymbol{\sigma}}_{TR}(t) = \tilde{\boldsymbol{\sigma}}_{SP}(t) = \eta_\alpha D^{\alpha_{TR}}(\tilde{\boldsymbol{\varepsilon}}_{TR}(t)), \tag{2.16}$$

where $\tilde{\boldsymbol{\sigma}}_{TR}(t)$ is the stress and $\tilde{\boldsymbol{\varepsilon}}_{TR}(t)$ is the strain for the transition regime while $\alpha_{TR}$ is the damping coefficient equal to $\alpha_{TR} = 1/2$. The jamming regime is more damped and can be expressed as:

$$\tilde{\boldsymbol{\sigma}}_{JR}(t) = \tilde{\boldsymbol{\sigma}}_{SP}(t) = \eta_\alpha D^{\alpha_{JR}}\left(\tilde{\boldsymbol{\varepsilon}}_{JR}(t)\right), \tag{2.17}$$

where $\tilde{\boldsymbol{\sigma}}_{JR}(t)$ is the stress and $\tilde{\boldsymbol{\varepsilon}}_{JR}(t)$ is the strain for the jamming regime while $\alpha_{JR}$ is the damping coefficient which satisfies the condition $0 < \alpha_{JR} < 1/2$.

Both regimes described by Eqs. (2.16) and (2.17) satisfy the following conditions: strain change under constant stress condition $\tilde{\boldsymbol{\sigma}}_0$ for the initial condition $\tilde{\boldsymbol{\varepsilon}}_K(0) = 0$ is equal to $\tilde{\boldsymbol{\varepsilon}}_K(t) = \frac{\tilde{\boldsymbol{\sigma}}_0}{\eta_\alpha} \frac{t^\alpha}{\Gamma(\alpha + 1)}$ (where $K \equiv TR, JR$, $\Gamma(\alpha + 1)$ is a gamma function). Strain increases during the time period $\Delta t$ from $\tilde{\boldsymbol{\varepsilon}}_K(0) = 0$ to $\tilde{\boldsymbol{\varepsilon}}_K(\Delta t)$ without the ability to relax (i.e., the relaxation time tends to infinity). If the system undergoes free relaxation at $t = \Delta t$ such that $\tilde{\sigma} = 0$ and $\tilde{\varepsilon}(\Delta t) = \tilde{\varepsilon}_0$, the strain stays constant for $t > \Delta t$ and equal to $\tilde{\boldsymbol{\varepsilon}}_K(t) = \tilde{\boldsymbol{\varepsilon}}_0$. Consequently, a strain cannot relax under constant stress condition similarly as for a viscoelastic liquid described by Eq. (2.1). Eqs. (2.16) and (2.17) are transformed from the time domain into the frequency domain using the Fourier integral transform. Fourier transform of the fractional derivative of the component of strain $\tilde{\varepsilon}(t)$ is equal to $F[D^{\alpha}(\tilde{\varepsilon}(t))] =$ as: $(i\omega)^{\alpha}F[\tilde{\varepsilon}(t)]$. The storage and loss moduli for the transition regime and the jamming regime can be expressed as:

$$G'(\omega) = \eta_{\alpha}\omega^{\alpha}\cos\left(\frac{\pi\alpha}{2}\right)$$
$$G''(\omega) = \eta_{\alpha}\omega^{\alpha}\sin\left(\frac{\pi\alpha}{2}\right). \tag{2.18}$$

Storage and loss moduli do not have a crossover point as was shown in Fig. 2.5. Systems under the transition regime satisfy the condition $G'_{TR}(\omega) = G''_{TR}(\omega)$ obtained for $\alpha = \alpha_{TR} = 1/2$. However, systems under jamming state are more damped and satisfy the condition that $\frac{G'_{JR}(\omega)}{G''_{JR}(\omega)} = \text{const.} > 1$ obtained for $0 < \alpha = \alpha_{JR} < 1/2$. The corresponding functional dependence of storage and loss moduli versus angular velocity described by Eq. (2.18) has been rheologically confirmed for various soft matter systems under oscillator strain conditions.

Jamming state transition has been used in literature to describe migrating-to-resting cell state transition during collective cell migration without proper rheological confirmation [37–40]. This state has been discussed for two cases (1) the increase in cell packing density[45] and (2) the collision of velocity fronts [46]. The increase of cell packing density for healthy tissue is regulated by contact inhibition [47] and is possible up to the physiological level. However, pathological state such as cancer can induce significant increase in cell packing density, which could lead to cell death [48]. The collision of velocity fronts during collective cell migration leads to migrating-to-resting cell state transition as well as the increase of local stress. Generated compressive stress during a collision of velocity fronts can induce softening of cell–cell adhesion contacts [49]. They pointed out that p120-catenin molecules are released into the cytoplasm under stress, destabilizing E-cadherin complexes and leading to intensive endocytosis of the E-cadherin. Softening of adhesion contacts could induce intensive fluctuations of cells around their equilibrium states. On the contrary, the jamming state corresponds to significantly damped cell fluctuations.

Bi et al. [40] considered migration of confluent cell monolayers and pointed out that cell packing density frequently stays approximately constant. They developed the vertex model for describing jamming state transition under simplified conditions such as homogeneous and isotropic multicellular systems. Multicellular systems are inhomogeneous even in 2D [9]. The limitation of this model is that they account for neither internal dissipation (necessary for considering cells as the viscoelastic system) nor anisotropic active stresses in the tissue [47]. Bi et al. [40] proposed the critical cell shape parameter as the indicator of the jamming state transition. This parameter represents the product of various processes such as (1) cell bulk elasticity, (2) cell contractility, and (3) interfacial energy. All processes correspond to the cellular level obtained at a time scale of minutes. Collective cell migration corresponds to the long-time scale, that is, time scale of hours. They considered the jamming state as the elastic solid state. Relaxation time of the elastic solid is $\tau_R \to 0$ while for the jamming state $\tau_R \to \infty$.

The migrating-to-resting cell state transition causes local changes of the state of viscoelasticity of the multicellular system as well as its stiffness, but it is not necessary the jamming state transition. The stiffness of multicellular systems depends on (1) cell state and (2) state of cell—cell adhesion contacts. Migrating cells are much stiffer than resting ones due to an accumulation of contractile energy [50]. Additional rheological experiments are necessary in order to confirm the presence of the jamming state in cellular systems and describe the underlying biological mechanism at the supracellular level.

## 2.4 The main characteristics of various viscoelastic models

Constitutive rheological model of multicellular systems depends on cell type and the intensity and the way of applying stress. Consequently, multicellular systems can behave as a viscoelastic liquid or viscoelastic solid depending on energy dissipation relative to energy storage during their rheological response. I would like to extract the main rheological characteristics of viscoelastic liquids and viscoelastic solids by discussing the main characteristics of various linear constitutive models.

Considered linear constitutive models for viscoelastic liquid are: (1) the Maxwell model and (2) the Jeffreys model. The Maxwell model describes stress relaxation under constant strain rate condition by introducing one relaxation time for stress. Strain cannot relax under constant stress condition. The Jeffreys model also describes stress relaxation under constant strain rate condition by introducing one relaxation time for stress. This model is more complex than the Maxwell model. It describes the relaxation of strain rate under constant stress condition by introducing one relaxation time for the strain rate while strain itself cannot relax.

Considered linear constitutive models for viscoelastic solids are: (1) the Kelvin–Voigt model, (2) the Zener model, and (3) the Four-parameter model. The Kelvin–Voigt model describes strain relaxation under constant stress condition by introducing one relaxation time for the strain. Stress cannot relax under constant strain conditions. The Zener model describes stress relaxation under constant strain conditions and strain relaxation under constant stress conditions by introducing one relaxation time for strain and the other for stress. The Four-parameter model describes stress relaxation under constant strain conditions and strain relaxation under constant stress conditions by introducing one relaxation time for stress and two relaxation times for strain. The proposed models are arranged in the context of their liquid-like and solid-like behaviors quantified by the $G'/G''$ ratio and schematically presented in Fig. 2.7.

This modeling consideration points out that the main rheological characteristics of viscoelastic liquid in general are:

1. strain cannot relax under constant stress conditions (the Maxwell and the Jeffreys models),
2. strain rate cannot relax under constant stress conditions for the Maxwell model or can relax under constant stress conditions for the Jeffreys model,
3. stress can relax under constant strain rate conditions (the Maxwell and the Jeffreys models).
4. stress cannot relax under constant strain condition.

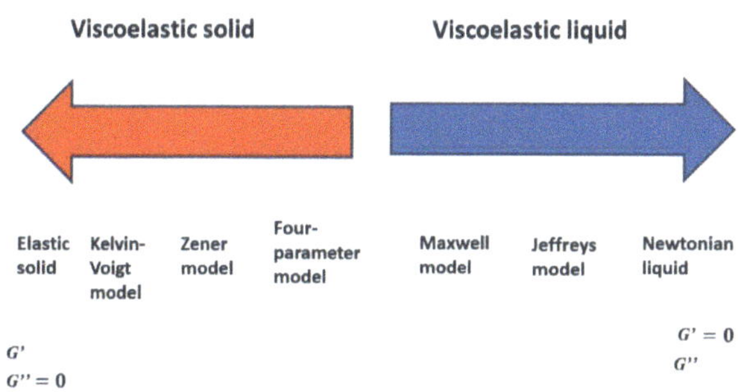


**Figure 2.7** Various viscoelastic solid and liquid models arranged based on the intensity of energy dissipation quantified by loss moduli $G''$.

The main characteristics of viscoelastic solids are:

1. Strain can relax under constant stress conditions (the Kelvin–Voigt model, the Zener model, the Four-parameter model);
2. Stress cannot relax under constant strain condition only for the Kelvin–Voigt model while for other model stress can relax.

## 2.5 Relaxation of multicellular systems under externally applied stress conditions

Spherical cellular aggregates are widely used as in vitro model systems to study the mechanical properties of tissues. Viscoelasticity of cellular aggregates and tumor spheroids is considered and compared under various stress conditions. Stress has been applied under (1) uniaxial compression between parallel plates [5,8], (2) hyperosmotic stress [25–27], (3) compression caused by hydrogel matrix rheological response during cell expansion [28,29,51], and (4) micropipette aspiration [30]. These systems were considered in the context of aggregate stress or strain relaxation ability obtained at various time scales.

### 2.5.1 Aggregate rounding after uniaxial compression

Stress relaxation under constant aggregate shape condition, cell aggregate shape relaxation under constant stress condition, or free relaxation after uniaxial compression have been considered by Mombach et al. [8] and Marmottant et al. [5]. The volume of single cells stays approximately constant during uniaxial compression and free relaxation. Marmottant et al. [5] proposed a single relaxation time for stress under constant aggregate shape condition equal to a few minutes. Stress relaxes exponentially from $\sim 27$ Pa to the residual stress value equal to $\sim 17$ Pa during 25 minutes. The phenomenon of stress relaxation has been related primarily to the adaptation of adhesion contacts [39]. Marmottant et al. [5] also considered the aggregates free relaxation (its rounding) after uniaxial compression driven by aggregate surface tension during 1 hour. They reported that the aggregate shape first relaxes quickly (the relaxation time corresponds to a few minutes) and then more slowly (the relaxation time corresponds to tens of minutes). Consequently, the shape relaxation occurs via two mechanisms related to (1) structural changes of adhesion complexes (which lead to the cell packing density relaxation) obtained at time scale of minutes and (2) collective cell migration obtained at the time scale of hours [10,52]. Mombach et al. [8] considered the free shape relaxation of 3D aggregates during 20 h. They pointed out to one average relaxation time that corresponds to several tens of minutes.

Aggregate shape and surface (i.e., surface strain) relaxation under compression and free relaxation after compression are characteristics of the viscoelastic solid rather than viscoelastic liquid. The main characteristic of viscoelastic liquid is that strain cannot relax under constant stress condition [1,53]. On the contrary, Flenner et al. [54] treated cell aggregate rounding as a viscoelastic liquid. They introduced two interconnected arguments: (1) cell aggregate rounding is driven by surface tension and (2) the surface tension represents the characteristic of a liquid. I agree that aggregate rounding is driven by tissue surface tension. However, surface tension is not necessarily the characteristic of liquid. Amorphous viscoelastic solids such as polymer hydrogels and foams also have surface tension [21,55].

The Four-parameter model [Eq. (2.12)] suitable for viscoelastic solid can be used for describing viscoelasticity of multicellular aggregate under compression between parallel plates and free relaxation after compression. This model introduces one relaxation time for stress and two relaxation times for the surface strain and/or aggregate shape. This model corresponds to experimental data reported by Marmottant et al. [5]. On the other hand, Mombach et al. [8] observed single relaxation time for the aggregate shape. The corresponding model can be the Zener model [Eq. (2.10)]. This model introduces one relaxation time for stress and one for the aggregate shape.

Pajic-Lijakovic and Milivojevic [53] reported that cell aggregate shape changes take place during successive relaxation cycles with various relaxation rates per cycle. A number of relaxation cycles vary from 7 to 15 from the experimental data by Mombach et al. [8]. Every relaxation rate $\kappa$ could be related to a corresponding configuration and volume fraction of migrating cells [53]. Configuration changes have feedback impact on dynamic surface tension and long-time viscoelasticity [19]. Experimental data by Mombach et al. [8] pointed out that the relaxation rates are not random but gather around two or three values indicating an organized cell migrating pattern. Three scenarios were considered: (1) $k_m$—most of the cells migrate, (2) $k_r$—some cell groups migrate while the others, at the same time, stay in resting state as reported by Pajic-Lijakovic and

Milivojevic [18,19,53], and (3) $k_r \sim 0$—most of the cells stay in the resting (passive) state.

**Case 1.** corresponds to an ordered trend of cell migration characterized by a minimum of biochemical and mechanical energy perturbation. The relaxation time for the aggregate shape is finite and equal $\tau_R = k_m^{-1}$ which indicates viscoelastic solid.

**Case 2.** corresponds to a disordered trend of cell migration characterized by a medium value of the energy perturbation. The relaxation time for the aggregate shape is finite and equal $\tau_R = k_t^{-1}$ which also indicates viscoelastic solid.

**Case 3.** corresponds to significantly suppressed cell migration caused by large energy perturbation. The relaxation time is $\tau_R = k_r^{-1} \to \infty$ while the aggregate shape parameter is $\varepsilon_d(t) = AR(t) - 1 \approx$ const. (where $AR(t)$ is the aggregate aspect ratio). If case 3 corresponds to viscoelastic solid, this rheological behavior could be treated as the jamming state described by Eq. (2.14) [19,53]. This result is not exact. Viscoelastic liquid described by the Maxwell model [Eq. (2.2)] also satisfies the conditions that: for $\sigma = 0$, the surface strain (or the shape deformation parameter) is constant, that is, $\varepsilon_d(t) \approx$ const., but it is not the jamming state. Consequently, confirmation of the jamming state in these systems needs additional rheological experiments.

## 2.5.2 Cell aggregate under hyperosmotic stress

Leroux et al. [56] considered cell aggregate volumetric strain relaxation under hyperosmotic stress in the range of 5−15 kPa for 10 minutes. Applied hyperosmotic stress induces (1) aggregate compression and (2) decrease of single-cell volumes, which leads to the accumulation of mechanical stress. Dolega et al. [26] reported that osmotic stress of 5 kPa induces mechanical stress at aggregate surface equal to ~1.5 kPa while the residual mechanical stress in the core region of the spheroid is three to four times higher. This stress induces isotopic compression of cells that leads to the decrease of single-cell volumes at a time scale of minutes [25]. At the end of the experiment ($t$ = 10 minutes), the volumetric compression of cells in the outer region is ~5%. The aggregate volume and volumetric strain relaxation time correspond to a few minutes and increases with stress. This time scale corresponds to the relaxation of single-cell volumes and the state of cell–cell adhesion contacts under stress. Delarue et al. [25] measured the cell–cell distances and pointed out their dependence on the local cell density and applied volumetric compression of the spheroid. Interestingly, the osmotic stress of 5 kPa significantly reduced the growth of mouse CT26 cell lines tumor spheroid at the time scale of days [27].

The volumetric strain relaxation under constant stress condition points to viscoelastic solid. The model equations suitable for viscoelastic solid that introduce single relaxation time for the volumetric strain relaxation are the Kelvin–Voigt model [Eq. (2.8)] and the Zener model [Eq. (2.10)].

### 2.5.3 Compression caused by hydrogel matrix rheological response

Microenvironmental restrictions around the cell aggregate include the following steps [51]: (1) entrapped aggregate induces radial strain of the surrounding hydrogel matrix during the cell expansion [57], (2) radial strain induces generation of the matrix resistance stress within the aggregate surrounding [58,59], (3) the generated matrix stress around the aggregate induces feedback generation of internal stress within the cell aggregate ([28,58–60]; 2017), and (4) the internal stress provokes a biological response of cells [28,29,60,61]. Cell rearrangement and growth and the resulted matrix anisotropic structural changes occur during multiscale relaxation cycles [58]. The cell–matrix mechanical interactions during cell migration and growth lead to an accumulation of residual stress within the cell aggregate. Consequently, the compressive strain and the resulted residual stress within an entrapped aggregate increase as a product of both, viscoelasticity of cell aggregate and viscoelasticity of the extracellular matrix.

### 2.5.4 Cell aggregate micropipette aspiration

Instead of cell aggregate uniaxial compression between parallel plates, aggregate micropipette aspiration is another widely used experimental system for considering a collective cell migration and its influence to viscoelasticity at supracellular level [30]. A part of surface cells is mechanically perturbed under external pressure and inserted into the pipette. The pipette length filled by cells depends on the magnitude of applied pressure. The lowest pressure of 0.5 kPa induces activation of perturbed cells and their entrance into the pipette via collective cell migration during 120 minutes. However, under higher pressure, one part of perturbed cells is instantaneously inserted into the pipette. These cells undergo collective cell migration within the pipette while the other part of already perturbed cells enters the pipette.

Guevorkian et al. [30] proposed the Maxwell model [Eq. (2.2)] for describing long-time cell flow inside the pipette under pressure at the time scale of several tens of minutes. These experimental conditions induced intensive energy dissipation during cell long-time rearrangement. The Maxwell model suitable for viscoelastic liquid points out that stress relaxes under constant strain rate conditions, while strain itself cannot relax under constant stress conditions. Consequently, this constitutive model represents a good choice as long as the equilibrium strain (which corresponds to the equilibrium pipette length filled by cells) is not observed. This condition was satisfied for higher pressure conditions, that is, $>0.5$ kPa during the experimental time period of 120 min.

## 2.6 Conclusions

Viscoelasticity of multicellular systems (such as cell aggregates or tumor spheroids) is related to their ability to relax under various stress and strain conditions, depending on (1) cell type and (2) intensity and way of applied stress. Stress has been applied in various ways, that is, under (1) uniaxial compression between parallel plates, (2) hyperosmotic stress, (3) compression caused by hydrogel matrix rheological response induced by cell expansion, and (4) micropipette aspiration. The corresponding relaxation of cellular systems occurs via two mechanisms. Short-time relaxation mechanism (the time scale of minutes) accounts for relaxation of (1) cell–cell adhesion contacts and (2) cell volumes (for the case of hyperosmotic stress). Long-time relaxation mechanism (the time scale of hours) accounts for collective cell migration. Type of viscoelasticity depends on energy dissipation relative to energy storage during cell rearrangement under various stress and strain conditions. Intensive energy dissipation points to viscoelastic liquid. This type of viscoelasticity is confirmed during cell aggregate micropipette aspiration under pressure higher than

0.5 kPa. The main characteristic of viscoelastic liquid is that strain cannot relax under constant stress condition. However, surface strain relaxation obtained during cell aggregate rounding after uniaxial compression and volumetric strain relaxation under constant hyperosmotic stress pointed to viscoelastic solid rather than viscoelastic liquid. The relaxation of entrapped cell aggregate within the hydrogel matrix is much complex. It accounts for permanent multitime relaxations of both viscoelastic systems, that is, the aggregate and surrounding matrix obtained at various time scales.

Additional rheological experiments are necessary in order to identify the type of viscoelasticity for particular loading conditions and to confirm the presence of jamming state transition caused by collective cell migration. Rheological analysis in the frequency domain is more suitable than that in the time domain for considering the relaxation phenomena.

## Acknowledgment

This work was supported by the Ministry of Education, Science and Technological Development of the Republic of Serbia (Contract No. 451-03-68/2020-14/200135).

## References

[1] I. Pajic-Lijakovic, M. Milivojevic, Viscoelasticity of multicellular surfaces, J. Biomech 60 (2017) 1–8.
[2] B. Fabry, G.N. Maksym, J.P. Butler, M. Glogauer, D. Navajas, J.J. Fredberg, Scaling the microrheology of living cells, Phys. Rev. Lett. 87 (14) (2001). 148102-1-4.
[3] J. Alcaraz, L. Buscemi, M. Grabulosa, X. Trepat, B. Fabry, R. Farre', et al., Microrheology of human lung epithelial cells measured by atomic force microscopy, Biophys. J 84 (2003) 2071–2079.
[4] I. Pajic-Lijakovic, M. Milivojevic, Actin cortex rearrangement caused by coupling with the lipid bilayer-modeling considerations, J. Membr. Biol. 248 (2) (2015) 337–347.
[5] P. Marmottant, A. Mgharbel, J. Kafer, B. Audren, J.P. Rieu, J.C. Vial, et al., The role of fluctuations and stress on the effective viscosity of cell aggregates, Proc. Natl. Acad. Sci. U. S. A. 106 (2009) 17271–17275.
[6] O. Thoumine, A. Ott, Time scale dependent viscoelastic and contractile regimes in fibroblasts probed by microplate manipulation, J. Cell Sci. 110 (1997) 2109–2116.
[7] F. Wottawah, S. Schinkinger, B. Lincoln, R. Ananthakrishnan, M. Romeyke, J. Guck, et al., Optical rheology of biological cells, Phys. Rev. Lett. 94 (2005) 098103.
[8] J.C.M. Mombach, D. Robert, F. Graner, G. Gillet, G.L. Thomas, M. Idiart, et al., Rounding of aggregates of biological cells: experiments and simulations, Phys. A 352 (2005) 525–534.
[9] J. Notbohm, S. Banerjee, K.J.C. Utuje, B. Gweon, H. Jang, Y. Park, et al., Cellular contraction and polarization drive collective cellular motion, Biophys. J. 110 (2016) (2016) 2729–2738.
[10] J.P. Rieu, A. Upadhyaya, J.A. Glazier, N.B. Ouchi, Y. Sawada, Diffusion and deformations of single hydra cells in cellular aggregates, Biophys. J. 79 (2000) 1903–1914.

[11] G. Petrungaro, L. Morelli, K. Uriu, Information flow in the presence of cell mixing and signalling delays during embryonic development, Sem. Cell Dev. Biol. 93 (2019) 26–35.
[12] A. Zemel, F. Rehfeldt, A.C.X. Brown, D.E. Disher, S.A. Safran, Optimal matrix rigidity for stress fiber polarization in stem cells, Nat. Phys 6 (6) (2010) 468–473.
[13] P. Lee, C.W. Wolgemuth, Wounds without purse tiring or signalling, PLoS Comp. Biol. 7 (3) (2011) e1002007.
[14] C. Mc Cann, P.W. Kriebel, C.A. Parent, W. Losert, Cell speed, persistence and information transmission during signal relay and collective migration, J. Cell Sci. 123 (10) (2010) 1724–1731.
[15] I. Pajic-Lijakovic, M. Milivojevic, Long-time viscoelasticity of multicellular surfaces caused by collective cell migration—multi-scale modeling considerations, Sem. Cell Dev. Biol. 93 (2019) 87–96.
[16] N. Özkaya, M. Nordin, Mechanical properties of biological tissues, in: N. Özkaya, M. Nordin, D. Goldsheyder, D. Leger (Eds.), Fundamentals of Biomechanics, Equilibrium, Motion and Deformation, Springer, 2012, pp. 195–218. ISBN 978-1-1-4614-1149-9.
[17] E.H. Barriga, K. Franze, G. Charras, R. Mayor, Tissue stiffening coordinates morphogenesis by triggering collective cell migration in vivo, Nature 554 (2018) 523–527.
[18] I. Pajic-Lijakovic, M. Milivojevic, Functional epithelium remodeling in response to applied stress under in vitro conditions, Appl. Bionics Biomech 2019 (2019) 4892709. Available from: https://doi.org/10.1155/2019/4892709.
[19] I. Pajic-Lijakovic, M. Milivojevic, Jamming state transition and collective cell migration, J. Biol. Eng. 13 (73) (2019) 1–12.
[20] E.H. Barriga, R. Mayor, Adjustable viscoelasticity allows for efficient collective cell migration, Sem. Cell Dev. Biol. 93 (2019) 55–68.
[21] G.B. Blanchard, A.G. Fletcher, L.J. Schumacher, The devil is in the mesoscale: mechanical and behavioural heterogeneity in collective cell movement, Sem. Cell Dev. Biol. 93 (2019) 46–54.
[22] R.M. Christensen, Theory of Viscoelasticity, Academic Press Inc, London, 1981. ISBN 0-12-174252-0.
[23] G.I. Hunter, E.R. Weeks, The physics of the colloidal glass transition, Rep. Prog. Phys. 75 (2012) 066501.
[24] I. Podlubny, Fractional Differential Equations, Mathematics in Science and Engineering, 198, London Academic Press, 1999, p. 78.
[25] M. Delarue, F. Montel, D. Vignjevic, J. Prost, J.F. Joanny, G. Cappello, Compressive stress inhibits proliferation in tumor spheroids through a volume limitation, Biophys. J. 107 (2014) 1821–1828.
[26] M.E. Dolega, M. Delarue, F. Ingremeau, J. Prost, A. Delon, G. Cappello, Cell-like pressure sensors reveal increase of mechanical stress towards the core of multicellular spheroids under compression, Nat. Commun 8 (2017) 14056.
[27] F. Montel, M. Delarue, J. Elgeti, D. Vignjevic, G. Cappello, J. Prost, Isotropic stress reduces cell proliferation in tumor spheroids, New J. Phys. 14 (2012) 055008.
[28] Q. Cheng, J. Tse, R.K. Jain, L.L. Munn, Micro-environmental mechanical stress controls tumor spheroid size and morphology by suppressing proliferation and inducing apoptosis in cancer cells, PLoS One 4 (2) (2009) e4632.
[29] Q. Chen, D. Yang, H. Zong, L. Zhu, L. Wang, X. Wang, et al., Growth-induced stress enhances epithelial-mesenchymal transition induced by IL-6 in clear cell renal cell carcinoma via the Akt/GSK-3β/β-catenin signalling pathway, Oncogenesis 6 (2017) e375.

[30] K. Guevorkian, D. Gonzalez-Rodriguez, C. Carlier, S. Dufour, F. Brochard-Wyart, Mechanosensitive shivering of model tissues under controlled aspiration, Proc. Natl. Acad. Sci. U. S. A. 108 (33) (2011) 13387–13392.
[31] C. Liu, J. He, E. van Ruymbeke, R. Keunings, C. Bailly, Evaluation of different methods for the determination of the plateau modulus and the entanglement molecular weight, Polymer 47 (2006) 4461–4479.
[32] Y. Liu, H.H. Winter, S.L. Perry, Linear viscoelasticity of complex coacervates, Adv. Coll. Int. Sci. 239 (2017) 46–60.
[33] S. Edwards, D.V. Grinev, Statistical mechanics of vibration-induced compaction of powders, Phys. Rev. E. 58 (4) (1988) 4758–4762.
[34] S.F. Edwards, The full canonical ensemble of a granular system, Phys. A 353 (2005) 114–118.
[35] M.P. Ciamarra, A. Coniglio, M. Nicodemi, Thermodynamics and statistical mechanics of dense granular media, Phys. Rev. Lett. 97 (2006) 158001.
[36] H.M. Hassanabadi, M. Wilhelm, D. Rodrigue, A rheological criterion to determine the percolation threshold in polymer nano-composites, Rheol. Acta 53 (2014) 869–882.
[37] L. Atia, D. Bi, Y. Sharma, J.A. Mitchel, B. Gweon, S. Koehler, et al., Universal geometric constraints during epithelial jamming, Nat. Phys (2018). Available from: https://doi.org/10.1038/s41567-018-0089-9.
[38] L. Oswald, S. Grosser, D.M. Smith, J.A. Kas, Jamming transition in cancer, J. Phys. D: Appl. Phys. 50 (2017) 483001.
[39] A.F. Pegoraro, J.J. Freddberg, J.A. Parkm, Problems in biology with many scales of length: cell-cell adhesion and cell jamming in collective cell migration, Exp. Cell Res. 343 (2016) 54–59.
[40] D. Bi, X. Yang, M.C. Marchetti, L.M. Manning, Motility driven glass and jamming transitions in biological tissues, Phys. Rev. X 6 (021011) (2015).
[41] P.G. De Gennes, Entangled polymers, Phys. Today 36 (6) (1983) 33.
[42] B.P. Tighe, Relaxations and rheology near jamming, Phys Rev Lett 107 (15) (2011) 1–4. 158303.
[43] S. Sastry, Critically jammed, Proc. Natl. Acad. Sci. U.S. A. 113 (35) (2016) 9673–9675.
[44] K. Baumgarten, B.P. Tighe, Viscous forces and bulk viscoelasticity near jamming, Soft Matter 13 (2017) 8368–8378.
[45] G. Duclos, S. Garcia, H.G. Yevick, P. Silberzan, Perfect nematic order in confined monolayers of spindle-shaped cells, Soft Matter 10 (14) (2014) 2346–2353.
[46] K.D. Nnetu, M. Knorr, J. Kas, M. Zink, The impact of jamming on boundaries of collectively moving weak-interacting cells, New J. Phys. 14 (2012) 115012.
[47] R. Alert, X. Trepat, Physical models of collective cell migration, Annu. Rev. Condens. Matter Phys. 11 (2020) 77–101.
[48] R. Levayer, Solid stress, competition for space and cancer: the opposing roles of mechanical cell competition in tumour initiation and growth, Semin. Cancer Biol. 63 (2019) 69–80. Available from: https://doi.org/10.1016/j.semcancer.2019.05.004.
[49] K.V. Iyer, R. Piscitello-Gómez, F. Jülicher, S. Eaton, Epithelial viscoelasticity is regulated by mechanosensitive E-cadherin turnover, Curr. Biol. 29 (2019) 1–14.
[50] J.R. Lange, B. Fabry, Cell and tissue mechanics in cell migration, Exp. Cell. Res. 319 (2013) 2418–2423.
[51] I. Pajic-Lijakovic, M. Milivojevic, S. Levic, K. Trifkovic, Z. Stevanovic-Dajic, R. Radosevic, et al., Matrix resistance stress: a key parameter for immobilized cell growth regulation, Proc. Biochem. 52 (2017) 30–43.
[52] J.C.M. Mombash, J.A. Glazier, Single cell motion in aggregates of embryonic cells, Phys. Rev. Lett. 76 (16) (1996) 3032–3035.

[53] I. Pajic-Lijakovic, M. Milivojevic, Successive relaxation cycles during long-time cell aggregate rounding after uni-axial compression, J. Biol. Phys. 43 (2017) 197–209.
[54] E. Flenner, L. Janosi, B. Barz, A. Neagu, G. Forgacs, I. Kosztin, Kinetic Monte Carlo and cellular particle dynamics simulations of multicellular systems, Phys. Rev. E 85 (2012) 031907.
[55] S. Mondal, M. Phukan, A. Ghatak, Estimation of solid–liquid interfacial tension using curved surface of a soft solid, PNAS 112 (41) (2015) 12563–12568.
[56] C.E. Leroux, J. Palmier, A.C. Boccara, G. Cappello, S. Monnier, Elastography of multicellular aggregates submitted to osmomechanical stress, New J. Phys. 17 (073035) (2015).
[57] A. Leal-Egana, U. Dietrich-Braumann, A. Diaz-Cuenca, M. Nowicki, A. Bader, Determination of pore size distribution at the cell-hydrogel interface, J. Nanobiotechnol. Open Access 9 (24) (2011) 1–7.
[58] I. Pajic-Lijakovic, S. Levic, V. Nedovic, B. Bugarski, Biointerface dynamics-multi scale modeling considerations, Col. Surf. B: Biointerf. 132 (2015) 236–245.
[59] I. Pajic-Lijakovic, Micro-environmentally restricted cell growth dynamics—modeling considerations, Critic. Rev. Biotechnol. 35 (3) (2015) 369–391.
[60] M. Kalli, T. Stylianopoulos, Defining the role of solid stress and matrix stiffness in cancer cell proliferation and metastasis, Front. Oncol. 8 (2018) 55. Available from: https://doi.org/10.3389/fonc.2018.00055.
[61] J.M. Tse, G. Cheng, J.A. Tyrrell, S.A. Wilcox-Adelman, Y. Boucher, R.K. Jain, et al., Mechanical compression drives cancer cells toward invasive phenotype, Proc. Natl. Acad. Sci. U. S. A. 109 (2012) 911–916.

CHAPTER THREE

# Biophysical origins of viscoelasticity during collective cell migration

**Andrew G. Clark**
Institut Curie, PSL Research University, CNRS, UMR 144 - Cell Biology and Cancer, Paris, France

## 3.1 Introduction

Cell migration is an inherently mechanical process driven by forces generated within the cell which are then transmitted to the substrate. How these mechanical forces drive cell translocation depends on the material properties of the cell and its substrate. The mechanics of the cell surface, where force transmission during migration takes place, are controlled largely by the actin cytoskeleton. The architecture and mechanical properties of actin networks can differ depending on the cell type, the mechanics of the substrate, and the mode of migration employed by cells. Mechanical forces generated through myosin motor activity on actin filaments are transmitted via receptors that span the plasma membrane and connect actin filaments to the substrate. In the case of collective migration, cell–cell adhesion proteins coordinate interactions between cells. Each of these individual functions of collective migration is mediated by different sets of proteins that act on different time- and lengthscales. This can give rise to viscoelastic behavior in cells and tissues that directly impacts migration dynamics.

In this chapter, I begin with basic principles of viscoelastic behavior and how viscoelastic properties can be measured in biology. I then describe the different biological components required for migration, including the actin cytoskeleton, cell–substrate adhesions, the extracellular matrix (ECM) and cell–cell adhesions, and how these systems behave viscoelastically. I finish by moving up in scale and discussing viscoelastic behavior in tissues and how viscoelastic properties can influence collective migration dynamics.

*Viscoelasticity and Collective Cell Migration.*
DOI: https://doi.org/10.1016/B978-0-12-820310-1.00007-0

## 3.2 Timescale-dependent behavior in viscoelastic materials

Materials that display characteristics of both elastic solids and viscous liquids are said to be viscoelastic. Elastic solids store energy when they are deformed by an external stress and return to their initial state once the stress is removed. If you were to press your finger on a piece of soft rubber (which is elastic), the rubber would deform, and you would feel a force pushing up on your finger until you remove it from the rubber. Viscous liquids, on the other hand, deform while the stress is being applied, but once the force is removed, they will not return to their initial state. If you were to plunge your finger into a pot of honey, the honey would simply flow around your finger and you would feel no restoring force pushing up on your finger.

Viscoelastic materials respond to force differently at different timescales. For example, a viscoelastic material may act like an elastic solid at short timescales and like a viscous liquid on long timescales. If you were to press your finger onto this viscoelastic material, you would initially feel a resisting force on your finger, but over time, this resisting force would decrease. When an external force is removed from a viscoelastic material, it may regain some, but not all, of its initial shape. While many biological systems exhibit viscoelastic behavior, viscoelasticity can also be observed on more everyday lengthscales.

One classical example of everyday viscoelasticity is the children's toy Silly Putty (Crayola LLC). Similar materials are also used for hand strength training or hand injury rehabilitation. These materials will bounce like a rubber ball when dropped onto a hard surface but will flow through a hole if left for longer times (Fig. 3.1A and B). Another common example of a viscoelastic material is a corn starch and water suspension, also called "Oobleck" after a Dr. Seuss book. Similar to Silly Putty, Oobleck behaves like a solid on subsecond timescales, resisting deformation and fracturing under high stress, but behaves like a liquid on second timescales and with much lower viscosity than Silly Putty (Fig. 3.1C). This makes the transition from solid to liquid behavior much more dramatic for Oobleck. Playing with such substances at home with one's own hands can be very helpful to understand how viscoelastic materials behave on different timescales and can lead to a better conception of viscoelasticity in biological materials.

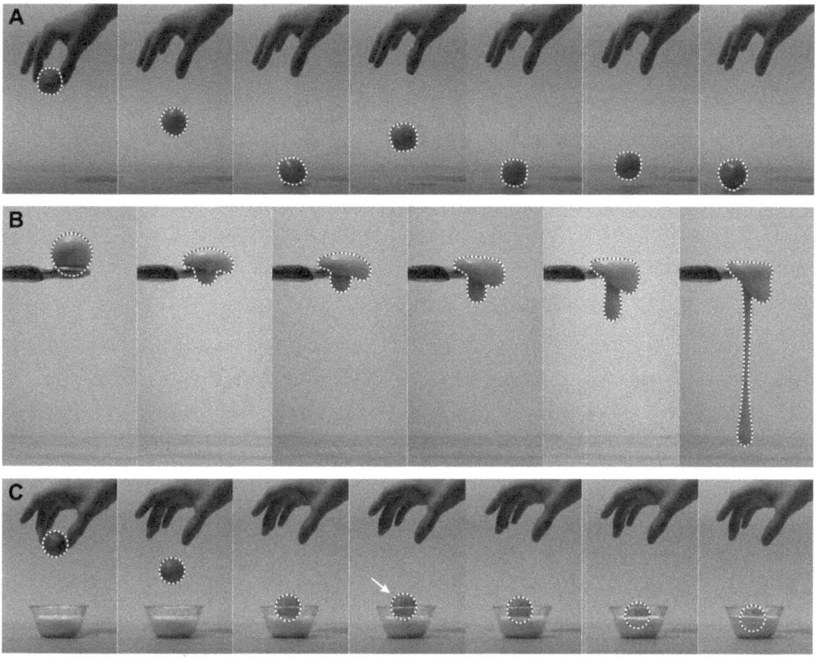


**Figure 3.1** Viscoelastic behavior in everyday materials. (A) When a putty for hand strength straining (similar to Silly Putty) is dropped onto a hard surface, it bounces several times before coming to rest, much like an elastic rubber ball. The putty therefore displays elastic behavior on subsecond timescales. Total time of montage: 0.8 seconds. (B) When the same putty is placed on a metal ring, it drips like honey after several minutes. The putty therefore behaves as a viscous liquid on the timescale of minutes. Total time of montage: 10 minutes. For panels *A* and *B*, the shape of the putty is outlined in black and white dotted lines. (C) A rubber ball is dropped onto a small bowl of Oobleck. Initially, the ball bounces on the surface of the Oobleck (white arrow) as if it were dropped onto a solid elastic surface. Once the rubber ball is at rest, it sinks into the Oobleck, as the Oobleck behaves as a viscous liquid. Total time of montage: 2 seconds. The outline of rubber ball is traced in black and white dotted lines.

## 3.3 Measuring viscoelastic behavior in biology

Cellular viscoelastic behavior can be measured by mechanically perturbing cells and assessing how they respond at different timescales. Most biological materials exhibit an elastic (solid-like) response at fast timescales and a viscous (liquid-like) response on longer timescales [1]. A number of tools exist to deform cells and assess their response, and each of these tools is tailored to measure viscoelastic behavior at different scales, from subcellular components to multicellular tissues (Fig. 3.2). Recently, the development of several less- or noninvasive techniques has provided new insight into viscoelastic behavior, especially in tissues in vivo, where perturbation-based techniques are often severely limited.

## 3.3.1 Experimental methods to measure biological viscoelasticity

### *Subcellular scale measurement tools*

A number of different tools exist for measuring subcellular viscoelastic properties. Atomic force microscopy (AFM) and active bead rheology, for example, can be used to measure the mechanics of submicron scale structures. Recent studies using AFM, in which a flexible cantilever is used to push down on the surface of the cell, have reported stiffness measurements across regions of the cell surface ranging from 10 to 100 kPa, indicating appreciable mechanical heterogeneities within single cells [2]. In active bead rheology, functionalized beads are attached to a biological material and a twisting or pulling force is applied via an optical or magnetic tweezers. This technique has also revealed heterogeneities in viscoelastic properties of the cell surface [3] and has been used to measure the mechanical properties of cellular cytoskeletal networks [4] and cell–cell junctions [5]. Microbead rheology can also be performed in a passive (noninvasive) manner. By embedding beads within a biological material such as a

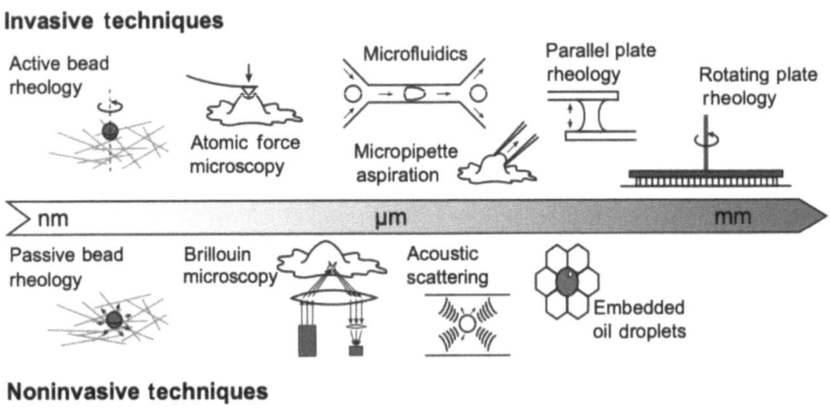


**Figure 3.2** Tools to measure biological viscoelasticity at different scales. Different invasive (top) and noninvasive (bottom) tools can be used to measure viscoelastic properties at the scale of subcellular components, single cells, and multicellular tissues.

cytoskeletal network, fluctuations in the position of the bead over time can be quantified and fit to a viscoelastic model to extract mechanical properties without applying an external force on the bead [6].

### *Whole-cell measurement tools*

Methods such as parallel plate rheometry and optical stretching, where whole cells are stretched using mechanical plates or using a focused dual laser beam, respectively, have been used to measure bulk viscoelastic properties of single cells [7,8]. Aspiration of cells into micropipettes can also be used to measure whole-cell viscoelastic properties [9]. Microfluidics-based approaches, where cells are deformed while being forced through micron-sized channels, offer possibilities for high-throughput measurements of whole-cell mechanical properties at rates greater than 100 cells/seconds [10]. In these cell-scale measurements, cell shape deformations are tracked over time. For such measurements, simple transmitted light images are typically sufficient to measure cell shape and thereby extract mechanical properties.

### *Tissue-scale measurement tools*

To measure viscoelastic properties in groups of cells and tissues, rotating plate rheology is commonly used. In this assay, a layer of cells or a tissue is placed within a circular chamber, and a rotating piston on top of the chamber applies a sheer force to the cells or tissue. Rotating plate rheology has been applied to measure viscoelastic properties of large groups of single (non-attached) cells [11] and more recently whole tissue samples [12]. It should be noted that rheological measurements of whole tissues are likely to reflect a combination of the viscoelastic properties of stromal ECM as well as the cells in the tissue (see Section 6). In addition, the rheological properties of the tissue depend on the timescale of the measurements. Parallel plate rotation performed at different measurement frequencies will provide information about the mechanics of the system at different levels: at subsecond to second timescales, measurements will correspond to the subcellular structures, while timescales of tens of seconds to minutes will reflect cellular behavior, and timescales of tens of minutes to hours will probe multicellular and tissue-level mechanics [13]. Rounding of cell aggregates following parallel plate compression and micropipette aspiration of cell aggregates has also been used to measure long timescale viscoelastic behavior in large groups of cells in vitro [14,15].

The recent development of suspended monolayer culture has allowed for measurements of viscoelastic properties of epithelia during 2D stretch or compression and to investigate the biological sources of viscoelasticity in these epithelia [16,17]. Laser ablation has recently become a popular and convenient technique to study viscoelastic behavior in biological materials ranging from cytoskeletal filaments [18] to epithelial junctions and tissues [19]. Following ablation, the deformation of the ablated material is imaged and tracked (typically by confocal fluorescence microscopy) to assess the forces acting on the material as well as its mechanical properties. Although this method is not typically used to measure absolute values of viscoelastic properties, when coupled with theoretical models, laser ablation experiments can also be used to extract real physical parameters [20]. Laser ablation, like many of the rheological tools described earlier, results in the destruction or significant perturbation of the biological sample and can be categorized as invasive techniques.

#### *Noninvasive measurement tools*

A number of new measurement techniques offer ways to measure viscoelastic properties of biological materials without significantly disturbing or destroying the sample. The major advantage of such noninvasive approaches is that measurements can be performed in live samples over time without affecting cell or tissue behavior. One such recent technique relies on embedding inert oil droplets into a developing epithelium. By analyzing the droplet shape, local forces and mechanical properties of living tissues can be measured. Oil droplets containing magnetic particles can be actuated with a magnet to produce cell-scale deformations, allowing an additional active rheological mode to this technique [21]. New imaging modalities, such as Brillouin microscopy and acoustic scattering rely on local scattering of light or acoustic waves by biological materials to reveal their viscoelastic properties. These approaches have been used to measure single-cell and tissue mechanics in a noninvasive manner [22–24].

### 3.3.2 Extraction of viscoelastic properties from experimental measurements

In rheology experiments, the mechanical properties of biological materials are extracted by relating the applied mechanical force (stress) and the resulting deformation (strain). Rheological assays can involve a single deformation (creep test), a deformation followed by removal of the stress (stress-relaxation test) or oscillatory deformations [25,26] (Fig. 3.3A and B).

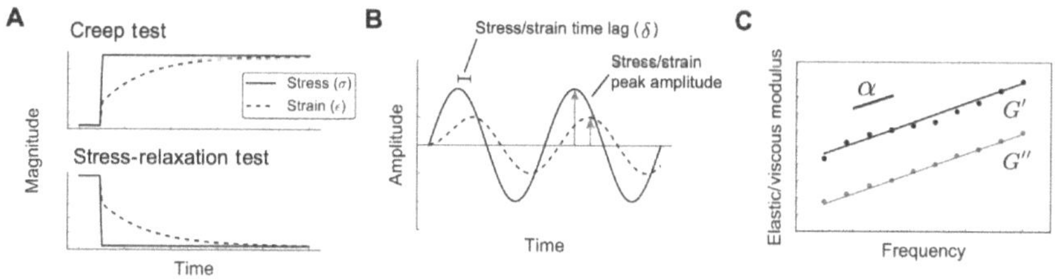


**Figure 3.3** Rheological assays to measure viscoelastic properties. (A) Single deformation tests include the creep test, where a stress is applied and sustained. Initially, many biological materials respond instantaneously in an elastic manner. Over long timescales, the strain response is more gradual and viscous-like. In stress-relaxation tests, the object is initially stressed over some time and then the stress is suddenly released. Again, the initial relaxation is instantaneous (elastic) followed by a longer timescale viscous relaxation. (B) In some rheological tests, oscillatory stress is applied. For viscoelastic materials, the deformation (strain) follows the same general pattern as the stress, but with a time delay ($\delta$). The viscous and elastic moduli can be extracted from the relationship between the stress and strain peak amplitudes and $\delta$. (C) For dynamic rheology experiments with oscillating stress, the viscous and elastic moduli ($G'$ and $G''$, respectively) can be extracted for each frequency of stress oscillation (dots). The dependence of $G'$ and $G''$ on the frequency often follows a power-law relationship (solid lines), and the exponent from this power-law fit gives an exponent, $\alpha$, which provides information about the relative viscous and elastic contributions.

By fitting the resulting stress–strain data to a theoretical model of the response, the respective elastic and viscous contributions can be measured independently. Depending on the experiment and the model used, different parameters can be extracted. In general, the key parameters that are extracted are the Storage Modulus, $G'$, which reflects the elastic component, and a Loss Modulus, $G''$, which reflects the viscous component. In many biological materials, $G'$ and $G''$ depend on the frequency of the oscillating stress, and this relationship (typically a power-law) reflects whether the stress response is predominantly elastic or predominantly viscous (Fig. 3.3C). A power-law exponent of $\alpha = 0$ reflects a purely elastic response, while $\alpha = 1$ indicates a purely viscous response [26].

Forces and mechanical properties can also be extracted by fitting live microscopic imaging data to theoretical models. Such *kinematic* approaches have been used to measure mechanical properties of biological polymers, such as cytoskeletal filaments [27]. More recently, kinematic approaches have been used to determine how different cellular behaviors in the tissue, such as cell divisions, cell rearrangements, or changes in cell shape, contribute to overall tissue mechanics and morphogenesis during *Drosophila* development [28,29].

Although the same methods for extracting viscoelastic parameters for inert objects can be applied to biological materials, cells and tissues are active materials. Through the conversion of chemical energy (e.g., ATP hydrolysis) cells themselves generate mechanical forces. One striking reality in cell mechanics is that the same cellular components that determine the mechanical properties of cells, such as elasticity and viscosity, are the same components that generate active forces [13]. Furthermore, these active forces can influence or control cell mechanics and must often also be considered in mechanical models of cell and tissue behavior [30–32].

Tissues and multicellular aggregates are inherently complex systems, and the origins of tissue viscoelasticity can arise from the combination of a number of underlying sources. In the remaining part of this chapter, I will discuss how different biophysical processes and components contribute to larger scale viscoelastic behavior, particularly during collective cell migration.

## 3.4 Viscoelasticity of the actin cytoskeleton

In most animal cell types, the actin cytoskeleton is the major regulator of cell mechanics. Actin monomers (G-actin) self-assemble into polymeric filaments (F-actin). Depending on G-actin concentrations, filaments can spontaneously assemble, disassemble, or exhibit treadmilling behavior [33]. Filament dynamics are also influenced by a number of actin-binding proteins (ABPs), which can serve to nucleate new filaments, bind monomers, or sever or cap existing filaments. Other ABPs cross-link or bundle F-actin, and myosin family motor proteins, particularly myosin-2, can bind one or more actin filaments and generate mechanical force. These individual molecular components of the actin cytoskeleton can be combined to form specialized structures with specific architectures that can drive processes like cell migration.

### 3.4.1 Elastic and viscous components in actin networks

Actin filaments themselves are semi-flexible polymers whose elasticity depends on the axis on which a stress acts. Actin filaments can more easily be twisted than bent, and more easily bent than stretched. For stresses along the longitudinal axis (stretch), filaments are nonlinearly elastic. F-actin is flexible enough so that its shape can be deformed by thermal fluctuations, meaning that filaments at biological temperatures are not straight, but wavy. For small amounts of stretching, filaments behave like an entropic spring, meaning that by applying stretch, filaments become straight, resisting bending deformations imposed by thermal fluctuations. Once the filaments are straight, they become extremely resistant to further stretch, acting at high stretch more like an enthalpic spring, meaning that the length of the filament increases during stretch. At sufficiently high force, the filaments eventually rupture. When filaments are compressed in an end-to-end manner, they undergo Euler buckling at low levels of force compared to the force required for stretching [34]. Buckling can also lead to increased filament severing, resulting in filament turnover within actin networks [35,36].

Viscoelastic behavior of the actin cytoskeleton arises from the elasticity of the filaments themselves and cross-linking ABPs bound to the filaments, combined with slower processes like turnover of filaments and cross-linkers. When an actin network is deformed, the network initially behaves in an elastic manner: filaments are bent, stretched, or twisted, and these mechanical perturbations result in storage of elastic free energy [34]. If the perturbation is prolonged, eventually the filaments will turn over, with old filaments depolymerizing, and new filaments repolymerizing. When the new filaments polymerize, they will not be under stress, as they have not experienced a perturbation in their lifetime. This results in a gradual loss of the stored free energy and a long timescale, viscous-like relaxation (Fig. 3.4). In addition to filament turnover, ABPs that serve to cross-link filaments, such as α-actinin, also turn over, and their turnover can also result in a long timescale relaxation in populations of cross-linked filaments. In cells, the turnover of cross-linkers has been demonstrated to be faster than actin turnover. Actin filaments in the cell cortex, a cross-linked F-actin network directly underlying the plasma membrane, turn over on timescales of tens of seconds, and separate subpopulations of actin in the cortex may have different dynamics and depend on the activity of different ABPs [37]. α-actinin has been shown to turn over on timescales of 5 seconds to tens of seconds, and its activity and regulation are thought to affect actin network mechanics, particularly during cytokinesis [37,38].

Although this simple example captures the origins of viscoelasticity in actin networks, it ignores the contribution of myosin motor activity. Myosin-2 activity in the cortex results in tensile stress on the actin filaments, which gives rise to cortical tension at the cell surface [39]. This surface tension results in a hydrostatic pressure inside the cell and causes

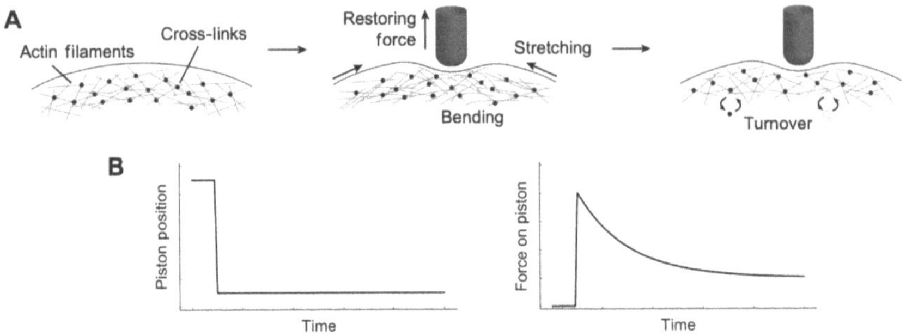


**Figure 3.4** Viscoelasticity in a pool of dynamic filaments in the absence of motor activity. (A) A pool of dynamic cross-linked filaments, such as F-actin, under a membrane. When a piston applies a stress on the population of filaments, filaments bend, twist, or stretch elastically. This elastic perturbation results in a restoring force back on the piston. Over time, the filaments and cross-linkers turn over, and the new filaments are no longer under elastic stress. This slower turnover gives rise to the viscous relaxation of the population of filaments. (B) Plots of the piston position and the restoring force on the piston over time from the experiment in *A*.

cells to prefer a spherical shape [30]. If a cell in suspension is deformed similarly to the simple example in Fig. 3.4, the restoring force would not go to zero even after all of the filaments and cross-links have turned over, due to cortical tension. Measurements of the resulting hydrostatic pressure have been used to determine cortical tension, for example, during rounding up in mitosis [40]. The tension-generation properties of actin networks depend not only on myosin but also on F-actin architecture and cross-linker dynamics [41–43]. How exactly turnover of actin filaments and ABPs influence the viscoelastic properties of actin networks remains an open question and a field of active research [44]. These viscoelastic properties, in turn, influence dynamic cellular processes such as cell migration.

### 3.4.2 Actin architecture during single-cell migration

Depending on their environment, single cells can migrate in ameboid or mesenchymal modes, which are associated with different actin organization [45,46]. During mesenchymal migration, cells typically form broad, flat actin-filled protrusions called lamellipodia. In lamellipodia, actin is assembled at the leading edge, pushing the membrane forward, and disassembled toward the cell body [33] (Fig. 3.5A). AFM-based rheology experiments have found that lamellipodia are viscoelastic, with predominantly elastic behavior; the elastic modulus is highest toward the leading edge, while cells are softer toward the cell body [47]. Recent theoretical work has used a

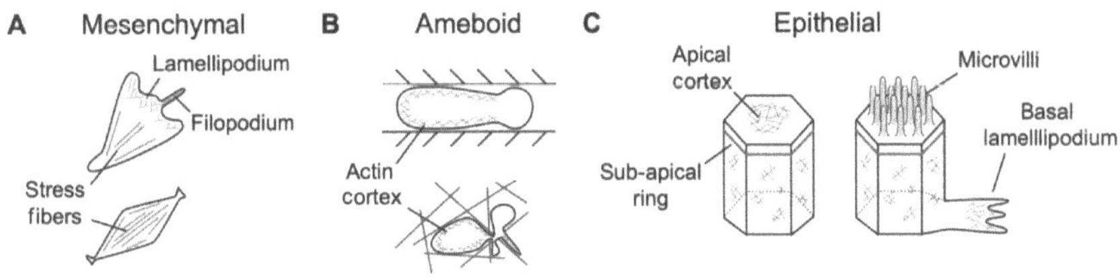


**Figure 3.5** Subcellular actin organization in different cell states. (A) In migratory single cells with a mesenchymal phenotype, actin is organized into lamellipodia and filopodia at the front that drive migration. In spindle-shaped mesenchymal cells, actin is organized primarily in stress fibers. (B) In cells migrating in microchannels, actin is organized in a peripheral cortex and interacts with the substrate in a nonspecific manner [51]. In more complex environments, many cell types take on a more lobulated form, using a combination of contractility-based squeezing using an actin cortex and leading-edge protrusions to navigate their environment [52]. (C) In epithelial cells, actin is organized into several substructures. At the apical surface, cells either display a loosely organized cortex or microvilli and a subapical ring of actin connected to the cell–cell junctions. Along the lateral surface, actin is organized in dynamic, loose clusters, and migratory epithelial cells display basal lamellipodia.

viscoelastic description of actin in the lamellipodium to faithfully describe actin flows and traction forces exerted during lamellipodia-based migration [48]. Mesenchymal cells also form finger-like protrusions called filopodia, which are filled with bundled actin fibers cross-linked by the protein fascin [49]. Filopodia exhibit a viscoelastic response following deflection by a magnetic tweezer [50]. Further toward the cell rear, actin is organized into tensile bundles of actin fibers called stress fibers. Stress fibers display viscoelastic recoil when they are cut by laser ablation, and myosin-2 activity is responsible for tensional prestress in stress fibers [18]. Mesenchymal cells can also adopt nonpolarized circular shapes, with lamellipodia in several directions and no net migration. Such circular lamellipodia drive substrate attachment during cell spreading. Alternatively, mesenchymal cells can adopt spindle-like cell shapes, with actin organized primarily in stress fibers (Fig. 3.5A).

During ameboid migration, the cell body is typically round, and cells extend actin-filled protrusions or make blebs at the leading edge. The actin cytoskeleton is typically organized in a cortical network during ameboid migration, and migration is not associated with well-defined cell–substrate adhesions. For cells migrating in more complex environments, actin is organized cortically in the cell rear and can also be present in protrusions (Fig. 3.5B). During ameboid migration in microfluidic channels, cortical actin flows rearward due to high concentrations of myosin at the cell rear. This results in contraction of the cell rear and expansion of the cell front that propels the cell forward [51,53]. Although the viscoelastic properties of the cortex have not been directly measured during such migratory flows, they have been measured in other systems with cortical flows, such as *Caenorhabditis elegans* zygotes [20,54]. The precise control of cortex viscoelasticity is crucial to ensure that the polarized cortical flows result in efficient migration.

### 3.4.3 Actin architecture in epithelia

In many tissues, cells migrate collectively in groups. Collective migration is crucial during developmental morphogenesis [55] and homeostatic migration in some adult tissues, such as the small intestine [56]. Collective migration has also been implicated in early stages of metastasis [45]. In many of these cases, cells display an epithelial morphology and migrate as small groups or sheets of cells. In epithelial cells, actin is typically organized in several distinct structures, each with its own architecture and accessory proteins. At the apical surface of some epithelial cells, a loosely overlapping meshwork of actin filaments form an apical cortex [57]. Just below the apical surface of epithelial cells, actin is organized into aligned bundles at the periphery of the cell at zonula adherens, where tight junction and adherens junction proteins are primarily localized [58] (Fig. 3.5C). These two actin structures, which are mechanically coupled, play a major role in developmental morphogenesis. During *Drosophila* gastrulation and dorsal closure, for example, apical and junctional actin structures undergo periodic, or pulsatile, contractions driven by bursts in myosin-2 motor activity, leading to the eventual constriction of the apical surface [57]. The cell shape changes associated with such periodic contractions can be described by viscoelastic models of the apical surface [5,59].

In some cell types, for example absorptive cells in the intestinal epithelium, the apical surface is covered with actin-rich microvilli. Microvilli are structurally similar to filopodia and share a number of common accessory proteins but are typically associated with polarized epithelial cells [60]. The mechanical properties of the microvilli themselves depend on the F-actin bundles inside of microvilli as well as actin-membrane attachments and adhesions between the microvilli themselves [61].

Actin is also present along the basolateral side of polarized epithelial cells. In cultured monolayers, actin accumulates into patches associated with cadherins along lateral junctions; similar to the apical cortex, bursts of myosin accumulation are associated with pulsatile contractions of the lateral actin network. It is thought that these contractions regulate clustering of cadherins along lateral junctions [62]. Actin networks can also be found on the very basal surface of epithelial cells. Recently, it was suggested that in epithelial cells in the small intestine, collective migration is driven by a basal network of actin that resembles a lamellipodium [56] (Fig. 3.5C). Such "cryptic" lamellipodia have also been observed to drive collective migration in cultured cells in vitro [63]. It is not currently clear how the basal actin network is coupled to the other actin structures in epithelial cells.

## 3.5 Cell–substrate adhesions and force transmission during migration

In order for cells to translocate during migration, the migration machinery inside of the cell (primarily the actin cytoskeleton) must be coupled to the external substrate. A number of different cell surface receptors, primarily those of the integrin family, link the cell with its substrate. Integrins are heterodimeric transmembrane proteins that bind specific ECM proteins, such as collagen, fibronectin, or laminin, on their extracellular N-terminus [64]. Individual integrins act as "catch bonds," meaning they bind for longer times with increasing force. At very high force, integrins act as "slip bonds," meaning binding time decreases with increasing force [65,66]. Integrins are coupled via their intracellular domain to the actin cytoskeleton by the protein talin. In response to mechanical force generated by the actin cytoskeleton, cryptic binding sites in talin are revealed, leading to the recruitment of additional adhesion proteins, integrin clustering, and focal adhesion maturation [67–69]. Focal adhesions have been observed in cells migrating in 2D and 3D in vitro as well as in vivo [70]. Magnetic bead experiments have shown that exerting force on single focal adhesions can induce increased recruitment of actin and that this actin recruitment increases the elastic moduli of focal adhesions but does not affect the viscous moduli [71]. Integrin binding in focal adhesions is highly dynamic, with pools of different integrin subtypes fluctuating between freely diffusing and immobilized states [72]. It is not currently well understood how local integrin dynamics affect nanoscale mechanical properties of focal adhesions.

Focal adhesion-based migration is often described by a "molecular clutch" model, where retrograde flow of actin is coupled to the substrate via elastic adhesions, resulting in rearward-facing traction forces. Simultaneously, actin polymerization at the leading edge opposes the retrograde flow and pushes the membrane forward. The forward movement of the cell is therefore determined by the relative speed of actin polymerization versus retrograde flow [73,74] (Fig. 3.6A). Clutch models can also include catch- and slip bond characteristics displayed by focal adhesions (Fig. 3.6B). Although the clutch is typically modeled as an elastic spring with characteristic binding and off rates, recent theoretical work has also explored the effects of clutch binding in a viscoelastic cell. The additional viscous component can effectively increase the binding rate of the clutch by allowing the receptor and ligand to stay closer together after unbinding, due to the slow, viscous-like recoil of the cell membrane [75] (Fig. 3.6C). In addition to the molecular clutch model, a number

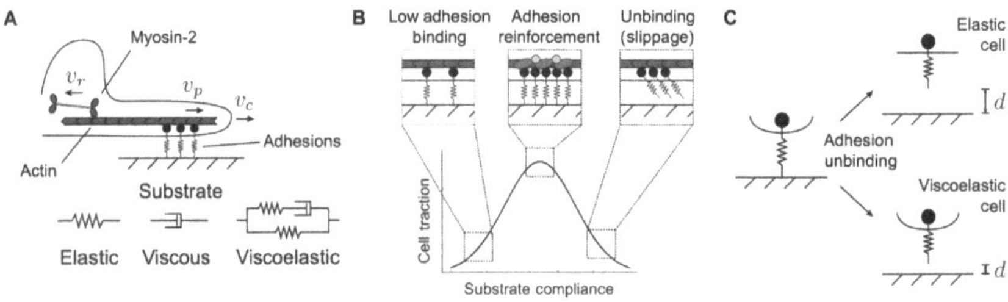


**Figure 3.6** Molecular clutch models with viscoelastic elements. (A) Cells spreading or migrating by lamellipodia are often described using a molecular clutch model, where retrograde flow of actin due to myosin-2 motor activity is opposed by actin polymerization at the leading edge. The molecular clutch (i.e., cell–substrate adhesions, most often integrins), couples the actin cytoskeleton and substrate when engaged, and thus retrograde flow depends on the deformation of the substrate. The substrate itself was first described as being purely elastic [73] and more recently as being purely viscous [77] or viscoelastic [78]. (B) The dynamics of the clutch are governed by binding rates and off rates, which are influenced by stress. Low levels of stress are insufficient to open talin, and focal adhesions cannot mature, resulting in low adhesion binding and low tractions and motility. At intermediate stress, focal adhesions mature via the recruitment of additional factors which reinforce adhesions. At high levels of stress, the bonds are destabilized, and unbinding (clutch slippage) occurs [79]. (C) The dynamics of the clutch can be influenced by cellular viscoelasticity. When an adhesion unbinds in a purely elastic cell, the cell membrane/actin immediately return to their resting position. However, long timescale relaxation for a viscoelastic cell results in a slower recoil of the membrane/actin. This results in integrins remaining closer to the substrate immediately after binding, increasing the probability of rebinding [75].

of other migration models exist, including "chimneying" and "flow-friction" models, which are used to describe cell migration in confined conditions [76].

During the migration of cells in confinement, focal adhesions often appear to be absent, and it has been proposed that nonspecific friction between the cell surface and the substrate gives rise to adhesive forces during migration. Although several candidate molecules have been proposed, including cadherins, cell surface receptors, and the glycocalyx matrix, the molecular origins of this friction are currently unknown [76]. Recently, clathrin, which is typically associated with endocytosis, has also been shown to play a role in cell–substrate adhesion during migration through collagen networks. Clathrin-coated structures associated with β1-integrin were shown to mediate the binding of cellular protrusions to collagen fibers in a focal adhesion-independent manner [80]. In addition to integrins, other cell surface receptors, such as syndecans, discoidal domain receptors, and CD44, have also been shown to bind ECM and support cell adhesion. However, the extent to which these receptors are involved in cell migration, and whether they are truly independent of integrin-ECM binding, is currently unclear [81]. In some epithelial cell types, most prominently skin keratinocytes, adhesion is mediated via integrins connected to intermediate filaments, in adhesive structures called hemidesmosomes. During wound repair in the skin, hemidesmosomes are rapidly disassembled in order to allow the tissue to reorganize and migrate using focal contacts in order to repair the wounded tissue [82].

## 3.6 Substrate mechanics during migration

Cellular viscoelastic properties can be influenced by the mechanical properties of the substrate to which cells are attached. In a number of different cell types, cells respond to increasing substrate stiffness by increasing cellular stiffness and viscosity, and these changes are associated with changes in actin organization [83–86]. However, recent work suggests that at least some of this apparent mechanosensory response may in fact be due to measurement error caused by deformation of soft substrates during AFM measurements [87]. Substrate stiffness has also been recently shown to be associated with transitions between epithelial and mesenchymal cell phenotypes [88].

The viscoelastic properties of the underlying substrate influence migration dynamics. Several cell types exhibit an optimal substrate stiffness for maximal migration speed, where substrates with elastic moduli above or below the optimum result in slower migration [89,90]. Molecular clutch models naturally predict an optimum substrate stiffness for efficient migration, and the magnitude of the optimal stiffness is predicted to depend on the number of motors and molecular clutch elements [91]. At stiffnesses below the optimum, force generation in focal adhesions is not sufficient to open talin, and focal adhesions cannot be reinforced [92]. At stiffnesses above the optimum, integrins are stabilized but bonds to the substrate break under the high traction force and focal adhesions cannot be stabilized (Fig. 3.6B). The mean size of focal adhesions within a cell has been shown to be a robust predictor of cell migration speed [93]. It has also been shown that cells can continue migrating at high velocity in stiff environments, provided that the cells are under confinement. It was suggested that enhanced polarization of traction forces allow cells to migrate efficiently in stiff confined environments [94]. However, it is unclear how adhesion stability is affected by stiffness in such confined conditions or how cells undergoing focal adhesion-free migration respond to substrate stiffness with or without confinement.

In addition to exhibiting stiffness-dependent migration dynamics, cells tend to migrate from soft substrates toward stiffer substrates, a process known as "durotaxis" [95]. Large groups of cells have also been shown to demonstrate durotactic behavior, and this behavior can be understood through a molecular clutch model. On stiffer substrates, retrograde flow of actin is low because the substrate is difficult to deform. This allows actin polymerization to dominate, promoting forward migration. On softer substrates, traction forces are similar compared to stiff substrates, but substrate deformation is higher. This results in faster rates of actin retrograde flow, which dominate over polymerization at the leading edge, leading to reduced migration toward softer substrates [96].

More recently, experiments combined with molecular clutch models have been applied to explore cell dynamics on purely viscous or viscoelastic substrates (Fig. 3.6A) (reviewed in Ref. [97]). Surface protein mobility (i.e., a viscous substrate component) has previously been shown to enhance cell spreading, though this effect may be cell-type dependent [98–100]. Viscous substrates with tuneable viscosity have been designed using supported lipid bilayers functionalized with ECM ligands. Similar to the differences observed between stiff and soft elastic substrates, on highly viscous bilayers, cells spread efficiently, while on less viscous substrates, spreading is reduced. This can again be explained by the molecular clutch:

highly viscous environments do not sufficiently deform on timescales relevant to actin and focal adhesion dynamics, meaning that actin retrograde flow is slow, so actin polymerization dominates. On substrates with low viscosity, focal adhesion formation is reduced and retrograde flow dominates, resulting in slower spreading [77]. Modeling of a molecular clutch on viscoelastic substrates suggests that substrate viscosity can impact cell spreading differently depending on the elasticity. For low elasticity regimes, cell spreading is maximum at intermediate viscosity, when the substrate relaxation time is similar to the clutch binding time. For stiff substrates, cell spreading is always high, regardless of viscosity [78]. Substrate viscosity has also been shown to influence collective migration dynamics. By tuning the viscoelastic properties of the underlying substrate, the relative viscous and elastic moduli can be tuned to optimize migration speed and directional correlation between cells [101]. Although such 2D viscoelastic substrates are useful to understand the importance of elastic and viscous properties of the substrate on migration, in biological environments, cells are often met with more complicated substrate topologies.

In the past several years, the use of biological and synthetic hydrogels has helped to shed light on the importance of substrate topology during migration. Collagen-I is an ECM protein that is a primary component of stroma, or connective tissue, in most organs. Collagen can self-assemble into fibrous networks that permit cell migration on the network surface or through the network. Collagen gels display viscoelastic behavior in stress-relaxation tests, and higher initial stresses result in more prominent long-term relaxation timescales. Chemical cross-linking of collagen networks results in faster relaxation times, meaning that cross-linked gels behave more like an elastic substrate, while non-cross-linked gels behave viscoelastically [102]. The mechanical and topological properties of the stroma have emerged as major factors in cell migration, particularly during early stages of metastasis [45]. Recent in vitro migration studies have argued that local topology, and not bulk viscoelastic properties, is the primary determinant of directionality during migration through collagen networks [103,104]. Ectopic alignment of collagen fibers can direct cell migration along the axis of alignment [105,106], and mechanical alignment of collagen by large groups of cells can predict the direction of collective migration and enhance invasion of single cells into collagen matrices [107–109]. The ability of cells to dynamically reorganize collagen networks will depend on the mechanical properties of the networks. For example, cells must exert more force to deform stiffer networks, and deformations of viscoelastic networks will be longer-lived than for purely elastic networks. Understanding the combinatorial effects of local network topology and viscoelastic properties of collagen gels remains an interesting, and largely unexplored, avenue of future research. New synthetic hydrogels offer more experimental control over network mechanics and biochemical functionalization and will no doubt be useful in addressing these questions [110,111].

## 3.7 Cell–cell adhesion dynamics

In tissues and multicellular structures, neighboring cells are attached via specific membrane receptors. The most prominent receptors involved in cell–cell adhesion are the cadherin superfamily proteins, and specifically the classical cadherins, which localize to adherens junctions, primarily in the zonula adherens. Cadherins are transmembrane proteins that bind via their extracellular N-terminal domains to form cell–cell adhesions. On their intracellular C-terminal domains, cadherins bind to proteins of the catenin family, which bind directly or indirectly to actin filaments [112].

Adherens junctions are closely integrated with the actin cytoskeleton, with bidirectional feedback between junctions and actomyosin. The actin cytoskeleton applies tensile stress on cadherins at the membrane, and tension increases upon cadherin–cadherin binding at cell–cell junctions [113]. Similar to focal adhesions, cadherin bonds act as mechanosensitive catch bonds; increased tension on cadherin junctions via myosin-2 contractility results in stronger cadherin bonding and increased cadherin stability up until the point of rupture [114–116]. Actomyosin prestress, which results in maturation of adherens junctions, allows tissues to maintain their integrity in the presence of large acute mechanical stresses [58]. Cadherin-based adhesions can also, in turn, stimulate actomyosin activity, both by recruiting actin nucleators and increasing myosin-2 activity [58,117–119]. Feedback between cadherins and actomyosin therefore tends to lead to reinforcement of cell–cell junctions. However, adherens junctions are dynamic structures that can also be remodeled in response to internal or external mechanical stress.

Cadherin dynamics can be viewed on two levels: (1) rearrangement of cadherin proteins along the membrane and (2) trafficking of cadherins between the cell membrane and intracellular membrane compartments (Fig. 3.7). Cadherins localized both to thc zonula adherens and the lateral surface of epithelial cells are mobile; clusters of cadherins can be assembled, disassembled, and translocated on timescales of tens of seconds to minutes, and their mobility is regulated by actomyosin activity (Fig. 3.7A) [62]. Junctional cadherins can also shuttle to and from the membrane via an exo/endocytic pathways, which are well-described and are thought to drive cell rearrangements during morphogenesis (Fig. 3.7B) (reviewed in Refs. [120,121]). Recent work has shown that cadherin turnover via membrane trafficking occurs on timescales of several minutes and is mechanosensitive. During morphogenesis in the *Drosophila* pupal wing, anisotropic tension stimulates endocytosis of cadherin, and this recycling event allows for cell rearrangements in the developing wing over the subsequent 8 hours [19]. During *Drosophila* germband extension, actomyosin contractility at junctions has been demonstrated to cluster cadherin and recruit the endocytotic machinery to junctions, thereby enhancing cadherin turnover. Cadherin recycling, in turn, facilitates junctional remodeling required for tissue extension during germband extension [122]. During collective migration in vitro, adherens junctions have been shown to form near the leading edge and disassemble near the trailing edge, creating a retrograde treadmill of junctions [123].

In addition to classical cadherins, desmosomal clathrins cluster to form cell–cell adhesions coupled to intermediate filaments in structures called desmosomes. Desmosomes play an important role in tissue integrity, predominantly in the skin and heart [124]. During development, desmosomes

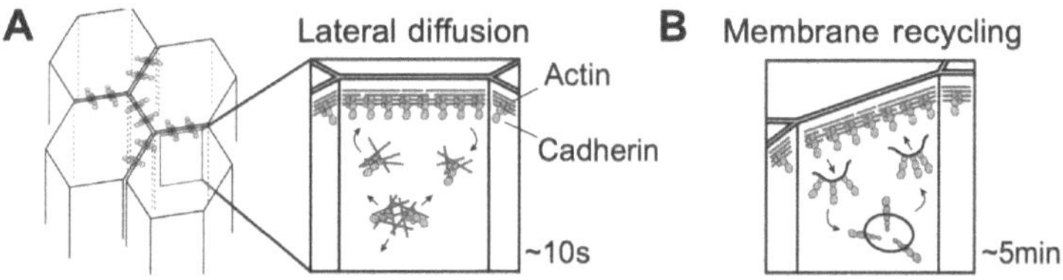


**Figure 3.7** Regulation of cell–cell junction dynamics. (A) Cell–cell junctions are mediated primarily by classical cadherins at the cell interfaces. In epithelial cells, junctions localize primarily to the subapical zonula adherens, where they colocalize with aligned actin filaments as well as to lateral junctions, where they are associated with actin patches. Cadherins can be rearranged on junctions by diffusion along lateral membranes on timescales of tens of seconds [62]. (B) Cadherins can also be exchanged on the lateral junctions by exo- and endocytosis of cadherins to and from the plasma membrane. Cadherin turnover by membrane recycling occurs on timescales of several minutes [19].

adopt a calcium-independent "hyperadhesive" state required for tissue integrity in adult organisms [125]. In response to tissue wounding, cells revert to the weaker, calcium-dependent desmosomes, which allows tissue reorganization required for wound closure [125,126]. Like classical cadherin bonds, desmosomal cadherin dynamics can also be regulated by endocytic recycling [127,128]. Cell–cell adhesion dynamics set the rules by which cells interact with each other and can facilitate cell rearrangements that give rise to large-scale tissue viscoelastic properties. Precise control of cell–cell junctions is crucial during dynamic processes like morphogenesis and collective cell migration.

## 3.8 Viscoelasticity in collective tissue migration

Migration of groups of cells, or collective migration, is a well-described phenomenon during embryonic tissue development [55]. During zebrafish gastrulation, for example, a sheet of cells called the prechordial plate migrates up toward the animal pole. At the leading edge, cells form protrusions to migrate forward, while cells behind the leading edge follow and maintain tight cell–cell junctions [129]. During the development of the lateral line, a sensory organ that runs along the length of the zebrafish body, clusters of cells migrate collectively along chemokine gradients, with leader cells sensing the gradients and follower cells being passively dragged behind [130]. In the *Drosophila* ovary, a small group of four to eight border cells migrate in a collective manner through the oocyte either using leading-edge protrusions or by collectively "rolling," while allowing for the exchange of neighboring cells during migration [131,132]. Recent work in the small intestine has highlighted the role of collective migration in adult tissues as well. Newly differentiated cells migrate from the intestinal crypt along the length of the villus, and during this collective migration, cells maintain tight boundaries and appear to migrate using cryptic basal lamellipodia [56]. These collective motions all rely on a combination of active migration mediated by the cytoskeleton, cell–substrate adhesions, substrate compliance, and cell–cell adhesions, each of which displays viscoelastic behavior that contributes to collective migration dynamics.

As cells at the leading edge exert traction forces on the environment and migrate forward, cells behind the leading edge will be pulled due to mechanical coupling via cell–cell junctions. This pulling generates tensile stress in the tissue, which has been measured during collective migration in border cells using fluorescent tension sensors [133] and in the small intestine using laser ablation [56]. Monolayer stress microscopy, in which stresses within a cell monolayer are inferred from traction force measurements, has also suggested that collectively migrating monolayers are under tension [134,135]. However, recent work suggests that the inherent assumptions in monolayer stress microscopy, namely that the monolayer behaves as a passive elastic or viscous material, may be insufficient to faithfully determine stresses in the monolayer [32]. In response to tissue tension during collective migration, cells can activate mechanosensory pathways to reinforce adhesions. On longer timescales, this tensile stress can be dissipated due to the dynamic nature of the cytoskeleton and cell–substrate and cell–cell adhesions, as described in the previous sections. In addition, oriented cell division and cell rearrangements can result in viscous relaxation at the tissue scale (reviewed in Ref. [13]). The combined effects of these numerous cell-scale and subcellular processes give rise to the viscoelastic properties of the tissue, which themselves be described as viscoelastic materials. Viscoelastic rheological properties of tissues can be measured experimentally [12], and theoretical models of tissues as continuous 2D viscous fluids have been successful in describing large-scale morphogenetic events during *Drosophila* and zebrafish development [136,137].

Tightly packed epithelia have been observed to rapidly transition between a fluid state, with high cell migration, to a solid-state, where migration is arrested. Such transitions are crucial to processes like epithelial wound healing, where sufficient tissue fluidity is required to rearrange cells as they repair a hole in the tissue [138]. Fluid-to-solid transitions are often described as cellular "jamming," using an analogy to glass transitions [139]. Jamming transitions have been observed in monolayers in vitro in response to changes in cell density; as density increases, migration speed decreases while directional correlation between cells increases, leading to large-scale flows of cells near the jamming transition [134,140,141] (Fig. 3.8A).

In tissues where cells are already densely packed, vertex models have been used to predict that jamming transitions can be governed by changes in cell shape [142,143]. Cell shape is often defined in this context by a shape parameter, which is the cell perimeter divided by the square root of the cell area ($p_0 = P/\sqrt{A}$). Whereas tissues containing cells with a high shape factor (elongated cells) are predicted to be in a motile, unjammed

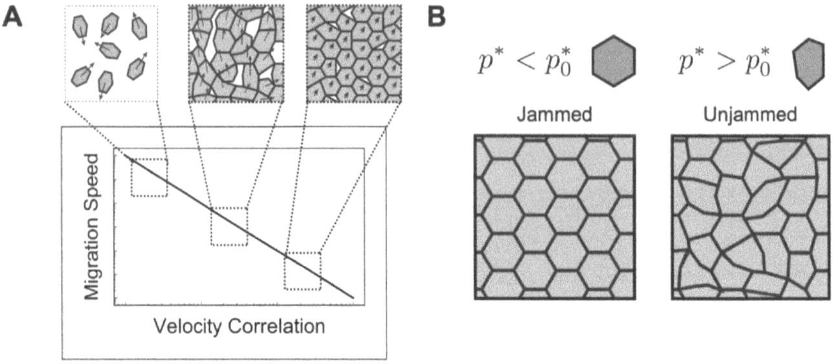


**Figure 3.8** Jamming transitions in monolayers and tissues. (A) As cell density increases in epithelial monolayers, cell migration slows down as directional correlation between cells increases. (B) In confluent epithelia, jamming transitions can be controlled by cell shape, described by the shape parameter, $p_0 = P/\sqrt{A}$, with the critical value $p_0^* = 3.81$. Below the critical value, the epithelium behaves like an ordered solid (jammed), while epithelia with cell shapes above the critical value, $p_0 > p_0^*$, behave like a disordered fluid.

state, tissues containing cells with a lower shape factor (less elongated) are predicted to be jammed (Fig. 3.8B). Because the shape is determined by cortical tension and cell–cell adhesion, this leads to the counter-intuitive prediction that increasing cell–cell adhesion will result in longer junctions and more elongated cells, leading to unjamming [142]. Cell shape was recently shown to be a good predictor of motility in primary lung tissues from healthy and asthmatic patients and during *Drosophila* embryogenesis [144]. Jamming transitions have been shown to regulate somitogenesis in zebrafish embryos [21] and have been proposed to govern cell migration capacity in models of invasive cancer [145,146]. Although jamming transitions have been useful in describing collective migration in different conditions, they do not necessarily reflect the rheological viscoelastic properties of the tissue. Vertex models, which have been used to study fluid–solid transitions, do not allow for dissipation of internal stress, which is required to describe a material as being viscoelastic, nor can it allow anisotropic active stresses, which could arise from cell polarization within a migrating tissue. A number of different theoretical models, each with its advantages and disadvantages, have been used to study collective cell dynamics [147]. Recent theoretical work has aimed at characterizing jamming transitions using viscoelastic rheological models in order to better understand how cellular jamming relates to tissue viscoelasticity [148].

## 3.9 Conclusion

The viscoelastic properties of migrating cells and tissues arise from a number of biological sources. A common theme in biological viscoelasticity is that while single elements may behave elastically, ensembles tend to behave in a viscoelastic manner due to turnover of the individual elements. Actin filaments, for example, exhibit nonlinear elastic properties in response to stress, but due to disassembly and assembly of individual filaments and turnover of cross-linkers, actin networks can also behave in a viscous manner at longer timescales. Ahdherens junctions, when placed under stress, both reinforce existing junctions and increase junction turnover, leading to a viscoelastic response of cell–cell junctions. At longer timescales, junctional remodeling can facilitate rearrangements of cells within a tissue, leading to viscoelastic behavior on a much larger scale.

Each of the biological processes discussed behaves according to specific length and timescales in response to perturbation. Millisecond timescales correspond to elastic effects of interactions between protein subunits, such as actin filament bending, while second timescales correspond to interactions between proteins, for example, tension at adhesion complexes or membrane–cortex interactions. On timescales of several tens of seconds, cell shape relaxes due to processes like turnover of actin filaments and cortical components, while on minute timescales cells polarize and recycle adhesion complexes. On longer timescales, from several tens of minutes to hours, cells in tissues rearrange and migrate. The combined effects of these different processes lead to complex viscoelastic behavior in groups of cells [13]. During collective migration, cells exert mechanical force on the substrate in order to move, and these movements generate stresses on neighboring cells. The collective viscoelastic response of the tissue to these stresses, in turn, affects migration dynamics. Although much is known about the individual components that contribute to cell and tissue viscoelasticity, understanding how these systems are coupled in space and time and how their concerted actions lead to viscoelastic behavior during collective migration remains an active field of research.

## Acknowledgments

I am grateful to D.M. Vignjevic, J. Barbazan, E.H. Barriga, and I. Pajic-Lijakovic for comments and suggestions on this manuscript as well as E.K. Paluch and G. Salbreux for countless discussions related to cellular viscoelasticity. I thank L.K. Clark-Liebermann and

M.C. Clark-Liebermann for their assistance with the putty and Oobleck experiments. I apologize to colleagues whose work I was not able to cite in this review. A.G. Clark is supported by the Institut National du Cancer (INCa grant PLBIO18-087 to D.M. Vignjevic).

## References

[1] G. Salbreux, G. Charras, E. Paluch, Actin cortex mechanics and cellular morphogenesis, Trends Cell Biol. 22 (10) (2012) 536–545.
[2] F.M. Hecht, J. Rheinlaender, N. Schierbaum, W.H. Goldmann, B. Fabry, T.E. Schäffer, Imaging viscoelastic properties of live cells by AFM: power-law rheology on the nanoscale, Soft Matter 11 (23) (2015) 4584–4591.
[3] A.R. Bausch, F. Ziemann, A.A. Boulbitch, K. Jacobson, E. Sackmann, Local measurements of viscoelastic parameters of adherent cell surfaces by magnetic bead microrheometry, Biophys. J. 75 (4) (1998) 2038–2049.
[4] B. Fabry, G.N. Maksym, J.P. Butler, M. Glogauer, D. Navajas, N.A. Taback, et al., Time scale and other invariants of integrative mechanical behavior in living cells, Phys. Rev. E 68 (4) (2003) 041914.
[5] R. Clément, B. Dehapiot, C. Collinet, T. Lecuit, P.F. Lenne, Viscoelastic dissipation stabilizes cell shape changes during tissue morphogenesis, Curr. Biol. 27 (20) (2017) 3132–3142. e4.
[6] M. Keller, R. Tharmann, M.A. Dichtl, A.R. Bausch, E. Sackmann, Slow filament dynamics and viscoelasticity in entangled and active actin networks, Philos. Trans. R. Soc. A 361 (1805) (2003) 699–712.
[7] F. Wottawah, S. Schinkinger, B. Lincoln, S. Ebert, K. Müller, F. Sauer, et al., Characterizing single suspended cells by optorheology, Acta Biomater. 1 (3) (2005) 263–271.
[8] N. Desprat, A. Richert, J. Simeon, A. Asnacios, Creep function of a single living cell, Biophys. J. 88 (3) (2005) 2224–2233.
[9] R.M. Hochmuth, Micropipette aspiration of living cells, J. Biomech. 33 (1) (2000) 15–22.
[10] O. Otto, P. Rosendahl, A. Mietke, S. Golfier, C. Herold, D. Klaue, et al., Real-time deformability cytometry: on-the-fly cell mechanical phenotyping, Nat. Methods 12 (3) (2015) 199–202.
[11] P. Fernández, L. Heymann, A. Ott, N. Aksel, P.A. Pullarkat, Shear rheology of a cell monolayer, N. J. Phys. 9 (11) (2007) 419.
[12] C.D. Madsen, T.R. Cox, Relative stiffness measurements of tumour tissues by shear rheology, Bio-protocol 7 (9) (2017) e2265.
[13] G. Charras, A.S. Yap, Tensile sorces and mechanotransduction at cell–cell junctions, Curr. Biol. 28 (8) (2018) R445–R457.
[14] K. Guevorkian, D. Gonzalez-Rodriguez, C. Carlier, S. Dufour, F. Brochard-Wyart, Mechanosensitive shivering of model tissues under controlled aspiration, Proc. Natl. Acad. Sci. U.S.A. 108 (33) (2011) 13387.
[15] P. Marmottant, A. Mgharbel, J. Käfer, B. Audren, J.P. Rieu, J.C. Vial, et al., The role of fluctuations and stress on the effective viscosity of cell aggregates, Proc. Natl. Acad. Sci. U.S.A. 106 (41) (2009) 17271.
[16] T.P.J. Wyatt, A.R. Harris, M. Lam, Q. Cheng, J. Bellis, A. Dimitracopoulos, et al., Emergence of homeostatic epithelial packing and stress dissipation through divisions oriented along the long cell axis, Proc. Natl. Acad. Sci. U.S.A. 112 (18) (2015) 5726–5731.

[17] A.R. Harris, J. Bellis, N. Khalilgharibi, T. Wyatt, B. Baum, A.J. Kabla, et al., Generating suspended cell monolayers for mechanobiological studies, Nat. Protoc. 8 (12) (2013) 2516–2530.
[18] S. Kumar, I.Z. Maxwell, A. Heisterkamp, T.R. Polte, T.P. Lele, M. Salanga, et al., Viscoelastic retraction of single living stress fibers and its impact on cell shape, cytoskeletal organization, and extracellular matrix mechanics, Biophys. J. 90 (10) (2006) 3762–3773.
[19] K.V. Iyer, R. Piscitello-Gómez, J. Paijmans, F. Jülicher, S. Eaton, Epithelial viscoelasticity is regulated by mechanosensitive E-cadherin turnover, Curr. Biol. 29 (4) (2019) 578–591.
[20] A. Saha, M. Nishikawa, M. Behrndt, C.P. Heisenberg, F. Jülicher, S.W. Grill, Determining physical properties of the cell cortex, Biophys. J. 110 (6) (2016) 1421–1429.
[21] A. Mongera, P. Rowghanian, H.J. Gustafson, E. Shelton, D.A. Kealhofer, E.K. Carn, et al., A fluid-to-solid jamming transition underlies vertebrate body axis elongation, Nature 561 (7723) (2018) 401–405.
[22] R. Prevedel, A. Diz-Muñoz, G. Ruocco, G. Antonacci, Brillouin microscopy: an emerging tool for mechanobiology, Nat. Methods 16 (10) (2019) 969–977.
[23] R. Schlüßler, S. Möllmert, S. Abuhattum, G. Cojoc, P. Müller, K. Kim, et al., Mechanical mapping of spinal cord growth and repair in living zebrafish larvae by Brillouin imaging, Biophys. J. 115 (5) (2018) 911–923.
[24] J.H. Kang, T.P. Miettinen, L. Chen, S. Olcum, G. Katsikis, P.S. Doyle, et al., Noninvasive monitoring of single-cell mechanics by acoustic scattering, Nat. Methods 16 (3) (2019) 263–269.
[25] Murata H. (2012). Rheology - Theory and Application to Biomaterials, chapter 17, pp. 403–426. Polymerization. IntechOpen.
[26] P.H. Wu, D.R.B. Aroush, A. Asnacios, W.C. Chen, M.E. Dokukin, B.L. Doss, et al., A comparison of methods to assess cell mechanical properties, Nat. Methods 15 (7) (2018) 491–498.
[27] J. Howard, Mechanics of Motor Proteins and the Cytoskeleton, Sinauer Associates, Sunderland, MA, 2001.
[28] R. Etournay, M. Popović, M. Merkel, A. Nandi, C. Blasse, B. Aigouy, et al., Interplay of cell dynamics and epithelial tension during morphogenesis of the Drosophila pupal wing, eLife 4 (2015) e07090.
[29] B. Guirao, S.U. Rigaud, F. Bosveld, A. Bailles, J. López-Gay, S. Ishihara, et al., Unified quantitative characterization of epithelial tissue development, eLife 4 (2015) e08519.
[30] A.G. Clark, E. Paluch, Mechanics and regulation of cell shape during the cell cycle, in: J.Z. Kubiak (Ed.), Cell Cycle in Development, Volume 53 of Results and Problems in Cell Differentiation, Chapter 3, Springer-Verlag, Berlin, Heidelberg, Germany, 2011.
[31] W.W. Ahmed, É. Fodor, T. Betz, Active cell mechanics: measurement and theory, BBA-Mol Cell Res. 1853 (2015) 3083–3094 (11, Part B).
[32] Y. Green, J.J. Fredberg, J.P. Butler, Relationship between velocities, tractions, and intercellular stresses in the migrating epithelial monolayer, Phys. Rev. E (2020). 16 April 2020.
[33] T.D. Pollard, G.G. Borisy, Cellular motility driven by assembly and disassembly of actin filaments, Cell 112 (4) (2003) 453–465.
[34] E.M. De La Cruz, M.L. Gardel, Actin mechanics and fragmentation, J. Biol. Chem. 290 (28) (2015) 17137–17144.

[35] M.P. Murrell, M.L. Gardel, F-actin buckling coordinates contractility and severing in a biomimetic actomyosin cortex, Proc. Natl. Acad. Sci. U.S.A. 109 (51) (2012) 20820.

[36] S.K. Vogel, Z. Petrasek, F. Heinemann, P. Schwille, Myosin motors fragment and compact membrane-bound actin filaments, eLife (2013) 2.

[37] M. Fritzsche, A. Lewalle, T. Duke, K. Kruse, G. Charras, Analysis of turnover dynamics of the submembranous actin cortex, Mol. Biol. Cell 24 (6) (2013) 757–767.

[38] S. Mukhina, Y.L. Wang, M. Murata-Hori, α-Actinin is required for tightly regulated remodeling of the actin cortical network during cytokinesis, Dev. Cell 13 (4) (2007) 554–565.

[39] A.G. Clark, O. Wartlick, G. Salbreux, E.K. Paluch, Stresses at the cell surface during animal cell morphogenesis, Curr. Biol. 24 (10) (2014) R484–R494.

[40] E. Fischer-Friedrich, A.A. Hyman, F. Jülicher, D.J. Müller, J. Helenius, Quantification of surface tension and internal pressure generated by single mitotic cells, Sci. Rep. 4 (2014) 6213.

[41] P. Chugh, A.G. Clark, M.B. Smith, D.A.D. Cassani, K. Dierkes, A. Ragab, et al., Actin cortex architecture regulates cell surface tension, Nat. Cell Biol. 19 (6) (2017) 689–697.

[42] H. Ennomani, G. Letort, C. Guérin, J.L. Martiel, W. Cao, F. Nédélec, et al., Architecture and connectivity govern actin network contractility, Curr. Biol. 26 (5) (2016) 616–626.

[43] T. Hiraiwa, G. Salbreux, Role of turnover in active stress generation in a filament network, Phys. Rev. Lett. 116 (18) (2016) 188101.

[44] P. Chugh, E.K. Paluch, The actin cortex at a glance, J. Cell Sci. 131 (14) (2018) jcs186254.

[45] A.G. Clark, D.M. Vignjevic, Modes of cancer cell invasion and the role of the microenvironment, Curr. Opin. Cell Biol. 36 (2015) 13–22.

[46] P. Friedl, K. Wolf, Plasticity of cell migration: a multiscale tuning model, J. Cell Biol. 188 (1) (2010) 11–19.

[47] S. Park, D. Koch, R. Cardenas, J. Käs, C.K. Shih, Cell motility and local viscoelasticity of fibroblasts, Biophys. J. 89 (6) (2005) 4330–4342.

[48] E.M. Craig, J. Stricker, M. Gardel, A. Mogilner, Model for adhesion clutch explains biphasic relationship between actin flow and traction at the cell leading edge, Phys. Biol. 12 (3) (2015) 035002.

[49] D. Vignjevic, S.I. Kojima, Y. Aratyn, O. Danciu, T. Svitkina, G.G. Borisy, Role of fascin in filopodial protrusion, J. Cell Biol. 174 (6) (2006) 863–875.

[50] A. Zidovska, E. Sackmann, On the mechanical stabilization of filopodia, Biophys. J. 100 (6) (2011) 1428–1437.

[51] M. Bergert, A. Erzberger, R.A. Desai, I.M. Aspalter, A.C. Oates, G. Charras, et al., Force transmission during adhesion-independent migration, Nat. Cell Biol. 17 (4) (2015) 524–529.

[52] M. Tozluoğlu, A.L. Tournier, R.P. Jenkins, S. Hooper, P.A. Bates, E. Sahai, Matrix geometry determines optimal cancer cell migration strategy and modulates response to interventions, Nat. Cell Biol. 15 (7) (2013) 751–762.

[53] R.J. Hawkins, R. Poincloux, O. Bénichou, M. Piel, P. Chavrier, R. Voituriez, Spontaneous contractility-mediated cortical flow generates cell migration in three-dimensional environments, Biophys. J. 101 (5) (2011) 1041–1045.

[54] M. Mayer, M. Depken, J.S. Bois, F. Julicher, S.W. Grill, Anisotropies in cortical tension reveal the physical basis of polarizing cortical flows, Nature 467 (7315) (2010) 617–621.

[55] E. Scarpa, R. Mayor, Collective cell migration in development, J. Cell Biol. 212 (2) (2016) 143–155.
[56] D. Krndija, F. El Marjou, B. Guirao, S. Richon, O. Leroy, Y. Bellaiche, et al., Active cell migration is critical for steady-state epithelial turnover in the gut, Science 365 (6454) (2019) 705–710.
[57] A.C. Martin, B. Goldstein, Apical constriction: themes and variations on a cellular mechanism driving morphogenesis, Development 141 (10) (2014) 1987–1998.
[58] B.R. Acharya, A. Nestor-Bergmann, X. Liang, S. Gupta, K. Duszyc, E. Gauquelin, et al., A Mechanosensitive RhoA pathway that protects epithelia against acute tensile stress, Dev. Cell 47 (4) (2018) 439–452.
[59] K. Dierkes, A. Sumi, J. Solon, G. Salbreux, Spontaneous oscillations of elastic contractile materials with turnover, Phys. Rev. Lett. 113 (14) (2014) 148102.
[60] S. Khurana, S.P. George, The role of actin bundling proteins in the assembly of filopodia in epithelial cells, Cell Adhes. Migr. 5 (5) (2011) 409–420.
[61] S.W. Crawley, M.S. Mooseker, M.J. Tyska, Shaping the intestinal brush border, J. Cell Biol. 207 (4) (2014) 441–451.
[62] S.K. Wu, S. Budnar, A.S. Yap, G.A. Gomez, Pulsatile contractility of actomyosin networks organizes the cellular cortex at lateral cadherin junctions, Eur. J. Cell Biol. 93 (10) (2014) 396–404.
[63] R. Farooqui, G. Fenteany, Multiple rows of cells behind an epithelial wound edge extend cryptic lamellipodia to collectively drive cell-sheet movement, J. Cell Sci. 118 (1) (2005) 51–63.
[64] R.O. Hynes, Integrins: bidirectional, allosteric signaling machines, Cell 110 (6) (2002) 673–687.
[65] M. Dembo, D.C. Torney, K. Saxman, D. Hammer, J.D. Murray, The reaction-limited kinetics of membrane-to-surface adhesion and detachment, Proc. R. Soc. B-Biol Sci. 234 (1274) (1988) 55–83.
[66] F. Kong, A.J. García, A.P. Mould, M.J. Humphries, C. Zhu, Demonstration of catch bonds between an integrin and its ligand, J. Cell Biol. 185 (7) (2009) 1275–1284.
[67] A. del Rio, R. Perez-Jimenez, R. Liu, P. Roca-Cusachs, J.M. Fernandez, M.P. Sheetz, Stretching single talin rod molecules activates vinculin binding, Science 323 (5914) (2009) 638–641.
[68] C.G. Galbraith, K.M. Yamada, M.P. Sheetz, The relationship between force and focal complex development, J. Cell Biol. 159 (4) (2002) 695–705.
[69] J.D. Humphries, P. Wang, C. Streuli, B. Geiger, M.J. Humphries, C. Ballestrem, Vinculin controls focal adhesion formation by direct interactions with talin and actin, J. Cell Biol. 179 (5) (2007) 1043–1057.
[70] S. Geraldo, A. Simon, N. Elkhatib, D. Louvard, L. Fetler, D.M. Vignjevic, Do cancer cells have distinct adhesions in 3D collagen matrices and in vivo? Eur. J. Cell Biol. 91 (11–12) (2012) 930–937.
[71] B.D. Matthews, D.R. Overby, F.J. Alenghat, J. Karavitis, Y. Numaguchi, P.G. Allen, et al., Mechanical properties of individual focal adhesions probed with a magnetic microneedle, Biochem. Biophys. Res. Comm. 313 (3) (2004) 758–764.
[72] O. Rossier, V. Octeau, J.B. Sibarita, C. Leduc, B. Tessier, D. Nair, et al., Integrins 1 and 3 exhibit distinct dynamic nanoscale organizations inside focal adhesions, Nat. Cell Biol. 14 (10) (2012) 1057–1067.
[73] C.E. Chan, D.J. Odde, Traction dynamics of filopodia on compliant substrates, Science 322 (5908) (2008) 1687–1691.
[74] T. Mitchison, M. Kirschner, Cytoskeletal dynamics and nerve growth, Neuron 1 (9) (1988) 761–772.
[75] L. Li, W. Zhang, J. Wang, A viscoelastic–stochastic model of the effects of cytoskeleton remodelling on cell adhesion, R. Soc. Open. Sci. 3 (10) (2016) 160539.

[76] E.K. Paluch, I.M. Aspalter, M. Sixt, Focal adhesion–independent cell migration, Ann. Rev. Cell Dev. Biol. 32 (1) (2016) 469–490.
[77] M. Bennett, M. Cantini, J. Reboud, J.M. Cooper, P. Roca-Cusachs, M. Salmeron-Sanchez, Molecular clutch drives cell response to surface viscosity, Proc. Natl. Acad. Sci. U.S.A. 115 (6) (2018) 1192–1197.
[78] Z. Gong, S.E. Szczesny, S.R. Caliari, E.E. Charrier, O. Chaudhuri, X. Cao, et al., Matching material and cellular timescales maximizes cell spreading on viscoelastic substrates, Proc. Natl. Acad. Sci. U.S.A. 115 (12) (2018) E2686–E2695.
[79] A. Elosegui-Artola, X. Trepat, P. Roca-Cusachs, Control of mechanotransduction by molecular clutch dynamics, Trends Cell Biol. 28 (5) (2018) 356–367.
[80] N. Elkhatib, E. Bresteau, F. Baschieri, A.L. Rioja, G. van Niel, S. Vassilopoulos, et al., Tubular clathrin/AP-2 lattices pinch collagen fibers to support 3D cell migration, Science 356 (6343) (2017) eaal4713.
[81] S. Schmidt, P. Friedl, Interstitial cell migration: integrin-dependent and alternative adhesion mechanisms, Cell Tissue Res. 339 (1) (2010) 83–92.
[82] S.B. Hopkinson, K.J. Hamill, Y. Wu, J.L. Eisenberg, S. Hiroyasu, J.C.R. Jones, Focal contact and hemidesmosomal proteins in keratinocyte migration and wound repair, Adv. Wound Care 3 (3) (2014) 247–263.
[83] J. Solon, I. Levental, K. Sengupta, P.C. Georges, P.A. Janmey, Fibroblast adaptation and stiffness matching to soft elastic substrates, Biophys. J. 93 (12) (2007) 4453–4461.
[84] Q. Xu, C. Li, Y. Kang, Y. Zhang, Long term effects of substrate stiffness on the development of hMSC mechanical properties, RSC Adv. 5 (128) (2015) 105651–105660.
[85] S. Jalali, M. Tafazzoli-Shadpour, N. Haghighipour, R. Omidvar, F. Safshekan, Regulation of endothelial cell adherence and elastic modulus by substrate stiffness, Cell Commun. Adhes. 22 (2-6) (2015) 79–89.
[86] A. Jannatbabaei, M. Tafazzoli-Shadpour, E. Seyedjafari, N. Fatouraee, Cytoskeletal remodeling induced by substrate rigidity regulates rheological behaviors in endothelial cells, J. Biomed. Mater. Res. A 107 (1) (2019) 71–80.
[87] J. Rheinlaender, A. Dimitracopoulos, B. Wallmeyer, N.M. Kronenberg, K.J. Chalut, M.C. Gather, T. Betz, G. Charras, K. Franze, Cortical cell stiffness is independent of substrate mechanics, Nat. Mater. 19 (2020) 1019–1025.
[88] A.J. Rice, E. Cortes, D. Lachowski, B.C.H. Cheung, S.A. Karim, J.P. Morton, et al., Matrix stiffness induces epithelial–mesenchymal transition and promotes chemoresistance in pancreatic cancer cells, Oncogenesis 6 (7) (2017) e352.
[89] S.R. Peyton, A.J. Putnam, Extracellular matrix rigidity governs smooth muscle cell motility in a biphasic fashion, J. Cell Physiol. 204 (1) (2005) 198–209.
[90] T.A. Ulrich, E.M. de Juan Pardo, S. Kumar, The mechanical rigidity of the extracellular matrix regulates the structure, motility, and proliferation of glioma cells, Cancer Res. 69 (10) (2009) 4167.
[91] B.L. Bangasser, G.A. Shamsan, C.E. Chan, K.N. Opoku, E. Tüzel, B.W. Schlichtmann, et al., Shifting the optimal stiffness for cell migration, Nat. Commun. 8 (1) (2017) 15313.
[92] A. Elosegui-Artola, R. Oria, Y. Chen, A. Kosmalska, C. Pérez-González, N. Castro, et al., Mechanical regulation of a molecular clutch defines force transmission and transduction in response to matrix rigidity, Nat. Cell Biol. 18 (5) (2016) 540–548.
[93] D.H. Kim, D. Wirtz, Focal adhesion size uniquely predicts cell migration, FASEB J. 27 (4) (2013) 1351–1361.
[94] A. Pathak, S. Kumar, Independent regulation of tumor cell migration by matrix stiffness and confinement, Proc. Natl. Acad. Sci. U.S.A. 109 (26) (2012) 10334–10339.

[95] C.M. Lo, H.B. Wang, M. Dembo, Y.L. Wang, Cell movement is guided by the rigidity of the substrate, Biophys. J. 79 (1) (2000) 144–152.
[96] R. Sunyer, V. Conte, J. Escribano, A. Elosegui-Artola, A. Labernadie, L. Valon, et al., Collective cell durotaxis emerges from long-range intercellular force transmission, Science 353 (6304) (2016) 1157–1161.
[97] M. Cantini, H. Donnelly, M.J. Dalby, M. Salmeron-Sanchez, The plot thickens: the emerging role of matrix viscosity in cell mechanotransduction, Adv. Healthc. Mater. (2019) 1901259.
[98] M. Cantini, P. Rico, D. Moratal, M. Salmerón-Sánchez, Controlled wettability, same chemistry: biological activity of plasma-polymerized coatings, Soft Matter 8 (20) (2012) 5575–5584.
[99] K. Mandal, Z. Gong, A. Rylander, V.B. Shenoy, P.A. Janmey, Opposite responses of normal hepatocytes and hepatocellular carcinoma cells to substrate viscoelasticity, Biomater. Sci. 8 (5) (2020) 1316–1328.
[100] H. Mirzadeh, F. Shokrolahi, M. Daliri, Effect of silicon rubber crosslink density on fibroblast cell behavior in vitro, J. Biomed. Mater. Res. A 67A (3) (2003) 727–732.
[101] M. Murrell, R. Kamm, P. Matsudaira, Substrate viscosity enhances correlation in epithelial sheet movement, Biophys. J. 101 (2) (2011) 297–306.
[102] B. Xu, H. Li, Y. Zhang, Understanding the viscoelastic behavior of collagen matrices through relaxation time distribution spectrum, Biomatter 3 (3) (2013) e24651.
[103] J. Sapudom, S. Rubner, S. Martin, T. Kurth, S. Riedel, C.T. Mierke, et al., The phenotype of cancer cell invasion controlled by fibril diameter and pore size of 3D collagen networks, Biomaterials 52 (2015) 367–375.
[104] S.I. Fraley, P.H. Wu, L. He, Y. Feng, R. Krisnamurthy, G.D. Longmore, et al., Three-dimensional matrix fiber alignment modulates cell migration and MT1-MMP utility by spatially and temporally directing protrusions, Sci. Rep. 5 (2015) 14580.
[105] A. Ray, Z.M. Slama, R.K. Morford, S.A. Madden, P.P. Provenzano, Enhanced directional migration of cancer stem cells in 3D aligned collagen matrices, Biophys. J. 112 (5) (2017) 1023–1036.
[106] K.M. Riching, B.L. Cox, M.R. Salick, C. Pehlke, A.S. Riching, S.M. Ponik, et al., 3D collagen alignment limits protrusions to enhance breast cancer cell persistence, Biophys. J. 107 (11) (2014) 2546–2558.
[107] Q. Shi, R.P. Ghosh, H. Engelke, C.H. Rycroft, L. Cassereau, J.A. Sethian, et al., Rapid disorganization of mechanically interacting systems of mammary acini, Proc. Natl. Acad. Sci. U.S.A. 111 (2) (2014) 658–663.
[108] R. Staneva, F. Burla, G.H. Koenderink, S. Descroix, D.M. Vignjevic, Y. Attieh, et al., A new biomimetic assay reveals the temporal role of matrix stiffening in cancer cell invasion, Mol. Biol. Cell 29 (25) (2018) 2979–2988.
[109] K.S. Kopanska, Y. Alcheikh, R. Staneva, D. Vignjevic, T. Betz, Tensile forces originating from cancer spheroids facilitate tumor invasion, PLoS One 11 (6) (2016) e0156442.
[110] E.E. Charrier, K. Pogoda, R.G. Wells, P.A. Janmey, Control of cell morphology and differentiation by substrates with independently tunable elasticity and viscous dissipation, Nat. Commun. 9 (1) (2018) 449.
[111] M. Dietrich, H. Le Roy, D.B. Brückner, H. Engelke, R. Zantl, J.O. Rädler, et al., Guiding 3D cell migration in deformed synthetic hydrogel microstructures, Soft Matter 14 (15) (2018) 2816–2826.
[112] D.E. Leckband, J. de Rooij, Cadherin adhesion and mechanotransduction, Ann. Rev. Cell Dev. Biol. 30 (1) (2014) 291–315.
[113] N. Borghi, M. Sorokina, O.G. Shcherbakova, W.I. Weis, B.L. Pruitt, W.J. Nelson, et al., E-cadherin is under constitutive actomyosin-generated tension that is

increased at cell–cell contacts upon externally applied stretch, Proc. Natl. Acad. Sci. U.S.A. 109 (31) (2012) 12568–12573.
[114] J. Jurado, J. de Navascués, N. Gorfinkiel, α-Catenin stabilises Cadherin-Catenin complexes and modulates actomyosin dynamics to allow pulsatile apical contraction, J. Cell Sci. 129 (24) (2016) 4496.
[115] B. Ladoux, E. Anon, M. Lambert, A. Rabodzey, P. Hersen, A. Buguin, et al., Strength dependence of cadherin-mediated adhesions, Biophys. J. 98 (4) (2010) 534–542.
[116] Q. le Duc, Q. Shi, I. Blonk, A. Sonnenberg, N. Wang, D. Leckband, et al., Vinculin potentiates E-cadherin mechanosensing and is recruited to actin-anchored sites within adherens junctions in a myosin II–dependent manner, J. Cell Biol. 189 (7) (2010) 1107–1115.
[117] L. Carramusa, C. Ballestrem, Y. Zilberman, A.D. Bershadsky, Mammalian diaphanous-related formin Dia1 controls the organization of E-cadherin-mediated cell-cell junctions, J. Cell Sci. 120 (21) (2007) 3870.
[118] B.R. Acharya, S.K. Wu, Z.Z. Lieu, R.G. Parton, S.W. Grill, A.D. Bershadsky, et al., Mammalian diaphanous 1 mediates a pathway for E-cadherin to stabilize epithelial barriers through junctional contractility, Cell Rep. 18 (12) (2017) 2854–2867.
[119] E.M. Kovacs, M. Goodwin, R.G. Ali, A.D. Paterson, A.S. Yap, Cadherin-directed actin assembly: E-cadherin physically associates with the Arp2/3 complex to direct actin assembly in nascent adhesive contacts, Curr. Biol. 12 (5) (2002) 379–382.
[120] C.M. Cadwell, W. Su, A.P. Kowalczyk, Cadherin tales: regulation of cadherin function by endocytic membrane trafficking, Traffic 17 (12) (2016) 1262–1271.
[121] H. Katsuno-Kambe, A.S. Yap, Endocytosis, cadherins and tissue dynamics, Traffic 21 (3) (2020) 268–273.
[122] R. Levayer, A. Pelissier-Monier, T. Lecuit, Spatial regulation of Dia and Myosin-II by RhoGEF2 controls initiation of E-cadherin endocytosis during epithelial morphogenesis, Nat. Cell Biol. 13 (5) (2011) 529–540.
[123] F. Peglion, F. Llense, S. Etienne-Manneville, Adherens junction treadmilling during collective migration, Nat. Cell Biol. 16 (7) (2014) 639–651.
[124] A.P. Kowalczyk, K.J. Green, Structure, function, and regulation of desmosomes, Prog. Mol. Biol. Transl. 116 (2013) 95–118.
[125] T.E. Kimura, A.J. Merritt, F.R. Lock, J.J. Eckert, T.P. Fleming, D.R. Garrod, Desmosomal adhesiveness is developmentally regulated in the mouse embryo and modulated during trophectoderm migration, Dev. Biol. 369 (2) (2012) 286–297.
[126] S. Wallis, S. Lloyd, I. Wise, G. Ireland, T.P. Fleming, D. Garrod, The isoform of protein kinase C Is involved in signaling the response of desmosomes to wounding in cultured epithelial cells, Mol. Biol. Cell 11 (3) (2000) 1077–1092.
[127] N. Resnik, K. Sepčić, A. Plemenitaš, R. Windoffer, R. Leube, P. Veranič, Desmosome assembly and cell-cell adhesion are membrane raft-dependent processes, J. Biol. Chem. 286 (2) (2011) 1499–1507.
[128] D. Brennan, S. Peltonen, A. Dowling, W. Medhat, K.J. Green, J.K. Wahl, et al., A role for caveolin-1 in desmoglein binding and desmosome dynamics, Oncogene 31 (13) (2012) 1636–1648.
[129] A. Diz-Muñoz, M. Krieg, M. Bergert, I. Ibarlucea-Benitez, D.J. Muller, E. Paluch, et al., Control of directed cell migration in vivo by membrane-to-cortex attachment, PLoS Biol. 8 (11) (2010) e1000544.
[130] T. Colak-Champollion, L. Lan, A.R. Jadhav, N. Yamaguchi, G. Venkiteswaran, H. Patel, et al., Cadherin-mediated cell coupling coordinates chemokine sensing across collectively migrating cells, Curr. Biol. 29 (15) (2019) 2570–2579.

[131] A. Cliffe, D.P. Doupé, H. Sung, I.K.H. Lim, K.H. Ong, L. Cheng, et al., Quantitative 3D analysis of complex single border cell behaviors in coordinated collective cell migration, Nat. Commun. 8 (1) (2017) 14905.

[132] D.J. Montell, W.H. Yoon, M. Starz-Gaiano, Group choreography: mechanisms orchestrating the collective movement of border cells, Nat. Rev. Mol. Cell Biol. 13 (10) (2012) 631–645.

[133] D. Cai, S.C. Chen, M. Prasad, L. He, X. Wang, V. Choesmel-Cadamuro, et al., Mechanical feedback through E-cadherin promotes direction sensing during collective cell migration, Cell 157 (5) (2014) 1146–1159.

[134] D.T. Tambe, C. Corey Hardin, T.E. Angelini, K. Rajendran, C.Y. Park, X. Serra-Picamal, et al., Collective cell guidance by cooperative intercellular forces, Nat. Mater. 10 (6) (2011) 469–475.

[135] M. Vishwakarma, J. Di Russo, D. Probst, U.S. Schwarz, T. Das, J.P. Spatz, Mechanical interactions among followers determine the emergence of leaders in migrating epithelial cell collectives, Nat. Commun. 9 (1) (2018) 3469.

[136] M. Smutny, Z. Ákos, S. Grigolon, S. Shamipour, V. Ruprecht, D. Čapek, et al., Friction forces position the neural anlage, Nat. Cell Biol. 19 (4) (2017) 306–317.

[137] B. He, K. Doubrovinski, O. Polyakov, E. Wieschaus, Apical constriction drives tissue-scale hydrodynamic flow to mediate cell elongation, Nature 508 (7496) (2014) 392–396.

[138] R.J. Tetley, M.F. Staddon, D. Heller, A. Hoppe, S. Banerjee, Y. Mao, Tissue fluidity promotes epithelial wound healing, Nat. Phys. 15 (11) (2019) 1195–1203.

[139] M. Sadati, N. Taheri Qazvini, R. Krishnan, C.Y. Park, J.J. Fredberg, Collective migration and cell jamming, Differentiation 86 (3) (2013) 121–125.

[140] T.E. Angelini, E. Hannezo, X. Trepat, M. Marquez, J.J. Fredberg, D.A. Weitz, Glass-like dynamics of collective cell migration, Proc. Natl. Acad. Sci. U.S.A. 108 (12) (2011) 4714–4719.

[141] S. Garcia, E. Hannezo, J. Elgeti, J.F. Joanny, P. Silberzan, N.S. Gov, Physics of active jamming during collective cellular motion in a monolayer, Proc. Natl. Acad. Sci. U.S.A. 112 (50) (2015) 15314–15319.

[142] D. Bi, J.H. Lopez, J.M. Schwarz, M.L. Manning, A density-independent rigidity transition in biological tissues, Nat. Phys. 11 (12) (2015) 1074–1079.

[143] D.B. Staple, R. Farhadifar, J.C. Röper, B. Aigouy, S. Eaton, F. Jülicher, Mechanics and remodelling of cell packings in epithelia, Eur. Phys. J. E 33 (2) (2010) 117–127.

[144] L. Atia, D. Bi, Y. Sharma, J.A. Mitchel, B. Gweon, S.A. Koehler, et al., Geometric constraints during epithelial jamming, Nat. Phys. 14 (6) (2018) 613–620.

[145] A. Palamidessi, C. Malinverno, E. Frittoli, S. Corallino, E. Barbieri, S. Sigismund, et al., Unjamming overcomes kinetic and proliferation arrest in terminally differentiated cells and promotes collective motility of carcinoma, Nat. Mater. 18 (2019) 1252–1263.

[146] R. Staneva, F. El Marjou, J. Barbazan, D. Krndija, S. Richon, A.G. Clark, et al., Cancer cells in the tumor core exhibit spatially coordinated migration patterns, J. Cell Sci. 132 (6) (2019) jcs220277.

[147] R. Alert, X. Trepat, Physical models of collective cell migration, Annu. Rev. Condens. Matter Phys. 11 (1) (2020) 77–101.

[148] I. Pajic-Lijakovic, M. Milivojevic, Jamming state transition and collective cell migration, J. Biol. Eng. 13 (1) (2019) 73.

CHAPTER FOUR

# Fine-tuning viscoelasticity: the key to collectively move in vivo

**Jaime A. Espina and Elias H. Barriga**
Mechanisms of Morphogenesis Lab, Gulbenkian Institute of Science (IGC), Oeiras, Portugal

## 4.1 Introduction

### 4.1.1 Collective cell migration

The coordinate movement of large groups of cells is essential for tissue rearrangements that drive biological processes such as embryogenesis, wound healing, and cancer. Migratory cell groups move in a coordinated manner, in the same direction and at similar speed. To do this, the group of cells synchronize their response to the environment, ensuring that the whole cluster behaves in a coordinated manner [1,2]. An advantage of moving as cell groups is the high degree of directionality featured by cell clusters compared to single cells, which can migrate faster but tend to change their direction quite often [1–3]. This high order of migratory efficiency that cellular clusters attain seems to emerge from the enormous level of coordination and communication that cells exhibit while migrating as a group. To achieve this coordination and successfully migrate, cells within the cluster require to finely tune their mechanical coupling between neighbors, fluently communicate, and integrate their cytoskeletal activities. This secures that environmental information that provides clusters with directional information is effectively internalized from the surroundings into each cell and transmitted across the migratory cluster. Hence, we can define collective cell migration (CCM) as a highly synchronized and cooperative mode of cellular motion [1–9]. A large body of evidence suggests that fine-tuning cell–cell adhesions within the cluster is essential for CCM to emerge [10]. Conversely, recent work shows that the viscoelastic properties of the environment can modulate cell–cell adhesions and with that the very own elastic properties of the cluster and

*Viscoelasticity and Collective Cell Migration.*
DOI: https://doi.org/10.1016/B978-0-12-820310-1.00003-3

in turn CCM [11]. This strongly suggests that CCM emerges as a context-dependent process that is in part determined by a feedback between the viscoelastic properties of the environment and those from the migratory cluster.

### 4.1.2 Collective cell migration in challenging environments

In order to interact with different chemical and mechanical properties of their environments, migratory cells require to modify their behavior [12,13]. Although the role of surrounding chemical cues on migratory cell behavior has been extensively studied [14–17], little is known about how migratory cells interact with the mechanical environment [15]. Some of the mechanical cues that cells encounter during in vivo migration are adhesion (friction), confinement, stiffness of migratory substrates, shear flow of extracellular fluids, topology, and density of the surrounding tissues or extracellular matrix (ECM) [18]. In order to confront these physical challenges in 3D environments, migratory cells respond by rapidly changing their viscoelastic properties, thus allowing the cells to resist deformation or to deform to extreme limits [18,19]. For single migratory cells, this only means changes of shape, but cells within a cluster have to respond as a coordinated group, changing their viscoelasticity as a supracellular unit [20] (Box 4.1, viscoelasticity). In this chapter our objective is to show that CCM in varying contexts is only achieved through the interplay of the environmental viscoelasticity with a fine regulation of viscoelastic properties of migrating clusters. To do this we will summarize and discuss some of the mechanisms identified in single and collective migrating cells to dynamically modify their viscoelasticity in response to different physical constraints.

## 4.2 Viscoelasticity of cellular components

Most of the tissues can reshape and adapt in order to respond to changes in their physical environment. For instance, cells cultured in 2D or 3D micropatterned surfaces tend to modify their shapes adjusting to the geometry of the surrounding pattern (extensively reviewed in Ref. [21]). On the other hand, in vivo migratory cells undergo enormous reshaping events, as they require to deform when moving across the

## BOX 4.1 Physical and biological terms glossary.

*Plastic deformation*: a permanent change of shape observed in rigid materials that are not reversible after removing the applied mechanical stress.

*Viscous deformation*: The resistance of a fluid to flow is easily explained by comparing water and syrup; water flows faster than syrup because it is less viscous, hence it opposes less resistance to flow.

*Elastic deformation*: The propensity of a material to return to its original position after a given load is removed. Elastic materials strain upon load and then load is removed they return to their original state, that is, elastic rubbers.

*Viscoelasticity*: A material that exhibit elastic and viscous characteristics when exposed to deformation is a viscoelastic material. Viscous or elastic response to deformation depends on the timescale, that is, when stress is removed the initial response is rapid and resembles the one of an elastic material, in a second phase the response will resemble more the one of a viscous material.

*Hookean behavior*: Linear response to deformation observed in a perfectly elastic material. Stress applied to deform a material and the strain are equal.

*Stress*: The term used to express the load in terms of force applied to the cross-sectional area of a given object. It can also be defined as the force required to deform a material. A material can experience two types of stress, normal or shear stress. When a perpendicular force is exerted to an object it is experiencing *normal stress*, but if the force acts parallel to the surface of that object, we talk about *shear stress*.

*Strain*: Deformations are normally observed when a material experiences stress and the ratio between its original shape and the deformation that the object suffered upon a given stress is the strain. As for the stress, a material can experience two types of strain, normal or shear strain. *Normal strain*: when a material is perpendicularly stressed. *Shear strain*: when stress acts parallel to the surface of a material. Blood flow on blood vessels for instance.

*Tension*: Is the reaction force applied by an object that has been stretched on the object that stretched it. Tension is a pulling force and as such it is active.

*Stiffness*: Is defined as the rigidity of an object or the extent of its resistance to deformation. The complementary property to stiffness is flexibility. To deform a stiff material, it takes a higher force than to deform a soft material.

*Epithelial-mesenchymal transition (EMT)*: Canonically EMT can be defined as the process where apico-basely polarized epithelial cells undergo a series of biochemical changes that leads to acquire front to rear polarity and motile mesenchymal phenotype. Nonetheless, this process is way more plastic and complex. The landscape of this complexity and a more up to date definition has been recently provided in Yang et al [182].

challenging geometries of their native extracellular environment [22,23]. Materials react differently when submitted to mechanical stress. For instance, some rigid materials deform in a plastic manner meaning that they reshape and cannot return to their original form (Fig. 4.1, curve C; Box 4.1, plastic deformation). On the other hand, the level of deformation that biological materials exhibit can differ depending on the

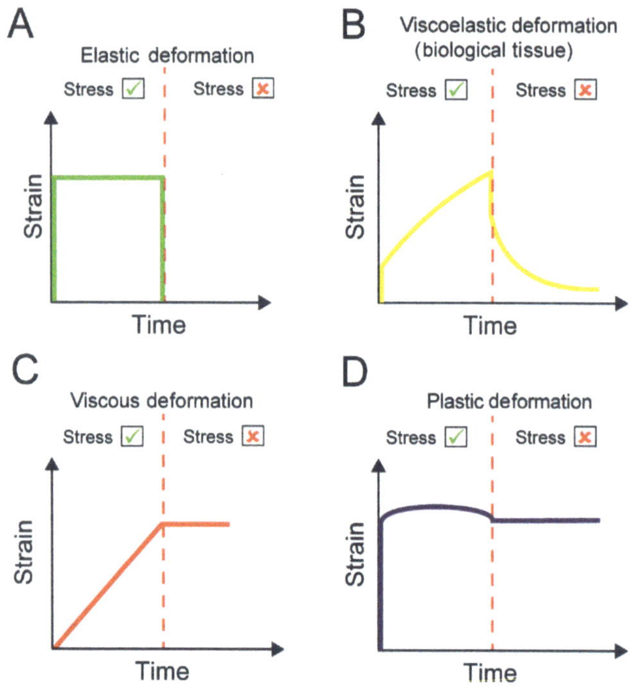


**Figure 4.1** ***Types of mechanical deformation in response to mechanical stress over time.*** (A–D) Drawings represent charts for the strain response and type of deformation of materials to mechanical stress along time in hypothetical stress relaxation experiments. (A) Materials with elastic deformation display proportional strain and stress. Elastic materials instantaneously strain upon stress and this deformation is fully and instantaneously reversed after removal of the applied stress. (B) Materials with viscoelastic deformation display time-dependent strain with an instantaneous elastic phase followed by time-dependent strain. Once stress is removed strain rapidly decreases as for elastic materials to then decrease, to some extent, in a time-dependent manner. Vertical dashed red lines indicate when mechanical stress is removed in the hypothetical stress relaxation experiment. (C) When a viscous material is stressed, deformation is not instantaneous and it strains linearly with time in an irreversible manner (once applied stress is removed strain remains the same). (D) Materials with plastic deformation display proportional strain and stress, but once stress is removed the material does not return to its form.

magnitude of force to which they are exposed, the time that they are exposed to this force, and the type of tissue that is being analyzed. Regardless, a common and distinctive feature of most biological tissues is that they respond as a nonlinear viscoelastic material when challenged against mechanical stress [24]. Viscoelastic biological materials can be defined as those tissues that exhibit both viscous and elastic characteristics when exposed to physical deformation (extensively discussed in Refs. [24–26] and depicted in Fig. 4.1; Box, 4.1, elastic deformation). Intriguingly, this definition of viscoelasticity is relatively recent, as based on pioneer studies, biological materials were initially assumed to have an elastic, also known as "Hookean behavior," in response to applied deformation tests (Box 4.1, Hookean behavior). The idea that tissues are "Hookean materials" predominated in the field until the second half of the 20th century. It was in these years when several groups started to report that if considering time as an independent parameter, the curves obtained after experimental deformation of a biological material would be rather nonlinear, when compared to the response of a known "Hookean material," used as experimental controls (for pioneer descriptions refer to Ref. [24]). In simple words, while if we analyze the immediate response of a biological material to deformation we will observe an elastic behavior, analyzing at a longer timescales will elicit a viscous recovery upon tissue deformation (Fig. 4.1, curve B). Accordingly, a common feature of biological tissues is that in order to resist deformation they stiffen in response to strain, by a process named strain stiffening [27,28] (Box 4.1, strain & stiffness). Due to its biomedical potential, this concept has been revisited and fully characterized by researchers that used purified and in vitro reconstituted cellular components—this simple experiment ultimately demonstrated that regardless of their initial nature, biological materials are intrinsically viscoelastic, and that they can undergo strain stiffening [29–31]. Currently, a clear framework where the viscoelastic response of a material is considered within three temporal regimes has been generated [26,32,33]. Briefly, (1) short-term regime for milliseconds to seconds, corresponding to a subcellular level and where, upon deformation it can be observed a linear relationship between stress–strain (linear regime); (2) mid-term for several seconds to minutes, corresponding to a cellular level; and (3) long-term viscoelastic regime, occurs in the range of several minutes to hours, corresponding to a supracellular level [26,32,33]. Of these three regimes, short and middle terms have been extensively studied, whereas long-term viscoelastic regimes are less clear [26]. Regardless, some explanations have been proposed to explain this long-term response displayed at the tissue level or by migrating cellular clusters—that is, when responding to stress, single cells can rearrange their cytoskeleton in a rapid manner [34], while group of cells require more time to do so, as they need to coordinate their cytoskeletal activities in a supracellular response [26]. This can also explain why collectively migrating cells migrate slow but with high degree of directionality and very persistently, while individual cells migrate with lower directionalities but with extremely high velocity. Hence, times matter when it comes to define the dual "viscous-elastic" behavior of biological materials [24], particularly when studying CCM [20].

Consequently, in the following paragraphs we discuss the collective movement of migrating cells that respond to the physical challenges of their environment in a rather long-term viscoelastic regime. Furthermore, we provide evidence of the cellular and molecular mechanisms that are thought to sense viscoelastic properties of the environment and transduce them into a mechanical cue that impact the molecular players involved in fine-tuning the viscoelasticity of migrating clusters.

## 4.3 Sensing and transducing environmental viscoelasticity

In order to adapt to the mechanical stress imposed by their migratory environment without compromising their integrity, migrating cells require to constantly adjust their shape. Intriguingly, cell shape can be instantaneously affected by rapid forces that act directly on cells, or they can use signaling mechanisms that allow cells to adapt to new mechanical environments in a longer timescale [20]. These signaling mechanisms involve sensing structures composed of molecules known as mechanosensors and adapter molecules known as mechanotransducers, which are in charge of transforming mechanical inputs into a cellular and eventual response [35]. While sensing occurs in short timescales, transduction of mechanical cues may work in mid-term or even long-term temporal scales, depending on the degree of response mediated by a given transducer. Thus, shape changes and adaptations of cells to their physical environments can be also categorized according to timescale. For instance, short- and mid-term responses correspond to cytoskeleton rearrangement mediated by posttranslational modifications of proteins related mainly to adhesion, polarity, and contractility, which can in turn rapidly affect cell behavior [36–38]. On the other hand, long-term responses to mechanical stimuli involve changes in gene expression profiles that can also affect cell behavior but often confer cells with new fates and specified functions [39,40]. Sensing and transduction of mechanical inputs are complex processes that require to integrate into the convoluted signaling pathways that operate in a cell. To simplify these complicated concepts, herein we just refer to the most studied sensing and transduction structures and their constituent molecules.

### 4.3.1 Focal adhesions

Focal adhesions (FAs) are structures that connect cells to the ECM of their migratory substrate and as such, they have been mostly studied in cells that migrate by performing cell-substrate adhesion-dependent migration [41]. FAs are composed, formed by dozens of proteins whose overall function is not only to provide adhesion to the substrate but also to transmit mechanical cues from the ECM by working as mechanotransduction hubs [42]. Owing to its complex and dynamical composition, it has been extremely complex to dissect the role of FAs-related proteins in mechanosensing and mechanotransduction. Nonetheless, thanks to recent efforts the field started to reach a consensus in which the FA itself can be divided into two functional domains, named mechanosensing and mechanosignaling modules that together compose the cell-substrate and mechanotransduction unit known as FAs [42]. One of the most important proteins in FAs is the Integrin receptors. These are transmembrane proteins that can bind to the ECM and recruit many adapters in order to strengthen the FA and link the FAs to the cytoskeleton [43], and they are common to both, mechanosensing and mechanosignaling modules [42]. This connection is mediated by Vinculin and Talin that allow cells to transfer force experienced at the FAs into the cytoskeleton [44–48]. In turn, the cytoskeleton, through myosin-II contractility, reinforce the FA dynamics, which leads to cell shape changes and migration of cells placed in stiff surfaces or substrate deformation when cells are placed on soft surfaces [49–51]. Based on this function, Vinculin, Talin, Tensin, and other FAs proteins are considered part of the mechanosensing FA module [42]. As stated, FAs do not only sense mechanical stimuli, but they are also a structure that allows cells to convert these mechanical stimuli into signals that promote diverse cellular activities and with that cell migration. For this, several proteins such as focal adhesion kinase, Paxilin, Src, among many others are in charge of promoting downstream signaling at the FAs, and owing to this activity they are part of the mechanosignaling module [42]. For instance, Twist1, an e-cadherin repressor and Epithelial-mesenchymal transition (EMT) inducer (Box 4.1, EMT), can upregulate the scaffold protein Nedd9 and Dock3, in turn, these two proteins can form a complex to activate Rac and thus, achieve mesenchymal movement [52]. Another example is the p130CAS protein. It has been proven that the CAS substrate domain submitted to mechanical extension in vitro present an increment of phosphorylation by Src kinases. Moreover, phosphorylated CAS is present in peripheral regions of cells, which is correlated to an extended CAS and higher traction forces, suggesting that CAS could be a sensor and force transducer within the cell [47].

### 4.3.2 Cytoskeleton

Another important structure mediating sensing and transductions of mechanical cues is the cytoskeleton. It is known that cytoskeleton undergoes significant changes during adaptation to physical stress and has important roles in many of the mechanisms proposed for mechanosensing and mechanotransduction [53]. Cytoskeleton is made by polymers that bind to form three types of fibers: intermediate filaments, actin filaments, and microtubules [54–56]. Viscoelastic properties of the cytoskeleton rely on both their passive mechanical properties and the dynamics of its fibers, these dynamics are thought to confer cytoskeletal fibers with a viscoelastic behavior [57–59]. Cytoskeleton dynamics depend on the rate of polymerization/depolymerization of its fibers as well as the degree of cross-linking between the monomer that composes them. Thus, it is intuitive to think that the combination of these parameters will determine if a cell maintains its form or reshape when exposed to mechanical stress [60]. For the passive mechanical properties of the cytoskeleton, research in many models have found that intermediate filaments are the softer of the three fibers, followed by the more rigid actin filament, and microtubules being the stiffer fiber of the cytoskeleton [54,56,61–64]. Mechanisms of modulation of cytoskeletal stiffness have been proposed, that is, particle flow analysis has allowed the measurement of mechanical strain contained in the cytoskeletal network of epithelial cells [65,66]. For the cytoskeleton dynamics, polymerization and depolymerization are regulated by many proteins. For instance, nucleation factors are in charge of initiating and extending the length of a fiber, capping factors can stabilize a fiber and terminate their growth, and finally, fibers can be disassembled by depolymerizing factors [43,55,67–72]. On the other hand, the elastic state of the cytoskeleton is also determined by its architecture, which depends on the cross-linking degree of its fibers [57,59,73]. Thus, cross-linking fibers as well as the activity of molecular motors like myosin-II have been proposed to play a role in the determination of cytoskeletal architecture and in consequence are related to the cytoskeletal viscoelastic properties [51,74]. For instance, it has been recently shown that the activity of the cortical nucleation promoting factor Spin90 coordinates the function of the actin nucleators Arp2/3 and mDia1 to control changes in cell mechanics, shape, and progression of the cell cycle [60]. Interestingly mechanical cues can reciprocally stimulate cross-linkers, motors, and filaments interaction. For example, real time live imaging of stretched individual cells revealed an actomyosin contractility response to environmental stiffness. Consequently, these authors also found that the binding dynamics of myosin to actin stress fibers correlate with an increment in applied force [34,49]. On the other hand, assays measuring actin assembly have shown that newly branched actin network can increase mechanical efficiency and density due to adaptation of force-generating actin networks to environmental mechanical cues. This study revealed a force-feedback loop between intra- and extracellular forces [75] and with that strongly suggested that this force-feedback between external and internal forces could help migratory cells to adjust their viscoelastic properties when exposed to a physically constrained microenvironment.

### 4.3.3 Nucleus

When migrating across their complex 3D native environments cells undergo major shape changes that in some cases can push their physical limits of integrity. Interestingly, in these confined environments, the nucleus has been found to be a factor that limits the extent to which cells deform [22,76]. The nuclear structure is limited by an outside lipid membrane, the nuclear membrane, covered on the inside with a dense network of proteins, named the nuclear lamina. The nuclear lamina is made mainly of type V intermediate filaments that are known to be formed, mostly, by a family of proteins known as nuclear Lamins [77,78]. Interestingly, it has been observed that inhibition of Lamins impairs DNA integrity as Lamin-deficient nuclei reach limits of deformation that induce a loss of chromatin architecture and DNA damage [79–82], cellular response to mechanical inputs, suggesting that these proteins are fundamental for the mechanical coupling between cytoskeleton and nucleus [79–82]. A cellular response to the physical environment, which is known to be modulated by Lamins, is cell differentiation. In a very elegant study, a group of authors showed that low stiffness values within undifferentiated cells are correlated with low levels of Lamins and differentiation to fat tissue, while accumulation of collagen would rather lead to high stiffness values that in turn correlates with increment of Lamin-A and cell differentiation to bone. These data suggest that Lamins has a role in mediating mechanically driven tissue differentiation [83]. As stated, during cell migration, the nucleus affects the degree of deformation of the cell by modulation of the density of the nuclear lamina, and in this way acts as a limiting factor for cells passing through confined spaces [22,82]. High levels of Lamins oppose to shape changes, while cells with low levels have a more flexible nucleus, however lowering Lamins levels to a minimum, have an effect on cell survival due to the above-mentioned effects on DNA damage [84]. More recently, the combination of quantitative biology and microfabrication techniques has allowed to dissect some of the mechanisms regulating nuclear lamina accumulation and nuclear viscoelastic properties [85]. In this chapter, destabilization of nuclear lamina of dendritic cells correlates with accumulation of the actin nucleator Arp2/3, allowing cells to reshape their nuclei in order to migrate over confined environments [85]. In the same line, the nucleus itself has been recently shown to work as a mechanosensor during cell migration across in vitro designed complex geometries [86]. In this study the authors show that leukocytes survey their environment to find the migratory path that offers the lowest resistance and that the nucleus is a key player in that surveying processes [86]. This may be relevant in vivo, where cells need to modify their viscoelasticity while migrating across confined spaces.

Sensing mechanisms described above are the most studied cases; however, during migration, the precise mechanism of sensing and adaptation to mechanical cues is starting to be studied. In vitro studies have shown that single cells can sense substrates with a depth of about 15 μm. Accordingly, in vivo Fibronectin-based ECM of Xenopus, chick, or zebrafish embryos (Barriga's lab unpublished) has a thickness of about 10 μm. Taken together, these data suggest that the models described above for mechanosensing fits with the sensing of the ECM for single migratory cells. However, it has been proposed that group of cells exhibit over 10-times larger forces into the substrate suggesting that cell clusters can sense environmental mechanical changes deeper than the one proposed for single cells [87]. How cells can achieve this deeper sensing ability? And also is this larger sensitivity also present and necessary *in vivo*? remains as open questions.

Regardless the mechanism that mediates sensing and transduction of mechanical inputs, the downstream effect of these understudied cues is known to directly feedback into the cytoskeleton and other cellular components. The outcome of this feedback is the cellular response to mechanics that involve changes in cell shape, activation of mechanically sensitive channels [88–90], or opening mechanically responsive nuclear pores allowing internalization and activation of specific transcription factors, such as Yap/Taz, member of the Hippo pathway—and with that to the modification of cell behavior by establishing a new genetic profile that will support cell migration and eventually differentiation [91–94].

### 4.3.4 Modifying cell–cell junctions to modulate viscoelasticity and synchronize CCM

As discussed previously, there is an increasing knowledge on how viscoelastic properties influence single-cell migration, and how these properties can be modulated to ensure efficient migration in complex 3D environments. However, migratory clusters not only have to synchronize their velocity and directionality but they also require reaching high levels of coordination to modify its viscoelastic properties as a supracellular entity. To achieve this coordination, cells within the cluster require to efficiently communicate molecular and mechanical information that individual cells within the cluster survey from the migratory environment [95,96]. Thus, in order to stay together and communicate to efficiently migrate cells within the cluster rely on their cell–cell junctions. For instance, it has been shown that resistance to shape changes [97], as well as the degree of fluidity of a migratory group depends on the modulation of the molecules that compose their cell–cell junctions [98]. The main types of cell–cell junctions are Adherens (AJ), Gap (GJ), and Tight junctions [99,100]. Interestingly, the constituent proteins of these junctional complexes are known to respond to mechanical cues. For example, organization and reinforcement of AJs at cell–cell contacts can be driven by actomyosin contractility [101,102]. Moreover, GJs were recently shown to coordinate force transmission between adjacent cells during collective migration

[103], and one of the main components of GJs, Connexins, has been shown to be mechanoresponsive proteins [104].

Out of the three types of cell–cell junctions, AJs seem to be the most relevant or the most studied during CCM. AJs maintain the integrity of the migrating cell group and promote cell coordination while allowing cellular rearrangements [105]. This means that through these contacts, the migratory cell cluster not only can mechanically couple but also AJs are an important source of signaling that helps to coordinate cell behavior. In this line, it has been shown that AJs can have a dual mechanical role during migration, on one side, allowing cells to exert tractions forces over the migratory surface, and on the other, coordinating the sensing and transmission of the force within the group. Thus, AJs are seen so far as the hinge mediating the force-feedback between the environment and the migratory cluster [100,106–109]. Therefore, it is clear that an efficient way to modulate the viscoelastic properties of a migratory cluster in a coordinated manner is through fine-tuning of their AJs.

### 4.3.5 Matching viscoelasticity in embryogenesis and wound healing

A good example of how collective migratory group of cells modulate their viscoelastic response at mid- and long-term timescales during embryogenesis is the *Xenopus* cranial neural crest. These cells are induced (specified) in dorsal territories, between the neural and nonneural ectoderm, from where they migrate as a cell collective to form most of the vertebrate head. When migrating in vivo, neural crest cells move across confined spaces by literally "bulldozing" their way between epidermis and mesoderm. To do so, neural crest cells evolved molecular mechanisms that allow them attaining group fluidity while staying together by fine-tuning their AJs [110–113]. First, in order to delaminate and migrate as a collective mesenchymal population, neural crest cells change the nature of their AJs, switching from E-cadherin-based junctions to N-cadherin [111]. Once delaminated, neural crest clusters maintain cohesivity and fluidity by dynamically fine-tuning the levels of N-cadherin at the cell–cell junction. This fine-tuning mechanism involves N-cadherin endocytosis that depends on lysophosphatidic acid and its receptor (LPAR2). Using phase map calculation and microfabrication experiments, these authors showed that this mechanism confers cells with higher fluidity rates while collectively migrating [98].

Another example of collective migratory cells that rely on the dynamic modulation of N-cadherin is astrocytes. In these cells, a fluid-like state during migration is achieved by a retrograde flow of N-cadherin that is essential to decrease AJs strength. The mechanism for this flow is determined by differential phosphorylation of p120-catenin within the cell, which in turn promotes N-cadherin internalization from the membrane, then N-cadherin translocates from the rear to the front of the cells to form new AJs [105,114]. Zebrafish blastodermal cells also modulate their cadherin dynamics to fluidly migrate as a collective. Here cells do not switch between cadherins like the other examples, but instead, they lose their polarity by increasing E-cadherin turnover dynamics via a process that involves EGF-driven E-cadherin endocytosis [115]. Another collective migratory tissue is the zebrafish lateral line primordia. This structure uses both E- and N-cadherin to migrate but differentially polarizes them over the structure leaving a mesenchymal front with high levels of N-cadherin and a more epithelial real with high levels of E-cadherin, and where cells will be deposited over the extent of the tail as neuromasts [116–118]. This mode of migration allows the migratory primordium to have a motile front of cells opening the space between epidermis and mesoderm, enabling the more static and "solid" rear cells to do both, efficiently deposit neuromasts, while resisting deformation maintaining this space opened. Even when the polarity in the cadherin expression and motility is clear [118,119], little is known about the mechanisms for establishing this polarity or whether the rear epithelial-like cells actually resist deformation.

The cases discussed above exemplify how cells modulate their AJs to acquire a more fluid-like configuration in order to move through confined spaces. In contrast, other cell groups require an increment in their resistance to deformation to "defeat" their challenging environments. These groups tend to maintain or increase the levels of E-cadherin in order to achieve a more rigid-like viscoelasticity that not only allows the cells to resist deformation when are exposed to environmental mechanical stress but also allows the cells to apply higher tractions forces during migration. One example of this case is the collective migration of Drosophila border cells. These cells are a group of six to eight cells that emerge from the follicular epithelium of the ovary and migrate to form the micropyle, a structure required for sperm entry [120,121]. Migrating border cells are surrounded by enormous nurse cells, so to migrate these cells have to resist the external pressure of the nurse cells while opening the space to pass through them. In this context, border cells exhibit an increment in the levels of E-cadherin from the moment of delamination that is maintained over the course of migration. This E-cadherin increment allows the cells to oppose the external pressing from the nurse cells, in this way maintaining their integrity [122–126].

During wound healing, E-cadherin also plays a role in coordinating the collective migration of epithelial. During in vitro wound healing assays, the front cells in contact with the damage form an actomyosin ring to close over site of the damage. The transmission of traction forces generated by this actomyosin ring is done through E-cadherin-based AJs [127,128]. Another morphogenetic event, mammary branching morphogenesis, also relies on epithelial collective migration with cadherin-dependent modulation [129]. In this case, activation of RAF and TGF-β activity promotes tissue rearrangement and E-cadherin endocytosis [130,131]. As discussed, AJs modulation through cadherin endocytosis has been proposed in many systems [132–136]. Noteworthy, the timescales of the mechanisms described above can be placed in the long-term regime of viscoelastic response proposed to act in collective migration [26]. For instance, cadherin internalization takes place in the order of minutes [137] and changes in gene expression can take even longer. Thus, mechanisms of cadherin modulation probably have a role during the long-term viscoelastic regime.

Finally, AJs protein levels have been shown to be controlled by mechanical cues in several in vitro and in vivo systems [11,138]. However, the level of this regulation still needs to be investigated, as it remains unknown whether it occurs at the mRNA or protein level. This is not trivial, as protein regulation occurs in shorter timescales than transcriptional regulations, this could shed light on how depending on their level of regulation AJs may act as sensors, transducers, or effectors of the viscoelastic response of cellular clusters to the mechanical properties of their environment. Thus, while neural crest cells represent a compelling model where fine-tuning of AJS has been linked to viscoelasticity, the CCM of all the examples discussed above has the potential to be explained as an emergent property of the fine-tuning of their viscoelastic properties via AJs regulation—an idea that would be interesting to explore by the researcher using these animal and cellular models of CCM.

### 4.3.6 Matching viscoelasticity in cancer cells

Cancer metastatic cells are highly migratory, and similarly to the examples discussed above they use multiple mechanisms of AJs control to confront their migratory. For example, invasive breast cancer cells secrete metalloprotease that initiates an EMT-like process. In this mechanism, MMP-3 metalloprotease secreted to the media cleaves E-cadherin extracellular domain, decreasing in this way the levels of cell adhesion strength. This reduction of E-cadherin is subsequently reinforced by the expression of the E-cadherin transcriptional repressor Snail, this together with activation of Rac1 at the leading edge of these cells leads to the acquisition of an invasive phenotype [139]. Moreover, it is widely accepted that a decrease of E-cadherin density in AJs reduces rigidity of the cell contacts allowing cancer cells to migrate in a fluid-like manner [140–142]. Furthermore, EMT is a reversible process where cell can transit back from a mesenchymal-like phenotype to an epithelial-like phenotype by undergoing mesenchymal-to-epithelial transition (MET)—a process that can be activated in part by mechanisms that are known to inhibit EMT [143,144]. MET has been proposed as one of the main mechanisms of secondary tumor formation from mesenchymal circulating tumor cells [145,146]. This indicates that MET can supply a mechanism for cells to stop migration, increasing adhesion, aggregation, proliferation, and as a result becoming more aggressive. We refer again to the regulation of MET by viscoelastic properties of the environment in Section 4.3.

## 4.4 Environmental viscoelasticity triggers and directs collective migration

Until here, we have discussed how a variety of migratory cell clusters modulate their viscoelastic properties to collectively migrate. However, CCM depends not only on the inner viscoelastic properties of the migrating clusters but it also depends on the viscoelastic properties of the migratory environment [20]. Indeed, it becomes every time clearer that environmental viscoelasticity does not only regulate CCM but it also regulates a wide range of processes, such as cell fate acquisition and tissue regeneration [40,147–152], and is extensively reviewed in Ref. [153]. In general, the CCM of a cellular cluster involves three main steps: First, for the onset of CCM, cells delaminate from their original position. Then, the cluster requires to directionally migrate to reach their target tissues or final anatomical positions. Finally, clusters require to stop migrating, aggregate into their new environments, and eventually differentiate into new cell types. Viscoelastic properties from the migratory environment have been proposed to participate in each one of these steps—in this part, we discuss in vitro and in vivo examples showing how the viscoelastic properties of the migratory environment can help clusters to delaminate, directionally migrate, and differentiate into new cell types.

### 4.4.1 Environmental viscoelasticity at the onset of CCM

To this date there is compelling evidence showing that stiffer surfaces increase the strength of ECM-Integrin and Integrin-cytoskeleton adhesion in cultured cells, leading to the generation of stronger traction forces and with that to efficient cell migration [154–161]. Also, cancer cells exhibit more aggressive phenotypes when plated on stiffer environments [162–165]. For instance, ECM increment in stiffness mediated by higher collagen cross-linking driven by the collagen cross-linked LOX can enhance metastatic behavior and aggressiveness of cultured mammary gland tumor cells [163]. Another process where substrate stiffness has been shown to be relevant for cell migration is during wound healing. For example, monolayers of mammary epithelial cells submitted to wound healing assays displayed more coordinated migratory behavior as well as faster propagation forces leading to more efficient migration when cultured on substrate of higher matrix rigidity, and the mechanism behind this behavior involves a feedback loop between force-dependent formation of AJs and myosin contractility [163,166]. These in vitro studies show that the amount of cadherin in AJs can be modulated by the changes in the viscoelastic properties of the microenvironment. These observations could have countless impact in the way we approach several pathologies, which are related to CCM but normally just addressed from a molecular perspective. Thus, whether these results are recapitulated in models that allow studying CCM in vivo is currently a hotspot of research, and great advances have been made in describing how mechanical cues can trigger and facilitate CCM in vivo. For example, it has been shown that during enteric system development an initial increase of stiffness promotes gut invasion by neural crest-derivative cells. However later in development, stiffness keeps increasing and the rate of gut colonization by neural crest-derivative cells is drastically reduced (Fig. 4.2B). The mechanism of stiffening in this case is collagen accumulation in the gut ECM. In response to this, migratory cells would normally secrete proteases to degrade the dense matrix creating space to migrate; however, in this case an excess of collagen in the environment precludes enteric neural crest cells to further colonize the gut [147]. Another example

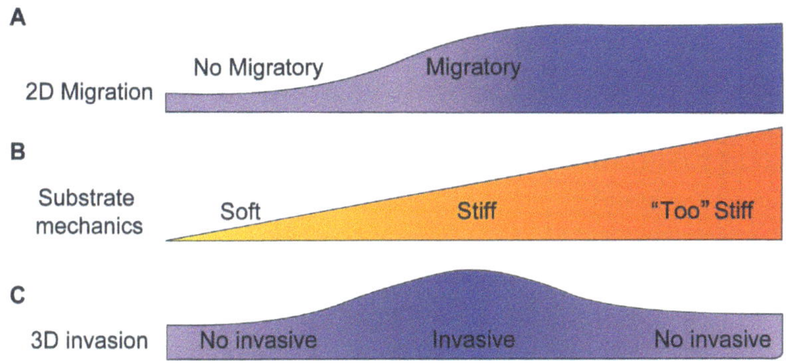


**Figure 4.2** ***Environmental mechanics modulate CCM behavior.*** (A) On 2D substrates, the migratory behavior is enhanced by increasing surface stiffness [11]. (B) Fine-tuning viscoelastic properties of the environment is essential to allow efficient 3D cell invasion while a soft matrix will inhibit migration and invasion; a stiffer matrix will favor invasion. Nonetheless, "too" stiff environments will inhibit collective cell invasion [147].

in vivo refers to *Xenopus* neural crest cells. In this model, the onset of neural crest migration is promoted by changes in the viscoelastic properties of the surface [11] (Fig. 4.2A). Here it was shown that prior to migration, cell already exhibits the expression of all the markers related to a migratory behavior but cells do not migrate, it is only when the mesoderm, the substrate of neural crest cells, increases its stiffness that the neural crest cluster starts its CCM. The mechanism underlying this increment rigidity depends on augment of cell density driven by conversion and extension movements of mesodermal cells starting on gastrulation. Thus, this example does not only show that viscoelastic properties of the environment trigger CCM in vivo, but that the long-range mechanical communication between neighboring tissue may be the key to synchronize morphogenetic events.

## 4.4.2 Environmental viscoelasticity and durotaxis

Environmental viscoelastic properties not only can trigger the onset of migration but also have been proposed to be the signaling cue regulating the directionality of the migratory cells. The process where the rigidity of the environment defines the direction of migration is called durotaxis and in vitro it has been shown to occur in single and collective migration. The first observation of how changes in the viscoelastic property of the media can promote directed cell migration was performed on fibroblasts plated in gradients of matrix rigidity. In this chapter, they observed that cells move from the soft part of the gel to the stiffer parts [167]. Latter, in a deeper characterization of this durotaxis behavior, it was found that polarization of cytoskeleton and myosin phosphorylation profiles was necessary for migration over these stiffness gradients [168]. Collective migratory groups also exhibit durotaxis. For instance, human mammary epithelial cells monolayers can follow stiffness gradients [168]. In this work, cell spreading of MCF10A monolayers plated on stiffness gradients is asymmetrical with cell collectives moving more efficiently toward the stiffer parts of the substrate. This behavior was explained not only through polarization at the monolayer edge but also with the long-range force transmission within the cluster through AJs to sense the gradients and orient the force to achieve a coordinated directional global response [169]. Another example where the viscoelastic properties of the environment are a relevant cue is vascular remodeling. In this tissue endothelial cells are known to migrate against the shear from the flow [170–172]. In one case, it was observed that the AJs of highly motile cells increased the vascular endothelial E-Cadherin when migrating opposite to the blood flow [170]. Finally, in a recent work, it was possible to measure the viscoelastic properties of a tissue in vivo. The researchers were able to detect stiffness gradients of mesoderm during mouse limb formation, matching these stiffness gradients to the cell movement observed by microscopy. These in vivo data suggest that durotaxis can be relevant in some in vivo processes. However, even when there is evidence suggesting that stiffness gradient exists and could have a role in vivo this idea needs to be further investigated. The examples described above of mechanical modulation of cell migration suggest that the viscoelastic properties of the surroundings are more than a barrier to be surpassed by migratory cells, but instead, they are an essential source of signaling that can modulate cell–cell adhesion to promote CCM in vivo and with that shape embryos or confer aggressive phenotypes to cancer cells.

### 4.4.3 Environmental viscoelasticity and the control of MET

As exemplified before, cell–cell contacts can be modified by transcriptional and translational mechanisms in a process that was initially described as EMT but that nowadays has undergone deep conceptual modifications, mostly in spite of new findings [173]. In this same manner, the MET transcription can also be regulated by the mechanical environment. For instance, mechanical properties of the target tissue of metastatic cancer cells can be an important factor to activate their programs as epithelial tumors [163,174]. During development, membranes surrounding early embryos, such as the zona pellucida in mammals, exert a mechanical force on blastomeres that contribute to compaction and constrain blastomere movements [175,176]. After this, trophectoderm development still requires tight junctions, actomyosin contractility, and cadherin-based adhesions [177]. Interaction between mesenchymal cells and their mechanical environment is not only required during early development. For instance, during kidney organogenesis pronephric ducts are formed by collective migration of kidney progenitors where follower cells undergo MET to form a caudal single epithelial tube [178]. Some studies have shown that this process is driven by the polarized traction generation from epithelial follower cells [179], using e-cadherin to transmit tensile forces in order to establish migration polarity [180]. There are many studies showing the importance of the interplay between the mechanical environment and MET during many processes like stem cell differentiation, development of cancer progression (reviewed in Ref. [145]). Noteworthy, EMT and MET are complex processes where it became every time clearer that the environment promotes changes from cellular to molecular levels in order to establish the ultimate cellular behavior. This indicates that the changes promoted by environmental viscoelasticity are not only in the scale of short and mid-term, but also the interaction with environmental mechanical cues can lead to modulation in gene expression profiles associated with long-term changes and cell differentiation [39,40].

## 4.5 Concluding remarks and future perspectives

In this chapter, we have illustrated that migratory cells have mechanisms that allow them to modulate their viscoelastic properties in order to face the different types of environments to which they are exposed in order to efficiently reach their targets. On the other hand, it is every time clearer that fine-tuning of the viscoelastic properties of the environment are signals that trigger but also serve as a guidance cue to direct migration in vitro and in vivo (Fig. 4.2). In this line, durotaxis, a process mainly observed in vitro for single and collective migratory cells, is starting to be demonstrated in vivo (Fig. 4.2A). However, how the viscoelasticity of the cells interplay with the viscoelastic properties of the environment in order to allow cells to efficiently migrate remains to be investigated for native contexts. As for durotaxis, it has been shown recently that cells can perform another type of migration called ratchetaxis, this is a mechanism of guiding cell migration through periodic asymmetries of the surrounding at cell scale when cells are confined in open channels [181]. However, how this newly described migratory can guide cell migration, how ratchetaxis modulates cell viscoelastic properties, and whether this kind of environmental architectures are present and are relevant in vivo remains as open questions.

Wound healing morphogenetic processes as well as pathologies rely on cell migration in order to progress, thus, understanding the mechanisms of cell migration is fundamental to develop more effective therapies when approaching effective regenerative treatments or to design better strategies against diseases such as cancer. Using in vivo rat models, it was observed that gastric tumors present poor prognosis correlated to molecular profiles activated by the mechanical stress exerted by the tumor to its microenvironment [39]. In this line, cells submitted to pressure from cancer proliferative adjacent cells can turn thus "converting" normal cells into tumor cells [39]. Finally, this chapter focused in general aspects of cell migration, but most of the cells and tissues are exposed to physical cues, and viscoelasticity is a property of most of the biological material. Interestingly, a recent work showed that kidney organoid morphogenesis requires a soft environment to properly form.

The study of how cells interact with their physical environment could allow us to develop new technologies for instance to form fully functional organs from stem cells or to develop new therapies to treat cancer progression. However, key in order to reach a comprehensive understanding of complex biological systems will require multidisciplinary approaches combining biochemical and biophysical approaches.

## Acknowledgments

EHB was supported by Fundacao Calouste Gulbenkian and by La Caixa Junior Leader Incoming (94978). JAE was supported by Fundacao Calouste Gulbenkian.

## References

[1] P. Friedl, D. Gilmour, Collective cell migration in morphogenesis, regeneration and cancer, Nat. Rev. Mol. Cell Biol. 10 (2009) 445–457. Available from: https://doi.org/10.1038/nrm2720.

[2] R. Mayor, S. Etienne-Manneville, The front and rear of collective cell migration, Nat. Rev. Mol. Cell Biol. 17 (2016) 97–109. Available from: https://doi.org/10.1038/nrm.2015.14.

[3] P. Rorth, Fellow travellers: emergent properties of collective cell migration, EMBO Rep. 13 (2012) 984–991. Available from: https://doi.org/10.1038/embor.2012.149.
[4] P. Rorth, Collective guidance of collective cell migration, Trends Cell Biol. 17 (2007) 575–579. Available from: https://doi.org/10.1016/j.tcb.2007.09.007.
[5] O. Ilina, P. Friedl, Mechanisms of collective cell migration at a glance, J. Cell Sci. 122 (2009) 3203–3208. Available from: https://doi.org/10.1242/jcs.036525.
[6] P. Rorth, Collective cell migration, Annu. Rev. Cell Dev. Biol. 25 (2009) 407–429. Available from: https://doi.org/10.1146/annurev.cellbio.042308.113231.
[7] E. Scarpa, R. Mayor, Collective cell migration in development, J. Cell Biol. 212 (2016) 143–155. Available from: https://doi.org/10.1083/jcb.201508047.
[8] P. Friedl, R. Mayor, Tuning collective cell migration by cell-cell junction regulation, Cold Spring Harb. Perspect. Biol. 9 (2017) a029199. Available from: https://doi.org/10.1101/cshperspect.a029199.
[9] E. Theveneau, C. Linker, Leaders in collective migration: are front cells really endowed with a particular set of skills? F1000Research 6 (2017) 1899. Available from: https://doi.org/10.12688/f1000research.11889.1.
[10] C. Collins, W.J. Nelson, Running with neighbors: coordinating cell migration and cell-cell adhesion, Curr. Opin. Cell Biol. 36 (2015) 62–70. Available from: https://doi.org/10.1016/j.ceb.2015.07.004.
[11] E.H. Barriga, K. Franze, G. Charras, R. Mayor, Tissue stiffening coordinates morphogenesis by triggering collective cell migration in vivo, Nature 554 (2018) 523–527. Available from: https://doi.org/10.1038/nature25742.
[12] P. Friedl, K. Wolf, Tumour-cell invasion and migration: diversity and escape mechanisms, Nat. Rev. Cancer 3 (2003) 362–374. Available from: https://doi.org/10.1038/nrc1075.
[13] C. Bonnans, J. Chou, Z. Werb, Remodelling the extracellular matrix in development and disease, Nat. Rev. Mol. Cell Biol. 15 (2014) 786–801. Available from: https://doi.org/10.1038/nrm3904.
[14] A. Dekanty, M. Milan, The interplay between morphogens and tissue growth, EMBO Rep. 12 (2011) 1003–1010. Available from: https://doi.org/10.1038/embor.2011.172.
[15] D. Gilmour, M. Rembold, M. Leptin, From morphogen to morphogenesis and back, Nature 541 (2017) 311–320. Available from: https://doi.org/10.1038/nature21348.
[16] A. Sagner, J. Briscoe, Morphogen interpretation: concentration, time, competence, and signaling dynamics, Wiley Interdiscip. Rev. Dev. Biol. 6 (2017) e271. Available from: https://doi.org/10.1002/wdev.271.
[17] T. Tabata, Y. Takei, Morphogens, their identification and regulation, Development 131 (2004) 703–712. Available from: https://doi.org/10.1242/dev.01043.
[18] G. Charras, E. Sahai, Physical influences of the extracellular environment on cell migration, Nat. Rev. Mol. Cell Biol. 15 (2014) 813–824. Available from: https://doi.org/10.1038/nrm3897.
[19] R.J. Petrie, K.M. Yamada, At the leading edge of three-dimensional cell migration, J. Cell Sci. 125 (2012) 5917–5926. Available from: https://doi.org/10.1242/jcs.093732.
[20] E.H. Barriga, R. Mayor, Adjustable viscoelasticity allows for efficient collective cell migration, Semin. Cell Dev. Biol. 93 (2019) 55–68. Available from: https://doi.org/10.1016/j.semcdb.2018.05.027.
[21] M. Thery, Micropatterning as a tool to decipher cell morphogenesis and functions, J. Cell Sci. 123 (2010) 4201–4213. Available from: https://doi.org/10.1242/jcs.075150.
[22] F.J. Calero-Cuenca, C.S. Janota, E.R. Gomes, Dealing with the nucleus during cell migration, Curr. Opin. Cell Biol. 50 (2018) 35–41. Available from: https://doi.org/10.1016/j.ceb.2018.01.014.

[23] J.R. Lange, B. Fabry, Cell and tissue mechanics in cell migration, Exp. Cell Res. 319 (2013) 2418–2423. Available from: https://doi.org/10.1016/j.yexcr.2013.04.023.
[24] A.H. Burstein, V.H. Frankel, The viscoelastic properties of some biological materials, Ann. N. Y. Acad. Sci. 146 (1968) 158–165. Available from: https://doi.org/10.1111/j.1749-6632.1968.tb20280.x.
[25] M. Kucharova, S. Doubal, P. Klemera, P. Rejchrt, M. Navratil, Viscoelasticity of biological materials - measurement and practical impact on biomedicine, Physiol. Res. 56 (Suppl 1) (2007) S33–S37.
[26] I. Pajic-Lijakovic, M. Milivojevic, Viscoelasticity of multicellular surfaces, J. Biomech. 60 (2017) 1–8. Available from: https://doi.org/10.1016/j.jbiomech.2017.06.035.
[27] R.K. Das, V. Gocheva, R. Hammink, O.F. Zouani, A.E. Rowan, Stress-stiffening-mediated stem-cell commitment switch in soft responsive hydrogels, Nat. Mater. 15 (2016) 318–325. Available from: https://doi.org/10.1038/nmat4483.
[28] C. Storm, J.J. Pastore, F.C. MacKintosh, T.C. Lubensky, P.A. Janmey, Nonlinear elasticity in biological gels, Nature 435 (2005) 191–194. Available from: https://doi.org/10.1038/nature03521.
[29] C. Goldenberg, I. Goldhirsch, Friction enhances elasticity in granular solids, Nature 435 (2005) 188–191. Available from: https://doi.org/10.1038/nature03497.
[30] K. Pogoda, L. Chin, P.C. Georges, F.J. Byfield, R. Bucki, R. Kim, et al., Compression stiffening of brain and its effect on mechanosensing by glioma cells, N. J. Phys. 16 (2014) 075002. Available from: https://doi.org/10.1088/1367-2630/16/7/075002.
[31] S. van Helvert, P. Friedl, Strain stiffening of fibrillar collagen during individual and collective cell migration identified by AFM nanoindentation, ACS Appl. Mater. Interfaces 8 (2016) 21946–21955. Available from: https://doi.org/10.1021/acsami.6b01755.
[32] A.R. Bausch, W. Moller, E. Sackmann, Measurement of local viscoelasticity and forces in living cells by magnetic tweezers, Biophys. J. 76 (1999) 573–579. Available from: https://doi.org/10.1016/S0006-3495(99)77225-5.
[33] O. Thoumine, A. Ott, Time scale dependent viscoelastic and contractile regimes in fibroblasts probed by microplate manipulation, J. Cell Sci. 110 (Pt 17) (1997) 2109–2116.
[34] D. Mitrossilis, J. Fouchard, D. Pereira, F. Postic, A. Richert, M. Saint-Jean, et al., Real-time single-cell response to stiffness, Proc. Natl Acad. Sci. U. S. A. 107 (2010) 16518–16523. Available from: https://doi.org/10.1073/pnas.1007940107.
[35] G. Charras, A.S. Yap, Tensile forces and mechanotransduction at cell-cell junctions, Curr. Biol. 28 (2018) R445–R457. Available from: https://doi.org/10.1016/j.cub.2018.02.003.
[36] S.W. Moore, P. Roca-Cusachs, M.P. Sheetz, Stretchy proteins on stretchy substrates: the important elements of integrin-mediated rigidity sensing, Dev. Cell 19 (2010) 194–206. Available from: https://doi.org/10.1016/j.devcel.2010.07.018.
[37] P. Pandya, J.L. Orgaz, V. Sanz-Moreno, Actomyosin contractility and collective migration: may the force be with you, Curr. Opin. Cell Biol. 48 (2017) 87–96. Available from: https://doi.org/10.1016/j.ceb.2017.06.006.
[38] Y. Ren, J.C. Effler, M. Norstrom, T. Luo, R.A. Firtel, P.A. Iglesias, et al., Mechanosensing through cooperative interactions between myosin II and the actin crosslinker cortexillin I, Curr. Biol. 19 (2009) 1421–1428. Available from: https://doi.org/10.1016/j.cub.2009.07.018.
[39] M.E. Fernandez-Sanchez, S. Barbier, J. Whitehead, G. Bealle, A. Michel, H. Latorre-Ossa, et al., Mechanical induction of the tumorigenic beta-catenin pathway by tumour growth pressure, Nature 523 (2015) 92–95. Available from: https://doi.org/10.1038/nature14329.

[40] A.E. Shyer, A.R. Rodrigues, G.G. Schroeder, E. Kassianidou, S. Kumar, R.M. Harland, Emergent cellular self-organization and mechanosensation initiate follicle pattern in the avian skin, Science 357 (2017) 811–815. Available from: https://doi.org/10.1126/science.aai7868.
[41] B. Geiger, J.P. Spatz, A.D. Bershadsky, Environmental sensing through focal adhesions, Nat. Rev. Mol. Cell Biol. 10 (2009) 21–33. Available from: https://doi.org/10.1038/nrm2593.
[42] B. Stutchbury, P. Atherton, R. Tsang, D.Y. Wang, C. Ballestrem, Distinct focal adhesion protein modules control different aspects of mechanotransduction, J. Cell Sci. 130 (2017) 1612–1624. Available from: https://doi.org/10.1242/jcs.195362.
[43] G. Salbreux, G. Charras, E. Paluch, Actin cortex mechanics and cellular morphogenesis, Trends Cell Biol. 22 (2012) 536–545. Available from: https://doi.org/10.1016/j.tcb.2012.07.001.
[44] C. Grashoff, B.D. Hoffman, M.D. Brenner, R. Zhou, M. Parsons, M.T. Yang, et al., Measuring mechanical tension across vinculin reveals regulation of focal adhesion dynamics, Nature 466 (2010) 263–266. Available from: https://doi.org/10.1038/nature09198.
[45] K. Honda, H. Yanai, H. Negishi, M. Asagiri, M. Sato, T. Mizutani, et al., IRF-7 is the master regulator of type-I interferon-dependent immune responses, Nature 434 (2005) 772–777. Available from: https://doi.org/10.1038/nature03464.
[46] A. Kumar, M. Ouyang, K. Van den Dries, E.J. McGhee, K. Tanaka, M.D. Anderson, et al., Talin tension sensor reveals novel features of focal adhesion force transmission and mechanosensitivity, J. Cell Biol. 213 (2016) 371–383. Available from: https://doi.org/10.1083/jcb.201510012.
[47] Y. Sawada, M. Tamada, B.J. Dubin-Thaler, O. Cherniavskaya, R. Sakai, S. Tanaka, et al., Force sensing by mechanical extension of the Src family kinase substrate p130Cas, Cell 127 (2006) 1015–1026. Available from: https://doi.org/10.1016/j.cell.2006.09.044.
[48] M. Yao, B.T. Goult, B. Klapholz, X. Hu, C.P. Toseland, Y. Guo, et al., The Mechanical response of talin, Nat. Commun. 7 (2016) 11966. Available from: https://doi.org/10.1038/ncomms11966.
[49] J. Fouchard, D. Mitrossilis, A. Asnacios, Acto-myosin based response to stiffness and rigidity sensing, Cell Adhes. Migr. 5 (2011) 16–19. Available from: 10.4161/cam.5.1.13281.
[50] E.K. Paluch, C.M. Nelson, N. Biais, B. Fabry, J. Moeller, B.L. Pruitt, et al., Mechanotransduction: use the force(s), BMC Biol. 13 (2015) 47. Available from: https://doi.org/10.1186/s12915-015-0150-4.
[51] R. Paul, P. Heil, J.P. Spatz, U.S. Schwarz, Propagation of mechanical stress through the actin cytoskeleton toward focal adhesions: model and experiment, Biophys. J. 94 (2008) 1470–1482. Available from: https://doi.org/10.1529/biophysj.107.108688.
[52] W.H. Yang, H.Y. Lan, C.H. Huang, S.K. Tai, C.H. Tzeng, S.Y. Kao, et al., RAC1 activation mediates Twist1-induced cancer cell migration, Nat. Cell Biol. 14 (2012) 366–374. Available from: https://doi.org/10.1038/ncb2455.
[53] Q. Wen, P.A. Janmey, Polymer physics of the cytoskeleton, Curr. Opin. Solid State Mater. Sci. 15 (2011) 177–182. Available from: https://doi.org/10.1016/j.cossms.2011.05.002.
[54] D.A. Fletcher, R.D. Mullins, Cell mechanics and the cytoskeleton, Nature 463 (2010) 485–492. Available from: https://doi.org/10.1038/nature08908.
[55] D. Stamenovic, N. Wang, Invited review: engineering approaches to cytoskeletal mechanics, J. Appl. Physiol. 89 (2000) 2085–2090. Available from: https://doi.org/10.1152/jappl.2000.89.5.2085.

[56] C.G. Tusan, Y.H. Man, H. Zarkoob, D.A. Johnston, O.G. Andriotis, P.J. Thurner, et al., Collective cell behavior in mechanosensing of substrate thickness, Biophys. J. 114 (2018) 2743–2755. Available from: https://doi.org/10.1016/j.bpj.2018.03.037.
[57] R. Brighenti, F.J. Vernerey, A simple statistical approach to model the time-dependent response of polymers with reversible cross-links, Compos. Part B, Eng. 115 (2017) 257–265. Available from: https://doi.org/10.1016/j.compositesb.2016.09.090.
[58] O. Lieleg, J. Kayser, G. Brambilla, L. Cipelletti, A.R. Bausch, Slow dynamics and internal stress relaxation in bundled cytoskeletal networks, Nat. Mater. 10 (2011) 236–242. Available from: https://doi.org/10.1038/nmat2939.
[59] O. Lieleg, K.M. Schmoller, M.M. Claessens, A.R. Bausch, Cytoskeletal polymer networks: viscoelastic properties are determined by the microscopic interaction potential of cross-links, Biophys. J. 96 (2009) 4725–4732. Available from: https://doi.org/10.1016/j.bpj.2009.03.038.
[60] L. Cao, A. Yonis, M. Vaghela, E.H. Barriga, P. Chugh, M.B. Smith, et al., SPIN90 associates with mDia1 and the Arp2/3 complex to regulate cortical actin organization, Nat. Cell Biol. (2020). Available from: https://doi.org/10.1038/s41556-020-0531-y.
[61] E.E. Charrier, P.A. Janmey, Mechanical properties of intermediate filament proteins, Methods Enzymol. 568 (2016) 35–57. Available from: https://doi.org/10.1016/bs.mie.2015.09.009.
[62] T. Hawkins, M. Mirigian, M. Selcuk Yasar, J.L. Ross, Mechanics of microtubules, J. Biomech. 43 (2010) 23–30. Available from: https://doi.org/10.1016/j.jbiomech.2009.09.005.
[63] B.J. Lopez, M.T. Valentine, Molecular control of stress transmission in the microtubule cytoskeleton, Biochim. Biophys. Acta 1853 (2015) 3015–3024. Available from: https://doi.org/10.1016/j.bbamcr.2015.07.016.
[64] J. Stricker, T. Falzone, M.L. Gardel, Mechanics of the F-actin cytoskeleton, J. Biomech. 43 (2010) 9–14. Available from: https://doi.org/10.1016/j.jbiomech.2009.09.003.
[65] B.P. Helmke, A.B. Rosen, P.F. Davies, Mapping mechanical strain of an endogenous cytoskeletal network in living endothelial cells, Biophys. J. 84 (2003) 2691–2699. Available from: https://doi.org/10.1016/S0006-3495(03)75074-7.
[66] G. Zagar, P.R. Onck, E. van der Giessen, Two fundamental mechanisms govern the stiffening of cross-linked networks, Biophys. J. 108 (2015) 1470–1479. Available from: https://doi.org/10.1016/j.bpj.2015.02.015.
[67] A.E. Carlsson, Actin dynamics: from nanoscale to microscale, Annu. Rev. Biophys. 39 (2010) 91–110. Available from: https://doi.org/10.1146/annurev.biophys.093008.131207.
[68] I. Dang, R. Gorelik, C. Sousa-Blin, E. Derivery, C. Guerin, J. Linkner, et al., Inhibitory signalling to the Arp2/3 complex steers cell migration, Nature 503 (2013) 281–284. Available from: https://doi.org/10.1038/nature12611.
[69] M. Krause, A. Gautreau, Steering cell migration: lamellipodium dynamics and the regulation of directional persistence, Nat. Rev. Mol. Cell Biol. 15 (2014) 577–590. Available from: https://doi.org/10.1038/nrm3861.
[70] A.C. Reymann, R. Boujemaa-Paterski, J.L. Martiel, C. Guerin, W. Cao, H.F. Chin, et al., Actin network architecture can determine myosin motor activity, Science 336 (2012) 1310–1314. Available from: https://doi.org/10.1126/science.1221708.
[71] D. Veltman, Actin dynamics: cell migration takes a new turn with arpin, Curr. Biol. 24 (2014) R31–R33. Available from: https://doi.org/10.1016/j.cub.2013.11.022.
[72] M.A. Wear, D.A. Schafer, J.A. Cooper, Actin dynamics: assembly and disassembly of actin networks, Curr. Biol. 10 (2000) R891–R895. Available from: https://doi.org/10.1016/s0960-9822(00)00845-9.
[73] M. Kirschner, T. Mitchison, Beyond self-assembly: from microtubules to morphogenesis, Cell 45 (1986) 329–342. Available from: https://doi.org/10.1016/0092-8674(86)90318-1.

[74] M. Murrell, P.W. Oakes, M. Lenz, M.L. Gardel, Forcing cells into shape: the mechanics of actomyosin contractility, Nat. Rev. Mol. Cell Biol. 16 (2015) 486–498. Available from: https://doi.org/10.1038/nrm4012.
[75] P. Bieling, T.D. Li, J. Weichsel, R. McGorty, P. Jreij, B. Huang, et al., Force feedback controls motor activity and mechanical properties of self-assembling branched actin networks, Cell 164 (2016) 115–127. Available from: https://doi.org/10.1016/j.cell.2015.11.057.
[76] P. Friedl, K. Wolf, J. Lammerding, Nuclear mechanics during cell migration, Curr. Opin. Cell Biol. 23 (2011) 55–64. Available from: https://doi.org/10.1016/j.ceb.2010.10.015.
[77] P.M. Davidson, J. Lammerding, Broken nuclei–lamins, nuclear mechanics, and disease, Trends Cell Biol. 24 (2014) 247–256. Available from: https://doi.org/10.1016/j.tcb.2013.11.004.
[78] G.R. Fedorchak, A. Kaminski, J. Lammerding, Cellular mechanosensing: getting to the nucleus of it all, Prog. Biophys. Mol. Biol. 115 (2014) 76–92. Available from: https://doi.org/10.1016/j.pbiomolbio.2014.06.009.
[79] M. Cupesi, J. Yoshioka, J. Gannon, A. Kudinova, C.L. Stewart, J. Lammerding, Attenuated hypertrophic response to pressure overload in a lamin A/C haploinsufficiency mouse, J. Mol. Cell. Cardiol. 48 (2010) 1290–1297. Available from: https://doi.org/10.1016/j.yjmcc.2009.10.024.
[80] J. Lammerding, J. Hsiao, P.C. Schulze, S. Kozlov, C.L. Stewart, R.T. Lee, Abnormal nuclear shape and impaired mechanotransduction in emerin-deficient cells, J. Cell Biol. 170 (2005) 781–791. Available from: https://doi.org/10.1083/jcb.200502148.
[81] J. Lammerding, P.C. Schulze, T. Takahashi, S. Kozlov, T. Sullivan, R.D. Kamm, et al., Lamin A/C deficiency causes defective nuclear mechanics and mechanotransduction, J. Clin. Invest. 113 (2004) 370–378. Available from: https://doi.org/10.1172/JCI19670.
[82] K. Wolf, M. Te Lindert, M. Krause, S. Alexander, J. Te Riet, A.L. Willis, et al., Physical limits of cell migration: control by ECM space and nuclear deformation and tuning by proteolysis and traction force, J. Cell Biol. 201 (2013) 1069–1084. Available from: https://doi.org/10.1083/jcb.201210152.
[83] R. Bainer, V. Weaver, Cell biology. Strength under tension, Science 341 (2013) 965–966. Available from: https://doi.org/10.1126/science.1243643.
[84] C.Y. Ho, J. Lammerding, Lamins at a glance, J. Cell Sci. 125 (2012) 2087–2093. Available from: https://doi.org/10.1242/jcs.087288.
[85] H.R. Thiam, P. Vargas, N. Carpi, C.L. Crespo, M. Raab, E. Terriac, et al., Perinuclear Arp2/3-driven actin polymerization enables nuclear deformation to facilitate cell migration through complex environments, Nat. Commun. 7 (2016) 10997. Available from: https://doi.org/10.1038/ncomms10997.
[86] J. Renkawitz, A. Kopf, J. Stopp, I. de Vries, M.K. Driscoll, J. Merrin, et al., Nuclear positioning facilitates amoeboid migration along the path of least resistance, Nature 568 (2019) 546–550. Available from: https://doi.org/10.1038/s41586-019-1087-5.
[87] H. Karcher, J. Lammerding, H. Huang, R.T. Lee, R.D. Kamm, M.R. Kaazempur-Mofrad, A three-dimensional viscoelastic model for cell deformation with experimental verification, Biophys. J. 85 (2003) 3336–3349. Available from: https://doi.org/10.1016/S0006-3495(03)74753-5.
[88] S.A. Gudipaty, J. Lindblom, P.D. Loftus, M.J. Redd, K. Edes, C.F. Davey, et al., Mechanical stretch triggers rapid epithelial cell division through Piezo1, Nature 543 (2017) 118–121. Available from: https://doi.org/10.1038/nature21407.
[89] W.C. Hung, J.R. Yang, C.L. Yankaskas, B.S. Wong, P.H. Wu, C. Pardo-Pastor, et al., Confinement sensing and signal optimization via Piezo1/PKA and Myosin II pathways, Cell Rep. 15 (2016) 1430–1441. Available from: https://doi.org/10.1016/j.celrep.2016.04.035.

[90] J. Wu, A.H. Lewis, J. Grandl, Touch, tension, and transduction - the function and regulation of piezo ion channels, Trends Biochem. Sci. 42 (2017) 57–71. Available from: https://doi.org/10.1016/j.tibs.2016.09.004.
[91] A. Elosegui-Artola, I. Andreu, A.E.M. Beedle, A. Lezamiz, M. Uroz, A.J. Kosmalska, et al., Force triggers YAP nuclear entry by regulating transport across nuclear pores, Cell 171 (2017) 1397–1410. Available from: https://doi.org/10.1016/j.cell.2017.10.008.
[92] T.J. Kirby, J. Lammerding, Emerging views of the nucleus as a cellular mechanosensor, Nat. Cell Biol. 20 (2018) 373–381. Available from: https://doi.org/10.1038/s41556-018-0038-y.
[93] S. Otsuka, J. Ellenberg, Mechanisms of nuclear pore complex assembly - two different ways of building one molecular machine, FEBS Lett. 592 (2018) 475–488. Available from: https://doi.org/10.1002/1873-3468.12905.
[94] G.J. Stanley, A. Fassati, B.W. Hoogenboom, Biomechanics of the transport barrier in the nuclear pore complex, Semin. Cell Dev. Biol. 68 (2017) 42–51. Available from: https://doi.org/10.1016/j.semcdb.2017.05.007.
[95] S. Etienne-Manneville, Control of polarized cell morphology and motility by adherens junctions, Semin. Cell Dev. Biol. 22 (2011) 850–857. Available from: https://doi.org/10.1016/j.semcdb.2011.07.023.
[96] M. Inaki, S. Vishnu, A. Cliffe, P. Rorth, Effective guidance of collective migration based on differences in cell states, Proc. Natl Acad. Sci. U. S. A. 109 (2012) 2027–22032. Available from: https://doi.org/10.1073/pnas.1115260109.
[97] D. Cai, W. Dai, M. Prasad, J. Luo, N.S. Gov, D.J. Montell, Modeling and analysis of collective cell migration in an in vivo three-dimensional environment, Proc. Natl Acad. Sci. U. S. A. 113 (2016) E2134–E2141. Available from: https://doi.org/10.1073/pnas.1522656113.
[98] S. Kuriyama, E. Theveneau, A. Benedetto, M. Parsons, M. Tanaka, G. Charras, et al., In vivo collective cell migration requires an LPAR2-dependent increase in tissue fluidity, J. Cell Biol. 206 (2014) 113–127. Available from: https://doi.org/10.1083/jcb.201402093.
[99] Z. Liu, J.L. Tan, D.M. Cohen, M.T. Yang, N.J. Sniadecki, S.A. Ruiz, et al., Mechanical tugging force regulates the size of cell-cell junctions, Proc. Natl. Acad. Sci. U. S. A. 107 (2010) 9944–9949. Available from: https://doi.org/10.1073/pnas.0914547107.
[100] S. Sluysmans, E. Vasileva, D. Spadaro, J. Shah, F. Rouaud, S. Citi, The role of apical cell-cell junctions and associated cytoskeleton in mechanotransduction, Biol. Cell 109 (2017) 139–161. Available from: https://doi.org/10.1111/boc.201600075.
[101] M. Krendel, N.A. Gloushankova, E.M. Bonder, H.H. Feder, J.M. Vasiliev, I.M. Gelfand, Myosin-dependent contractile activity of the actin cytoskeleton modulates the spatial organization of cell-cell contacts in cultured epitheliocytes, Proc. Natl Acad. Sci. U. S. A. 96 (1999) 9666–9670. Available from: https://doi.org/10.1073/pnas.96.17.9666.
[102] D.E. Leckband, J. de Rooij, Cadherin adhesion and mechanotransduction, Annu. Rev. Cell Dev. Biol. 30 (2014) 291–315. Available from: https://doi.org/10.1146/annurev-cellbio-100913-013212.
[103] E. Bazellieres, V. Conte, A. Elosegui-Artola, X. Serra-Picamal, M. Bintanel-Morcillo, P. Roca-Cusachs, et al., Control of cell-cell forces and collective cell dynamics by the intercellular adhesome, Nat. Cell Biol. 17 (2015) 409–420. Available from: https://doi.org/10.1038/ncb3135.
[104] L. Bao, F. Sachs, G. Dahl, Connexins are mechanosensitive, Am. J. Physiol. Cell Physiol. 287 (2004) C1389–C1395. Available from: https://doi.org/10.1152/ajpcell.00220.2004.

[105] F. Peglion, F. Llense, S. Etienne-Manneville, Adherens junction treadmilling during collective migration, Nat. Cell Biol. 16 (2014) 639–651. Available from: https://doi.org/10.1038/ncb2985.
[106] M. Cavey, T. Lecuit, Molecular bases of cell-cell junctions stability and dynamics, Cold Spring Harb. Perspect. Biol. 1 (2009) a002998. Available from: https://doi.org/10.1101/cshperspect.a002998.
[107] A.R. Harris, A. Daeden, G.T. Charras, Formation of adherens junctions leads to the emergence of a tissue-level tension in epithelial monolayers, J. Cell Sci. 127 (2014) 2507–2517. Available from: https://doi.org/10.1242/jcs.142349.
[108] M. Kishikawa, A. Suzuki, S. Ohno, aPKC enables development of zonula adherens by antagonizing centripetal contraction of the circumferential actomyosin cables, J. Cell Sci. 121 (2008) 2481–2492. Available from: https://doi.org/10.1242/jcs.024109.
[109] S. Tsukita, M. Furuse, M. Itoh, Multifunctional strands in tight junctions, Nat. Rev. Mol. Cell Biol. 2 (2001) 285–293. Available from: https://doi.org/10.1038/35067088.
[110] I. Bahm, E.H. Barriga, A. Frolov, E. Theveneau, P. Frankel, R. Mayor, PDGF controls contact inhibition of locomotion by regulating N-cadherin during neural crest migration, Development 144 (2017) 2456–2468. Available from: https://doi.org/10.1242/dev.147926.
[111] E.H. Barriga, P.H. Maxwell, A.E. Reyes, R. Mayor, The hypoxia factor Hif-1alpha controls neural crest chemotaxis and epithelial to mesenchymal transition, J. Cell Biol. 201 (2013) 759–776. Available from: https://doi.org/10.1083/jcb.201212100.
[112] E.H. Barriga, R. Mayor, Embryonic cell-cell adhesion: a key player in collective neural crest migration, Curr. Top. Dev. Biol. 112 (2015) 301–323. Available from: https://doi.org/10.1016/bs.ctdb.2014.11.023.
[113] E. Theveneau, R. Mayor, Neural crest delamination and migration: from epithelium-to-mesenchyme transition to collective cell migration, Dev. Biol. 366 (2012) 34–54. Available from: https://doi.org/10.1016/j.ydbio.2011.12.041.
[114] I. Dupin, E. Camand, S. Etienne-Manneville, Classical cadherins control nucleus and centrosome position and cell polarity, J. Cell Biol. 185 (2009) 779–786. Available from: https://doi.org/10.1083/jcb.200812034.
[115] S. Song, S. Eckerle, D. Onichtchouk, J.A. Marrs, R. Nitschke, W. Driever, Pou5f1-dependent EGF expression controls E-cadherin endocytosis, cell adhesion, and zebrafish epiboly movements, Dev. Cell 24 (2013) 486–501. Available from: https://doi.org/10.1016/j.devcel.2013.01.016.
[116] S. Durdu, M. Iskar, C. Revenu, N. Schieber, A. Kunze, P. Bork, et al., Luminal signalling links cell communication to tissue architecture during organogenesis, Nature 515 (2014) 120–124. Available from: https://doi.org/10.1038/nature13852.
[117] A. Nechiporuk, D.W. Raible, FGF-dependent mechanosensory organ patterning in zebrafish, Science 320 (2008) 1774–1777. Available from: https://doi.org/10.1126/science.1156547.
[118] C. Revenu, S. Streichan, E. Dona, V. Lecaudey, L. Hufnagel, D. Gilmour, Quantitative cell polarity imaging defines leader-to-follower transitions during collective migration and the key role of microtubule-dependent adherens junction formation, Development 141 (2014) 1282–1291. Available from: https://doi.org/10.1242/dev.101675.
[119] D. Dalle Nogare, A.B. Chitnis, A framework for understanding morphogenesis and migration of the zebrafish posterior Lateral Line primordium, Mech. Dev. 148 (2017) 69–78. Available from: https://doi.org/10.1016/j.mod.2017.04.005.

[120] D.J. Montell, Border-cell migration: the race is on, Nat. Rev. Mol. Cell Biol. 4 (2003) 13–24. Available from: https://doi.org/10.1038/nrm1006.
[121] M. Prasad, X. Wang, L. He, D. Cai, D.J. Montell, Border cell migration: a model system for live imaging and genetic analysis of collective cell movement, Methods Mol. Biol. 1328 (2015) 89–97. Available from: https://doi.org/10.1007/978-1-4939-2851-4_6.
[122] D. Cai, S.C. Chen, M. Prasad, L. He, X. Wang, V. Choesmel-Cadamuro, et al., Mechanical feedback through E-cadherin promotes direction sensing during collective cell migration, Cell 157 (2014) 1146–1159. Available from: https://doi.org/10.1016/j.cell.2014.03.045.
[123] K. Campbell, J. Casanova, A role for E-cadherin in ensuring cohesive migration of a heterogeneous population of non-epithelial cells, Nat. Commun. 6 (2015) 7998. Available from: https://doi.org/10.1038/ncomms8998.
[124] E.R. Geisbrecht, D.J. Montell, Myosin VI is required for E-cadherin-mediated border cell migration, Nat. Cell Biol. 4 (2002) 616–620. Available from: https://doi.org/10.1038/ncb830.
[125] D.J. Montell, W.H. Yoon, M. Starz-Gaiano, Group choreography: mechanisms orchestrating the collective movement of border cells, Nat. Rev. Mol. Cell Biol. 13 (2012) 631–645. Available from: https://doi.org/10.1038/nrm3433.
[126] P. Niewiadomska, D. Godt, U. Tepass, DE-Cadherin is required for intercellular motility during Drosophila oogenesis, J. Cell Biol. 144 (1999) 533–547. Available from: https://doi.org/10.1083/jcb.144.3.533.
[127] A. Brugues, E. Anon, V. Conte, J.H. Veldhuis, M. Gupta, J. Colombelli, et al., Forces driving epithelial wound healing, Nat. Phys. 10 (2014) 683–690. Available from: https://doi.org/10.1038/nphys3040.
[128] T. Zulueta-Coarasa, R. Fernandez-Gonzalez, Tension (re)builds: biophysical mechanisms of embryonic wound repair, Mech. Dev. 144 (2017) 43–52. Available from: https://doi.org/10.1016/j.mod.2016.11.004.
[129] E.R. Shamir, A.J. Ewald, Adhesion in mammary development: novel roles for E-cadherin in individual and collective cell migration, Curr. Top. Dev. Biol. 112 (2015) 353–382. Available from: https://doi.org/10.1016/bs.ctdb.2014.12.001.
[130] E. Janda, M. Nevolo, K. Lehmann, J. Downward, H. Beug, M. Grieco, Raf plus TGFbeta-dependent EMT is initiated by endocytosis and lysosomal degradation of E-cadherin, Oncogene 25 (2006) 7117–7130. Available from: https://doi.org/10.1038/sj.onc.1209701.
[131] A.P. Kowalczyk, B.A. Nanes, Adherens junction turnover: regulating adhesion through cadherin endocytosis, degradation, and recycling, Sub-cellular Biochem. 60 (2012) 197–222. Available from: https://doi.org/10.1007/978-94-007-4186-7_9.
[132] B. Baum, M. Georgiou, Dynamics of adherens junctions in epithelial establishment, maintenance, and remodeling, J. Cell Biol. 192 (2011) 907–917. Available from: https://doi.org/10.1083/jcb.201009141.
[133] T.L. Le, A.S. Yap, J.L. Stow, Recycling of E-cadherin: a potential mechanism for regulating cadherin dynamics, J. Cell Biol. 146 (1999) 219–232.
[134] W. Shih, S. Yamada, N-cadherin-mediated cell-cell adhesion promotes cell migration in a three-dimensional matrix, J. Cell Sci. 125 (2012) 3661–3670. Available from: https://doi.org/10.1242/jcs.103861.
[135] C.Y. Tai, S.P. Mysore, C. Chiu, E.M. Schuman, Activity-regulated N-cadherin endocytosis, Neuron 54 (2007) 771–785. Available from: https://doi.org/10.1016/j.neuron.2007.05.013.
[136] F. Ulrich, C.P. Heisenberg, Trafficking and cell migration, Traffic 10 (2009) 811–818. Available from: https://doi.org/10.1111/j.1600-0854.2009.00929.x.

[137] C.M. Niessen, D. Leckband, A.S. Yap, Tissue organization by cadherin adhesion molecules: dynamic molecular and cellular mechanisms of morphogenetic regulation, Physiol. Rev. 91 (2011) 691–731. Available from: https://doi.org/10.1152/physrev.00004.2010.

[138] B.D. Cosgrove, K.L. Mui, T.P. Driscoll, S.R. Caliari, K.D. Mehta, R.K. Assoian, et al., N-cadherin adhesive interactions modulate matrix mechanosensing and fate commitment of mesenchymal stem cells, Nat. Mater. 15 (2016) 1297–1306. Available from: https://doi.org/10.1038/nmat4725.

[139] D.C. Radisky, D.D. Levy, L.E. Littlepage, H. Liu, C.M. Nelson, J.E. Fata, et al., Rac1b and reactive oxygen species mediate MMP-3-induced EMT and genomic instability, Nature 436 (2005) 123–127. Available from: https://doi.org/10.1038/nature03688.

[140] N. Gavert, M. Sheffer, S. Raveh, S. Spaderna, M. Shtutman, T. Brabletz, et al., Expression of L1-CAM and ADAM10 in human colon cancer cells induces metastasis, Cancer Res. 67 (2007) 7703–7712. Available from: https://doi.org/10.1158/0008-5472.CAN-07-0991.

[141] M.H. Yang, M.Z. Wu, S.H. Chiou, P.M. Chen, S.Y. Chang, C.J. Liu, et al., Direct regulation of TWIST by HIF-1alpha promotes metastasis, Nat. Cell Biol. 10 (2008) 295–305. Available from: https://doi.org/10.1038/ncb1691.

[142] H. Yano, Y. Mazaki, K. Kurokawa, S.K. Hanks, M. Matsuda, H. Sabe, Roles played by a subset of integrin signaling molecules in cadherin-based cell-cell adhesion, J. Cell Biol. 166 (2004) 283–295. Available from: https://doi.org/10.1083/jcb.200312013.

[143] O.H. Ocana, R. Corcoles, A. Fabra, G. Moreno-Bueno, H. Acloque, S. Vega, et al., Metastatic colonization requires the repression of the epithelial-mesenchymal transition inducer Prrx1, Cancer Cell 22 (2012) 709–724. Available from: https://doi.org/10.1016/j.ccr.2012.10.012.

[144] J.H. Tsai, J.L. Donaher, D.A. Murphy, S. Chau, J. Yang, Spatiotemporal regulation of epithelial-mesenchymal transition is essential for squamous cell carcinoma metastasis, Cancer Cell 22 (2012) 725–736. Available from: https://doi.org/10.1016/j.ccr.2012.09.022.

[145] H.Y. Kim, T.R. Jackson, L.A. Davidson, On the role of mechanics in driving mesenchymal-to-epithelial transitions, Semin. Cell Dev. Biol. 67 (2017) 113–122. Available from: https://doi.org/10.1016/j.semcdb.2016.05.011.

[146] R. Li, J. Liang, S. Ni, T. Zhou, X. Qing, H. Li, et al., A mesenchymal-to-epithelial transition initiates and is required for the nuclear reprogramming of mouse fibroblasts, Cell Stem Cell 7 (2010) 51–63. Available from: https://doi.org/10.1016/j.stem.2010.04.014.

[147] N.R. Chevalier, E. Gazguez, L. Bidault, T. Guilbert, C. Vias, E. Vian, et al., How tissue mechanical properties affect enteric neural crest cell migration, Sci. Rep. 6 (2016) 20927. Available from: https://doi.org/10.1038/srep20927.

[148] R.C. Hardie, K. Franze, Photomechanical responses in Drosophila photoreceptors, Science 338 (2012) 260–263. Available from: https://doi.org/10.1126/science.1222376.

[149] M. Iwashita, N. Kataoka, K. Toida, Y. Kosodo, Systematic profiling of spatiotemporal tissue and cellular stiffness in the developing brain, Development 141 (2014) 3793–3798. Available from: https://doi.org/10.1242/dev.109637.

[150] D.E. Koser, E. Moeendarbary, J. Hanne, S. Kuerten, K. Franze, CNS cell distribution and axon orientation determine local spinal cord mechanical properties, Biophys. J. 108 (2015) 2137–2147. Available from: https://doi.org/10.1016/j.bpj.2015.03.039.

[151] D.E. Koser, A.J. Thompson, S.K. Foster, A. Dwivedy, E.K. Pillai, G.K. Sheridan, et al., Mechanosensing is critical for axon growth in the developing brain, Nat. Neurosci. 19 (2016) 1592–1598. Available from: https://doi.org/10.1038/nn.4394.

[152] E. Moeendarbary, I.P. Weber, G.K. Sheridan, D.E. Koser, S. Soleman, B. Haenzi, et al., The soft mechanical signature of glial scars in the central nervous system, Nat. Commun. 8 (2017) 14787. Available from: https://doi.org/10.1038/ncomms14787.
[153] A. Kindberg, J.K. Hu, J.O. Bush, Forced to communicate: integration of mechanical and biochemical signaling in morphogenesis, Curr. Opin. Cell Biol. 66 (2020) 59−68. Available from: https://doi.org/10.1016/j.ceb.2020.05.004.
[154] D. Choquet, D.P. Felsenfeld, M.P. Sheetz, Extracellular matrix rigidity causes strengthening of integrin-cytoskeleton linkages, Cell 88 (1997) 39−48. Available from: https://doi.org/10.1016/s0092-8674(00)81856-5.
[155] A. Elosegui-Artola, R. Oria, Y. Chen, A. Kosmalska, C. Perez-Gonzalez, N. Castro, et al., Mechanical regulation of a molecular clutch defines force transmission and transduction in response to matrix rigidity, Nat. Cell Biol. 18 (2016) 540−548. Available from: https://doi.org/10.1038/ncb3336.
[156] M.H. Kopf, Collective cell migration induced by mechanical stress and substrate adhesiveness, Phys. Rev. E Stat. Nonlinear Soft Matter Phys. 91 (2015) 012712. Available from: https://doi.org/10.1103/PhysRevE.91.012712.
[157] S.V. Plotnikov, A.M. Pasapera, B. Sabass, C.M. Waterman, Force fluctuations within focal adhesions mediate ECM-rigidity sensing to guide directed cell migration, Cell 151 (2012) 1513−1527. Available from: https://doi.org/10.1016/j.cell.2012.11.034.
[158] N. Saxena, P. Mogha, S. Dash, A. Majumder, S. Jadhav, S. Sen, Matrix elasticity regulates mesenchymal stem cell chemotaxis, J. Cell Sci. 131 (2018). Available from: https://doi.org/10.1242/jcs.211391.
[159] M.P. Sheetz, D.P. Felsenfeld, C.G. Galbraith, Cell migration: regulation of force on extracellular-matrix-integrin complexes, Trends Cell Biol. 8 (1998) 51−54. Available from: https://doi.org/10.1016/s0962-8924(98)80005-6.
[160] L. Trichet, J. Le Digabel, R.J. Hawkins, S.R. Vedula, M. Gupta, C. Ribrault, et al., Evidence of a large-scale mechanosensing mechanism for cellular adaptation to substrate stiffness, Proc. Natl Acad. Sci. U. S. A. 109 (2012) 6933−6938. Available from: https://doi.org/10.1073/pnas.1117810109.
[161] C. Wu, S.B. Asokan, M.E. Berginski, E.M. Haynes, N.E. Sharpless, J.D. Griffith, et al., Arp2/3 is critical for lamellipodia and response to extracellular matrix cues but is dispensable for chemotaxis, Cell 148 (2012) 973−987. Available from: https://doi.org/10.1016/j.cell.2011.12.034.
[162] A.G. Clark, D.M. Vignjevic, Modes of cancer cell invasion and the role of the microenvironment, Curr. Opin. Cell Biol. 36 (2015) 13−22. Available from: https://doi.org/10.1016/j.ceb.2015.06.004.
[163] T.R. Cox, D. Bird, A.M. Baker, H.E. Barker, M.W. Ho, G. Lang, et al., LOX-mediated collagen crosslinking is responsible for fibrosis-enhanced metastasis, Cancer Res. 73 (2013) 1721−1732. Available from: https://doi.org/10.1158/0008-5472.CAN-12-2233.
[164] P. Friedl, S. Alexander, Cancer invasion and the microenvironment: plasticity and reciprocity, Cell 147 (2011) 992−1009. Available from: https://doi.org/10.1016/j.cell.2011.11.016.
[165] K.R. Levental, H. Yu, L. Kass, J.N. Lakins, M. Egeblad, J.T. Erler, et al., Matrix crosslinking forces tumor progression by enhancing integrin signaling, Cell 139 (2009) 891−906. Available from: https://doi.org/10.1016/j.cell.2009.10.027.
[166] M.R. Ng, A. Besser, G. Danuser, J.S. Brugge, Substrate stiffness regulates cadherin-dependent collective migration through myosin-II contractility, J. Cell Biol. 199 (2012) 545−563. Available from: https://doi.org/10.1083/jcb.201207148.
[167] C.M. Lo, H.B. Wang, M. Dembo, Y.L. Wang, Cell movement is guided by the rigidity of the substrate, Biophys. J. 79 (2000) 144−152. Available from: https://doi.org/10.1016/S0006-3495(00)76279-5.

[168] M. Raab, J. Swift, P.C. Dingal, P. Shah, J.W. Shin, D.E. Discher, Crawling from soft to stiff matrix polarizes the cytoskeleton and phosphoregulates myosin-II heavy chain, J. Cell Biol. 199 (2012) 669–683. Available from: https://doi.org/10.1083/jcb.201205056.
[169] R. Sunyer, V. Conte, J. Escribano, A. Elosegui-Artola, A. Labernadie, L. Valon, et al., Collective cell durotaxis emerges from long-range intercellular force transmission, Science 353 (2016) 1157–1161. Available from: https://doi.org/10.1126/science.aaf7119.
[170] C.A. Franco, M.L. Jones, M.O. Bernabeu, I. Geudens, T. Mathivet, A. Rosa, et al., Dynamic endothelial cell rearrangements drive developmental vessel regression, PLoS Biol. 13 (2015) e1002125. Available from: https://doi.org/10.1371/journal.pbio.1002125.
[171] C.A. Franco, M.L. Jones, M.O. Bernabeu, A.C. Vion, P. Barbacena, J. Fan, et al., Non-canonical Wnt signalling modulates the endothelial shear stress flow sensor in vascular remodelling, eLife 5 (2016) e07727. Available from: https://doi.org/10.7554/eLife.07727.
[172] R.S. Udan, T.J. Vadakkan, M.E. Dickinson, Dynamic responses of endothelial cells to changes in blood flow during vascular remodeling of the mouse yolk sac, Development 140 (2013) 4041–4050. Available from: https://doi.org/10.1242/dev.096255.
[173] J. Yang, P. Antin, G. Berx, C. Blanpain, T. Brabletz, M. Bronner, et al., Guidelines and definitions for research on epithelial-mesenchymal transition, Nat. Rev. Mol. Cell Biol. 21 (2020) 341–352. Available from: https://doi.org/10.1038/s41580-020-0237-9.
[174] M.S. Samuel, J.I. Lopez, E.J. McGhee, D.R. Croft, D. Strachan, P. Timpson, et al., Actomyosin-mediated cellular tension drives increased tissue stiffness and beta-catenin activation to induce epidermal hyperplasia and tumor growth, Cancer Cell 19 (2011) 776–791. Available from: https://doi.org/10.1016/j.ccr.2011.05.008.
[175] W.N. De Vries, A.V. Evsikov, B.E. Haac, K.S. Fancher, A.E. Holbrook, R. Kemler, et al., Maternal beta-catenin and E-cadherin in mouse development, Development 131 (2004) 4435–4445. Available from: https://doi.org/10.1242/dev.01316.
[176] T.M. Grana, E.A. Cox, A.M. Lynch, J. Hardin, SAX-7/L1CAM and HMR-1/cadherin function redundantly in blastomere compaction and non-muscle myosin accumulation during *Caenorhabditis elegans* gastrulation, Dev. Biol. 344 (2010) 731–744. Available from: https://doi.org/10.1016/j.ydbio.2010.05.507.
[177] T.P. Fleming, J. McConnell, M.H. Johnson, B.R. Stevenson, Development of tight junctions de novo in the mouse early embryo: control of assembly of the tight junction-specific protein, ZO-1, J. Cell Biol. 108 (1989) 1407–1418. Available from: https://doi.org/10.1083/jcb.108.4.1407.
[178] Y. Atsuta, Y. Takahashi, FGF8 coordinates tissue elongation and cell epithelialization during early kidney tubulogenesis, Development 142 (2015) 2329–2337. Available from: https://doi.org/10.1242/dev.122408.
[179] X. Trepat, M.R. Wasserman, T.E. Angelini, E. Millet, D.A. Weitz, J.P. Butler, et al., Physical forces during collective cell migration, Nat. Phys. 5 (2009) 426–430. Available from: https://doi.org/10.1038/nphys1269.
[180] G.F. Weber, M.A. Bjerke, D.W. DeSimone, A mechanoresponsive cadherin-keratin complex directs polarized protrusive behavior and collective cell migration, Dev. Cell 22 (2012) 104–115. Available from: https://doi.org/10.1016/j.devcel.2011.10.013.
[181] D. Caballero, J. Comelles, M. Piel, R. Voituriez, D. Riveline, Ratchetaxis: long-range directed cell migration by local cues, Trends Cell Biol. 25 (2015) 815–827. Available from: https://doi.org/10.1016/j.tcb.2015.10.009.
[182] J. Yang, P. Antin, G. Berx, C. Blanpain, T. Brabletz, M. Bronner, et al. EMT International Association (TEMTIA), Guidelines and definitions for research on epithelial-mesenchymal transition., Nat. Rev. Mol. Cell Biol. 21 (6) (2020) 341–352. Available from: https://doi.org/10.1038/s41580-020-0237-9.

CHAPTER FIVE

# Effects of time delays and viscoelastic parameters in oscillatory response of cell monolayers

Cristian Borja[1], Elena Moral[2] and Jose J. Muñoz[3]
[1]Barcelona University (UB), Spain
[2]Polytechnic University of Catalonia (UPC), Barcelona, Spain. Currently at Technical University of Munich (TUM), Germany
[3]Laboratory of Numerical Analysis (LaCàN). Polytechnic University of Catalonia (UPC), Barcelona, Spain

## 5.1 Introduction

The analysis of the oscillatory behavior of cells has attracted the attention of experimental and theoretical researchers in recent years. They are ubiquitous and may give rise to drastic morphogenetic shape changes. For instance, sustained planar oscillations of monolayers have been observed during embryonic germband extension in Drosophila embryo [1], or during dorsal closure [2,3] or ventral invagination [4].

These oscillations have been associated with the coupling between the chemical and mechanical signaling [5,6], or the relation between cell polarity and force transmission [7,8], with a potential dependence on the confinement [9]. More recently, it has been observed that the relocation of actomyosin components from medial to junctional areas of the cell may trigger these pulsatile response, suggesting that this biochemical delay between motor molecules is responsible of the observed oscillations [10].

Some recent works have analyzed the dynamics of the chemo-mechanical coupling through diffusion reaction equations, with applications in apical constriction in epithelia [5], monolayers on substrates [7,8] or wound healing [11]. In order to analyze the stability of such systems, we here focus on the delay and model it explicitly by resorting to a delayed cell rheology and the resulting delay differential equations, in

*Viscoelasticity and Collective Cell Migration.*
DOI: https://doi.org/10.1016/B978-0-12-820310-1.00005-7

similar manner to the stress analysis in yeast [12], neural axons conduction [13], transport in respiratory systems [14], or in cell maturation [15].

We here analyze the stability and oscillatory response of a planar deformation mode that preserves cell areas. We include a viscous like rheological law based on an adaptive rest-length [16], where the latter varies according to a delayed measure of strains. This work extends previous unidimensional analysis [17] to a planar and non-linear mode of deformation.

We note that we focus our analysis on cell oscillations due to the cytoskeletal reorganization, but without considering cell intercalation and rearrangements. In the former cases, the time scale of the oscillations is of the orders of several minutes [2], in contrast to longer periods observed in cell migration of monolayers. Consequently, our resulting stresses do not include any contribution due to the intercalation events observed during cell migration, whose contribution on the induced residual stresses must not be discarded.

This chapter is structured as follows. We first present the periodic planar deformation mode and the resulting equilibrium equations for are preserving planar deformations of a tissue in Section 5.2. We then analyze the linearized equations and the resulting delay differential equations, and provide bounds for the delay and material parameters in Section 5.2.3. We also study the stability and oscillatory bounds of the non-linear system (Section 5.2.4). We also present the equilibrium equations for the cross-section of the tissue in Section 5.3, and analyze the stability for on-plane (Section 5.3.1) and out-of-plane (Section 5.3.2) deformations. We finally give some concluding remarks in Section 5.4.

## 5.2 Planar deformations

### 5.2.1 Planar deformation mode and equilibrium equations

We will simplify our mechanical analysis by assuming that the epithelium can be represented by a two-dimensional patch of regular hexagons, as shown in Fig. 5.1A. We will additionally suppose that the cells deform in an area preserving mode, and that this deformation has the same periodicity of the whole pattern. The tissue deformation is thus fully described by the vertex displacement given in the grayed area of Fig. 5.1B. Although this periodicity is not exactly observed experimentally, cells locally share similar distortions. The simplification employed here also allows us to

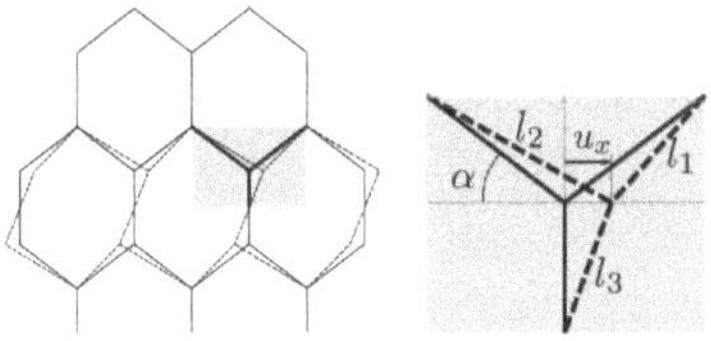


**Figure 5.1** (A) Scheme of distribution of cells as a regular pattern of hexagons. (B) Detail of the vertex displacement in the shaded area of the left.

parametrize the whole deformation through the displacement of a single vertex, denoted by $\mathbf{u} = \{u_x \ u_y\}^T$, and to obtain analytical results on the stability of the resulting oscillations.

According to these assumptions, the planar deformations of the two-dimensional patch are fully described with the three elements depicted in Fig. 5.1B, which represent the junctions of three generic cells. We will henceforth study the motion of the cells in the system by analyzing the equilibrium equations and the rheology of these 3 elements.

The equilibrium equations of the three element system are written in terms of an elastic potential function $W(\boldsymbol{u})$ given by,

$$W(\mathbf{u}) = \frac{1}{2}\sum_{i=1}^{3} k(l_i(\mathbf{u}) - L_i)^2 \tag{5.1}$$

where $k$ is the element stiffness, assumed as constant and equal for all the elements, $l_i(\mathbf{u})$ is the apparent (observed) length of each element $i = 1, 2, 3$, and $L_i$ its *rest-length*. We point out that the latter is assumed to be not necessarily constant. Its evolution will be detailed in the next section.

The equilibrium equations are deduced from the stationary condition of $W(\boldsymbol{u})$ as,

$$\frac{dW(\mathbf{u})}{d\mathbf{u}}\bigg|_{L_i=const} = \mathbf{0}$$

which from the expression of $W(\boldsymbol{u})$ in (5.1) yields,

$$\sum_{i=1}^{3}(l_i - L_i)\frac{dl_i}{d\mathbf{u}} = \mathbf{0} \tag{5.2}$$

Note that the equation above includes two scalar equations, one for $u_x$ (horizontal equilibrium) and another for $u_y$ (vertical equilibrium). For simplicity, we will assume that $u_y = 0$, and denote by $u$ the horizontal displacement $u_x$. The general case with $u_y \neq 0$ will be discussed in Section 5.2.4. It will be shown that a very similar analysis to the one presented here when $u_y = 0$ also applies for non-zero vertical displacements.

The expressions of $l_i$ in terms of $u$ are found using elemental geometry,

$$l_1 = \sqrt{(l_0 sin\alpha)^2 + (l_0 cos\alpha - u)^2} \tag{5.3}$$

$$l_2 = \sqrt{(l_0 sin\alpha)^2 + (l_0 cos\alpha + u)^2} \tag{5.4}$$

$$l_3 = \sqrt{l_0^2 + u^2} \tag{5.5}$$

where $\alpha$ is the initial angle that elements 1 and 3 form with the horizontal axis, and $l_0$ is the length of the undeformed configuration, assumed equal for all the elements (see Fig. 5.1B). Using the expressions above, the horizontal mechanical equilibrium is deduced from (5.2) as,

$$(l_1 - L_1)\frac{u - l_0 cos\alpha}{l_1} + (l_2 - L_2)\frac{l_0 cos\alpha + u}{l_2} + (l_3 - L_3)\frac{u}{l_3} = 0 \tag{5.6}$$

In the subsequent sections, we will use the compact notation $\mathbf{L}(t) = \{L_1(t)\ L_2(t)\ L_3(t)\}^T$ as the vector of unknown rest-lengths. Consequently, by replacing the geometrical relations in (5.3)–(5.5) into the previous non-linear equation, it can be re-casted as

$$G(u, \mathbf{L}) = 0 \tag{5.7}$$

### 5.2.2 Cortex rheology and delay differential equations

We assume that each element has the ability to adapt to its current length $l_i$ by changing its resting length $L_i$ according to the following evolution law,

$$\dot{L}_i(t) = \gamma_i(l_i(t-\tau) - L_i(t-\tau)), \quad i = 1, 2, 3 \tag{5.8}$$

where $\gamma_i > 0$ is a material parameter called the *remodeling rate*, and $\tau > 0$ is a delay that measures the retarding time between the measured strain $l_i - L_i$ and the cell active rest-length changes. Parameters $\gamma_i$ represents the rate at which cells adapt their length. It has been proved that when $\tau = 0$, the evolution law in (5.8) mimics Maxwell viscoelastic response [16]. Such model has been also employed in embryogenesis [18–21] or relaxation tests on suspended monolayers [22]. We remark that for $\gamma_i = 0$ the resting length becomes constant, in which case we recover purely elastic rheology.

The presence of delay $\tau$ is motivated by the observed oscillatory response and the measured delay between myosin concentration and area oscillation in Drosophila fly dorsal closure [2]. The stability of single elements with the evolution law in (5.8) has been analyzed in [17]. We are here interested in extending these results to multicellular systems.

The presence of the the delay $\tau$ in (5.8) turns this differential equation into a so-called Delay Differential Equation (DDE) [23], which jointly with the equilibrium equation in (5.6) and the geometrical relations in (5.3)–(5.5) will allow to analyze the stability of our primary unknowns $u(t)$ and $L_i(t), i = 1, 2, 3$. More precisely, by replacing the expression of $l_i$ in (5.8) by the relations in (5.3)–(5.5), the time evolution of $u$ and $\mathbf{L}(t)$ are given by the following system of algebraic-DDE,

$$\begin{cases} \dot{\mathbf{L}}(t) = \mathbf{f}(u(t-\tau), \mathbf{L}(t-\tau)), \\ G(u(t), \mathbf{L}(t)) = 0, \end{cases} \tag{5.9}$$

with the initial conditions

$$\begin{matrix} \mathbf{L}(t) & = \mathbf{L}_0 \\ u(t) & = u_0 \end{matrix} \} \forall t \in [-\tau, 0]. \tag{5.10}$$

The explicit expression of function $\mathbf{f}(u(t), \mathbf{L}(t))$ in (5.9) is

$$\mathbf{f}(u(t), \mathbf{L}(t)) = \begin{Bmatrix} \gamma_1 \left( \sqrt{(l_0 \sin\alpha)^2 + (l_0 \cos\alpha - u(t))^2} - L_1(t) \right) \\ \gamma_2 \left( \sqrt{(l_0 \sin\alpha)^2 + (l_0 \cos\alpha + u(t))^2} - L_2(t) \right) \\ \gamma_3 \left( \sqrt{l_0{}^2 + u(t)^2} - L_3(t) \right) \end{Bmatrix}. \tag{5.11}$$

Note that due to the presence of the DDE in (5.9), the initial conditions are defined on the interval $[-\tau, 0]$, and that we have assumed for simplicity that $u_0$ and $\mathbf{L}_0 = \{L_{10} \;\; L_{20} \;\; L_{30}\}^T$ are constant. For consistency, the values of $u_0$ and $\mathbf{L}(0)$ should satisfy $G(u_0, \mathbf{L}_0) = 0$. We will next study the linearized form of the initial value problem in (5.9)–(5.11).

## 5.2.3 Linearized equations of planar deformation

### *5.2.3.1 Linear delay-differential equations*

The algebraic-DDE system is non-linear due to the expression of $\mathbf{f}(u(t), \mathbf{L}(t))$ and $G(u(t), \mathbf{L}(t))$. We will henceforth assume that $u(t) \ll 1$, and that the apparent lengths $l_i(t)$ can be approximated by the linear terms,

$$\begin{aligned} l_1(t) &\approx l_0 - u(t)cos\alpha, \\ l_2(t) &\approx l_0 + u(t)cos\alpha, \\ l_3(t) &\approx l_0, \end{aligned} \tag{5.12}$$

which are valid at all times. By inserting these linearized forms into the DDEs in (5.8) we obtain,

$$\begin{cases} \dot{L}_1(t) &= -\gamma_1 u(t-\tau)cos\alpha - \gamma_1 L_1(t-\tau) + \gamma_1 l_0, \\ \dot{L}_2(t) &= \gamma_2 u(t-\tau)cos\alpha - \gamma_2 L_2(t-\tau) + \gamma_2 l_0, \\ \dot{L}_3(t) &= -\gamma_3 L_3(t-\tau) + \gamma_3 l_0, \end{cases} \tag{5.13}$$

Furthermore, the linearized form of the equilibrium equation in (5.6) at $(u, \mathbf{L}) = (0, \mathbf{L}_0)$ becomes:

$$\left(3 - \frac{L_{10}}{l_0} sin^2\alpha - \frac{L_{20}}{l_0} sin^2\alpha - \frac{L_{30}}{l_0}\right) u(t) + cos\alpha(L_1(t) - L_2(t)) = 0$$

which after setting

$$\lambda_0 = 3 - \frac{L_{10}}{l_0} sin^2\alpha - \frac{L_{20}}{l_0} sin^2\alpha - \frac{L_{30}}{l_0} \tag{5.14}$$

may be written at time $t - \tau$ as,

$$u(t-\tau) = \frac{\cos\alpha}{\lambda_0}(L_2(t-\tau) - L_1(t-\tau)). \tag{5.15}$$

Replacing this expression of $u(t)$ into (5.13), our system of algebraic-DDE in (5.9)–(5.11) is approximated by the following linear system of DDEs:

$$\begin{cases} \dot{\mathbf{L}}(t) &= \boldsymbol{A}\mathbf{L}(t-\tau) + l_0\gamma, \\ \mathbf{L}(t) &= \mathbf{L}_0, \qquad \forall t \in [-\tau, 0]. \end{cases} \tag{5.16}$$

with

$$\mathbf{A} = \begin{bmatrix} \gamma_1\left(\frac{cos^2\alpha}{\lambda_0} - 1\right) & -\gamma_1\frac{cos^2\alpha}{\lambda_0} & 0 \\ -\gamma_2\frac{cos^2\alpha}{\lambda_0} & \gamma_2\left(\frac{cos^2\alpha}{\lambda_0} - 1\right) & 0 \\ 0 & 0 & -\gamma_3 \end{bmatrix}, \quad \gamma = \begin{Bmatrix} \gamma_1 \\ \gamma_2 \\ \gamma_3 \end{Bmatrix}. \tag{5.17}$$

### 5.2.3.2 Solution of linear delay differential equation system

For simplicity we will study the stability of the DDE systems in (5.16)–(5.17) assuming that

$$\gamma_1 = \gamma_2 = \gamma_3 = \gamma. \tag{5.18}$$

In this case, since matrix $\mathbf{A}$ is a real symmetric matrix, it accepts real eigenvalues $\lambda_1, \lambda_2, \lambda_3$. Its eigenvalue decomposition may then be written as,

$$\mathbf{A} = \mathbf{V}^T \mathbf{D} \mathbf{V} \tag{5.19}$$

with $\mathbf{D} = diag(\lambda_1, \lambda_2, \lambda_3)$ and $V = [\mathbf{v}_1, \ \mathbf{v}_2, \mathbf{v}_3]$ the matrix of normalized eigenvectors such that $\mathbf{V}^T\mathbf{V} = \mathbf{I}$. In our case, from the expression of $\mathbf{A}$ in (5.17) we have that,

$$\mathbf{D} = \begin{bmatrix} -\gamma & 0 & 0 \\ 0 & -\gamma & 0 \\ 0 & 0 & -\gamma\left(1 - \dfrac{2cos^2\alpha}{\lambda_0}\right) \end{bmatrix}, \mathbf{V} = \frac{1}{\sqrt{2}}\begin{bmatrix} 1 & 0 & 1 \\ 1 & 0 & -1 \\ 0 & \sqrt{2} & 0 \end{bmatrix}. \tag{5.20}$$

It follows that the DDE system in (5.16)–(5.17) is equivalent to the following *diagonal* system,

$$\begin{cases} \dot{\mathbf{y}}(t) & = \boldsymbol{D}\mathbf{y}(t-\tau) + l_0\gamma\mathbf{v}, \\ \mathbf{y}(t) & = \boldsymbol{V}\mathbf{L}_0, \qquad \forall t \in [-\tau, 0]. \end{cases} \tag{5.21}$$

with $\mathbf{y}(t) = \boldsymbol{V}\mathbf{L}(t)$ and $\mathbf{v} = \sum_i \mathbf{v}_i$. It is worth pointing out though that the linear system in (5.16) can also be written as,

$$\begin{cases} \dot{L}_2(t) - \dot{L}_1(t) & = \gamma\left(1 - \dfrac{2\cos^2\alpha}{\lambda_0}\right)(L_2(t-\tau) - L_1(t-\tau)), \\ \dot{L}_3(t) & = -\gamma L_3(t-\tau) + \gamma l_0, \end{cases} \tag{5.22}$$

with initial conditions

$$\begin{cases} L_2(t) - L_1(t) & = L_{20} - L_{10}, \quad t \in [-\tau, 0] \\ L_3(t) & = L_{30}, \quad t \in [-\tau, 0] \end{cases}$$

**Remark 5.1.** *The latter forms prompts us to consider the following particular situation:*

*(a) When $L_{10} = L_{20}$, due to the first DDE in (5.22), we have that*

$$L_2(t) - L_1(t) = 0 \quad \forall t \in [-\tau, \infty)$$

*This corresponds to reducing the dynamics to solely mode* $\mathbf{v}_3$ *in* (5.20). *Also, due to the equilibrium condition in* (5.15), *and when* $\lambda_0 \neq 0$, *the displacement* $u(t) \equiv 0$. *In this case, the delay equations in* (5.13) *are simplified, and matrix* $\mathbf{A}$ *reduces to*

$$\mathbf{A} = -\gamma \mathbf{I} \tag{5.23}$$

That is, $\lambda_1 = \lambda_2 = \lambda_3 = -\gamma$.

*(b) Additionally, let us assume that* $L_{10} = L_{20} = L_{30}$. *Then, due to the form of the system matrix* $\mathbf{A}$ *in (5.23) and initial conditions, all rest-lengths have the same evolution:*

$$L_1(t) = L_2(t) = L_3(t) \quad \forall t \in [-\tau, \infty).$$

In order to analyze the stability of the DDE, let us introduce the general form of the solutions to the DDE:

**Proposition 5.1.** *The solution of the DDE in (5.21) is given by function* $\mathbf{y}_m = \{y_{m1}, \; y_{m2}, \; y_{m3}\}^T$, *with components,*

$$y_{mi}(t) = e^{m_i t} y_{0i} - \frac{l_0 \gamma}{\lambda_i}, \quad i = 1, 2, 3, \tag{5.24}$$

if $m_i$ *is solution of the characteristic equations,*

$$\tau \lambda_i = m_i e^{m_i}, \quad i = 1, 2, 3. \tag{5.25}$$

The proposition can be proved by simply inserting the expression in (5.24) into (5.21), which gives rise to the condition

$$\lambda_i = m_i e^{m_i \tau}, \quad i = 1, 2, 3. \tag{5.26}$$

By using the change of variable $t' = t/\tau$, the DDE in (5.21) reads $d\mathbf{y}/dt' = \tau \mathbf{D} \mathbf{y}(t' - 1) - l_0 \gamma \mathbf{v}$, and thus condition in (5.26) turns into condition in (5.25). The latter equation can be written in terms of the Lambert function $W(z)$, which is defined as the inverse function such that $z = W(ze^z)$, $\forall z \in \mathbb{C}$, or equivalently, the solution of [23–25],

$$z = W(z) e^{W(z)}. \tag{5.27}$$

Alternatively, functions $y_{mi}(t)$ in (5.24) are solution of the DDE if

$$m_i = W(\lambda_i \tau), \quad i = 1, 2, 3. \tag{5.28}$$

**Remark 5.2.** *Lambert function can be extended to matrices by using the definition* [26,27],

$$\mathbf{W}(\mathbf{A})e^{\mathbf{W}(\mathbf{A})} = \mathbf{A}$$

*In our case though, since* $\mathbf{A}$ *admits the diagonalisation* $\mathbf{A} = \mathbf{V}\mathbf{D}\mathbf{V}^T$, *the Lambert matrix expression reads* $\mathbf{W}(\mathbf{A}) = \mathbf{V}^T\mathbf{W}(\mathbf{D})\mathbf{V}^T$, *and thus the analysis of the values of* $\mathbf{W}(\mathbf{A})$ *reduces to the simpler scalar definition in* (5.27).

Lambert function is multivalued, giving rise to a set of solutions (or branches) $m_{ik}, k\in\mathbf{Z}$ that satisfy Eq. (5.27) [24]. Due to the linearity of the DDE, component $y_i(t)$ of the general solution $\mathbf{y}(t)$ can be written as

$$y_i(t) = \sum_{k=-\infty}^{+\infty} e^{m_{ik}t} c_{ik} - l_0 \frac{\gamma}{\lambda_i}, \quad i = 1, 2, 3 \tag{5.29}$$

where the coefficients $c_{ik}$, $i = 1, 2, 3$, $k\in\mathbf{Z}$ depend on the initial conditions.

#### 5.2.3.3 Stability and oscillatory analysis of linear delay differential equation

We first recall standard results of the Lambert function and the characteristic equation [23,25]:

**Proposition 5.2.** *Lambert function* $W(z)$ *satisfies,*

$$i) Re(W(z)) > 0 \text{ if } Re(z) < -\frac{\pi}{2} \text{ or } Re(z) > 0$$

$$ii) Im(W(z)) \neq 0 \text{ if } Re(z) < -\frac{1}{e}$$

The bounds in Proposition 5.2 can be verified by searching the values for which the real part or the imaginary part of $me^m$ vanish [23]. Indeed, setting $m = a + ib$ and $z = \xi + i\eta$, equation $z = me^m$ implies

$$\xi = e^a(a\cos b - b\sin b)$$

$$\eta = e^a(b\cos b + a\sin b).$$

When $a = 0$, we have from the first equation that $\xi = b = \pi/2 \pm k\pi$, $k\in\mathbf{Z}$. When $b\rightarrow 0$, the second equation gives $a = -\lim_{b\rightarrow 0} b\cot b = -1$, which inserted into the first equation yields $\xi = -1/e$.

Since the solutions $y_{mi}(t)$ contain the term $e^{m_i t}$, we know that stable solution will be obtained whenever $Re(m_i) < 0$, and oscillations will appear whenever $Im(m_i) \neq 0$. Therefore, from Proposition 5.2 and the eigenvalues $\lambda_i$ in (5.20), we conclude our main result for the linearized problem:

**Proposition 5.3.** *The DDE in* (5.16)–(5.17), *with* $\gamma_1 = \gamma_2 = \gamma_3 = \gamma > 0$ *and* $\tau > 0$ will have unstable solutions if

(ia) $L_{10} \neq L_{20}$:

$$\gamma\tau\max\left(1, 1 - \frac{2\cos^2\alpha}{\lambda_0}\right) > \frac{\pi}{2}, \quad \text{or,} \quad 1 - \frac{2\cos^2\alpha}{\lambda_0} < 0,$$

(ib) $L_{10} = L_{20}$

$$\gamma\tau > \frac{\pi}{2},$$

*and oscillatory solutions if*

(iia) $L_{10} \neq L_{20}$:

$$\gamma\tau\max\left(1, 1 - \frac{2\cos^2\alpha}{\lambda_0}\right) > \frac{1}{e}$$

(iib) $L_{10} = L_{20}$:

$$\gamma\tau > \frac{1}{e}$$

Consequently, we have that when $\lambda_0 < 0$ and $L_{10} \neq L_{20}$, the initial conditions $\mathbf{L}_0$ may determine the presence of oscillations, and that when $\lambda_0 < \lambda_{crit}$ with

$$\lambda_{crit} = 2cos^2\alpha, \tag{5.30}$$

and $L_{10} \neq L_{20}$, the stability conditions depend also on the initial conditions. For instance, for $\alpha = \pi/6$, and setting $L_{10} \approx L_{20} = L_{30} = m_L l_0$, with $m_L \geq 0$ a model constant, we have from (5.14) that $\lambda_0 \approx 3 - 2m_L$. Therefore, the initial conditions determine the stability of the DDE whenever $3 - 3m_L/2 < 2cos^2\alpha = 3/2$, that is, when

$$m_L > 1. \tag{5.31}$$

This corresponds to compressive initial states ($L_{i0} > l_0$), which in cellular materials is not a common situation. When $\lambda_0 > \lambda_{crit}$, the stability is determined by the value of $\gamma\tau$ because $0 < \left(1 - \frac{2cos^2\alpha}{\lambda_0}\right) < 1$, and for unstressed initial conditions ($L_{i0} = l_0$), $\lambda_0 = \lambda_{crit}$, and thus the stability is not affected by the initial conditions. A similar conclusion can be drawn when $L_{i0} = 0$ (bars are stretched to their maximum), which yields $\lambda_0 = 3$.

Fig. 5.2 shows the stability diagram on the $(\gamma, \tau)$ plane for $L_{10} = L_{20}$, which shows the same patterns as the one-for-one-element system [17]. Fig. 5.3 illustrates the dependence of the stability on $\lambda_0$. The figure shows

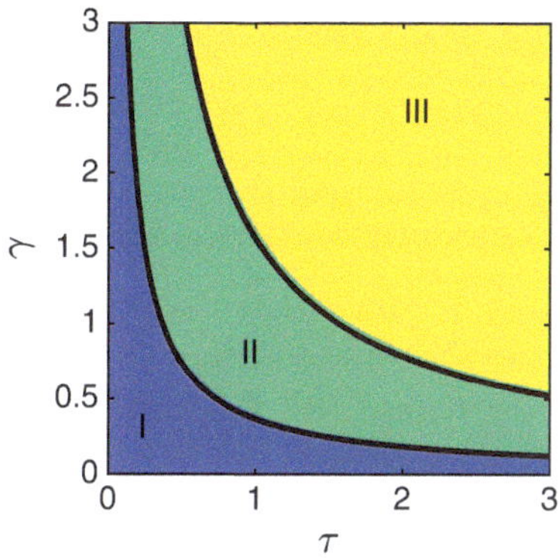


**Figure 5.2** Stability diagram on the $(\gamma, \tau)$ plane for $L_{10} = L_{20} = L_{30} = 0.9l_0$, $\alpha = \pi/6$, so that $\lambda_0 = 1.65$.

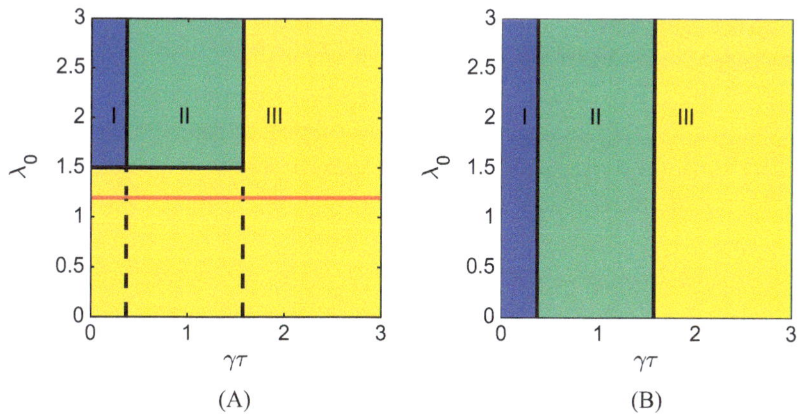


**Figure 5.3** Stability diagram of the linear system (A) and non-linear system (B) in the $(\lambda_0, \gamma\tau)$ plane, obtained with $l_0 = 1$, $\alpha = \frac{\pi}{6}$, $L_{10} = L_{30} = 1.2l_0$ and $L_{20}$ calculated with the expression of $\lambda_0$ in (14). The vertical black lines are the critical values of the stability and oscillatory, $\frac{1}{e}$ and $\frac{\pi}{2}$, respectively. The horizontal black line is the value of $\lambda_{crit} = 2\cos^2\alpha$. The red line is the value of $\lambda_0$ such that $L_{10} = L_{20}$ and corresponds to the value $\lambda_0 = 3 - \frac{L_{20}}{l_0}(2sin^2\alpha + 1)$. The sectors **I** are the stable and non-oscillatory solutions, **II** are the stable and oscillatory solutions and **III** are the unstable solutions.

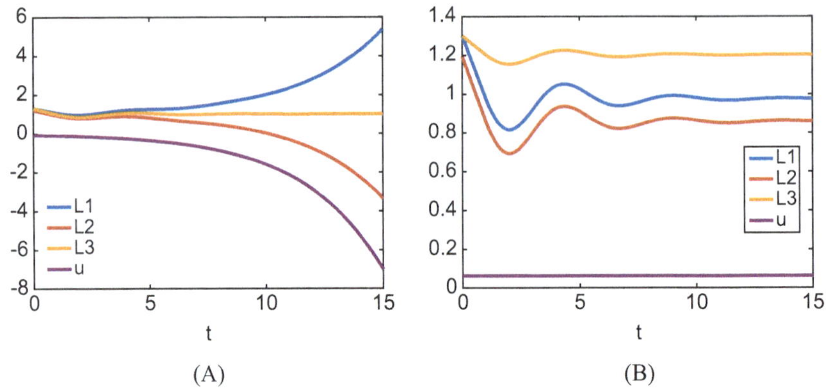


**Figure 5.4** Numerical simulations with $l_0 = \gamma = \tau = 1$, $\alpha = \frac{\pi}{6}$, $L_{10} = L_{30} = 1.3l_0$, $L_{20} = 1.2l_0$, $\lambda_0 = 1.075$. (A) Linear problem. (B) Non-linear problem.

the analytical results for the linear system (A) and also the numerical results for the non-linear system (B). The time evolution of $\mathbf{L}(t)$ is also shown for a value of $\lambda_0$ in Fig. 5.4. The fact that the non-linear system shows no dependence on the initial conditions has motivated the analysis of the following section.

### 5.2.4 Non-Linear system of planar deformations

We have numerically solved the non-linear system in (5.9) for $\gamma_1 = \gamma_2 = \gamma_3 = \gamma$, and verified that it exhibits no dependence on the initial conditions. This system uses the same strain function in (5.1) as in the previous section, but with no simplification (linearisation).

The non-linear system exhibits the same stability diagram shown in Fig. 5.2 for all values of $\mathbf{L}_0$. This fact will be justified in the next result:

**Proposition 5.4.** *The system of non-linear DDE in (9) admits the solution $\dot{u} = 0$ if $\gamma_1 = \gamma_2 = \gamma_3 = \gamma$ and the initial conditions preserve equilibrium, that is, $\mathbf{G}(u_0, \mathbf{L}_0) = \mathbf{0}$.*

**Proof 5.1.** *This can be proved by verifying that indeed the solution $\dot{u} = 0$ is compatible with the equilibrium balance in (5.2) and the associated DDE equations in (5.8). First note that when $u = const$, we have that $dl_i/du = c_i = const$ and $l_i(t) \equiv l_0$. Therefore, the equilibrium equations*

$$\sum_{i=1}^{3} \frac{dW}{dl_i} \frac{dl_i}{du} = 0 \tag{5.32}$$

*read*

$$\sum_{i=1}^{3}(l_0 - L_i(t))c_i = 0 \tag{5.33}$$

*The time variation of this equation yields,*

$$-\sum_{i=1}^{3}\dot{L}_i(t)c_i = 0, \tag{5.34}$$

*which must hold* $\forall t \in [-\tau, \infty)$. *By replacing* $\dot{L}_i(t)$ *by the expression of the three evolution laws in (5.8) we obtain an equivalent equation to the equilibrium in* (5.33),

$$\gamma\sum_{i=1}^{3}(l_0 - L_i(t-\tau))c_i = 0. \tag{5.35}$$

*If we denote by* $G(u(t), \mathbf{L}(t)) = 0$ *the equilibrium equation in* (5.33), *we have thus shown that when* $\dot{u} = 0$ *and* $\gamma_1 = \gamma_2 = \gamma_3$, *the evolution laws in* (5.8) *imply that,*

$$-\frac{d}{dt}G(u_0, \mathbf{L}(t)) = \gamma G(u_0, \mathbf{L}(t-\tau)). \tag{5.36}$$

*Consequently, if the initial conditions are such that* $G(u_0, \mathbf{L}_0) = 0$, *that is, equilibrium is preserved during* $t \in [-\tau, 0]$, *then equilibrium is preserved at all times due to the evolution laws.*

The stability bounds in the non-linear case can now be explained by pointing out that when $\dot{u} = 0$, each element has a constant length $l_i = l_0$, and therefore behaves as a single element, where the evolution laws in (5.8) become independent and linear. The stability can be then analyzed like in the unidimensional case [17], with stability bounds for $\gamma\tau$ given by $\pi/2$.

**Remark 5.3.** *The result of Proposition 5.4 for the non-linear systems also holds if vertical displacements are allowed, that is,* $u_y \neq 0$. *In this case, when* $\dot{\mathbf{u}} = 0$, the equilibrium equations are written for components $x$ and $y$ as,

$$\sum_{i=1}^{3}(l_0 - L_i(t))\mathbf{c}_i = \boldsymbol{0}$$

*with* $\mathbf{c}_i = \frac{dl}{d\mathbf{u}} = \{dl_i/du_x \ dl_i/du_y\}^T$. *If this equation is expressed as* $\mathbf{G}(\mathbf{u}_0, \mathbf{L}(t)) = \boldsymbol{0}$, *a similar relation to the one in (5.36) is derived, and thus the same conclusions apply.*

**Remark 5.4.** *Proposition 5.4 does not necessary hold when the parameters* $\gamma_i$ *are not equal, or for the linear system. Indeed, the linearized equilibrium equations at a point* $u = 0, L_i = L_{i0}$ are derived from the expression

$$G(0,\mathbf{L}_0) + \frac{dG(u,\mathbf{L})}{du}|_{(0,\mathbf{L}_0)}u + \sum_{i=1}^{3}\frac{dG(u,\mathbf{L})}{dL_i}|_{(0,\mathbf{L}_0)}(L_i - L_{i0}) = 0,$$

*which result into the equation written in* (5.15),

$$\lambda_0 u(t) + (L_1(t) - L_2(t))cos\alpha = 0, \tag{5.37}$$

*with* $\lambda_0$ *given in (5.14). This equation, which is the linearized version of (5.32), differs from the form in (5.33) and may not be compatible with the evolution laws in (5.8). In fact, the time differentiation of (5.37) reveals that if* $\dot{u} = 0$*, then* $\dot{L}_1 = \dot{L}_2$*. However, the evolution laws in (5.8) imply that* $\dot{L}_1 \neq \dot{L}_2$ *when* $L_{10} \neq L_{20}$*. Consequently, the solution* $\dot{u} = 0$ *is not admissible for this initial condition.*

## 5.3 Analysis of tissue cross-section

### 5.3.1 On-plane deformation

We consider here a strip of tissue where cells are adjacent to each other and a horizontal displacement $\mathbf{u}(t)$ is acting on the nodes. We will first analyze on-plane displacements that preserve cross-section area and that combine top and bottom surface, as indicated in Fig. 5.5.

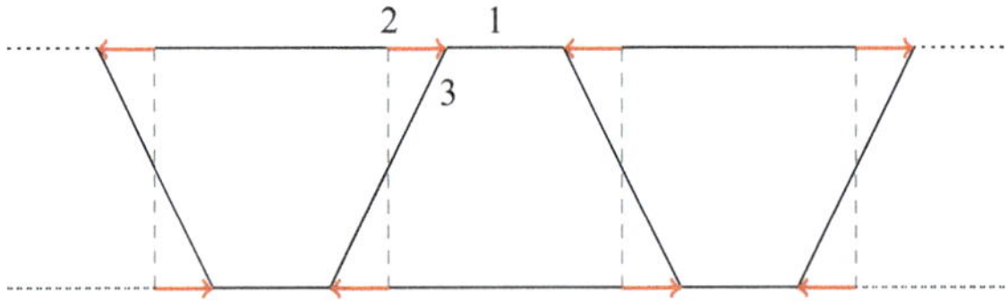


**Figure 5.5** A section of tissue under the displacement $\mathbf{u}(t)$ -in red-. Edges are enumerated.

By additionally considering a *constant height* assumption, the DDE for the rest-length and the *equilibrium* condition lead to a fully determined system,

$$\begin{cases} \dot{L}_1(t) = \gamma(l_0 - 2u(t-\tau) - L_1(t-\tau)) \\ \dot{L}_2(t) = \gamma(l_0 + 2u(t-\tau) - L_2(t-\tau)) \\ \dot{L}_3(t) = \gamma\left(\sqrt{l_0^2 + 4u^2(t-\tau)} - L_3(t-\tau)\right) \\ 0 = L_2(t) - L_1(t) - 6u(t) + \dfrac{2u(t)L_3(t)}{\sqrt{l_0^2 + 4u^2(t)}} \end{cases} \tag{5.38}$$

Let us consider $m_u$ and $m_L$ two dimensionless parameters such that:

$$L_1(0) = L_3(0) = m_L l_0, \qquad u(0) = m_u l_0$$

In order to guarantee the initial equilibrium, we introduce a small perturbation $\epsilon$ such that $L_2(0) = m_L l_0 + \epsilon$. The equilibrium equation provides an explicit expression for $\epsilon$:

$$\epsilon = \epsilon(l_0, m_L, m_u) = 2m_u l_0\left(3 - \frac{m_L}{\sqrt{1+4m_u^2}}\right) > 0$$

Note that in order to avoid overlapping, the dimensionless parameters must satisfy $m_u \in \left(-\frac{1}{2}, \frac{1}{2}\right)$ and $m_L \in (0,1)$.

After the linearisation of the system in (5.38) at the initial condition we obtain a system of DDEs that can be written as,

$$\begin{bmatrix} \dot{L}_1(t) \\ \dot{L}_2(t) \\ \dot{L}_3(t) \end{bmatrix} = \begin{bmatrix} A & B & C \\ B & A & -C \\ C & -C & D \end{bmatrix} \begin{bmatrix} L_1(t-\tau) \\ L_2(t-\tau) \\ L_3(t-\tau) \end{bmatrix} + \gamma l_0 \begin{bmatrix} 1 + \dfrac{2m_L m_u}{3\left(1+4m_u^2\right)^{\frac{3}{2}} - m_L} \\ 1 - \dfrac{2m_L m_u}{3\left(1+4m_u^2\right)^{\frac{3}{2}} - m_L} \\ \dfrac{1}{\sqrt{1+4m_u^2}}\left(1 - \dfrac{4m_u^2 m_L}{3\left(1+4m_u^2\right)^{\frac{3}{2}} - m_L}\right) \end{bmatrix} \tag{5.39}$$

where,

$$A = \gamma \frac{m_L - 2\left(1 + 4m_u^2\right)^{\frac{3}{2}}}{3\left(1 + 4m_u^2\right)^{\frac{3}{2}} - m_L} \qquad B = -\gamma \frac{\left(1 + 4m_u^2\right)^{\frac{3}{2}}}{3\left(1 + 4m_u^2\right)^{\frac{3}{2}} - m_L}$$

$$C = -\gamma \frac{2m_u\left(1 + 4m_u^2\right)}{3\left(1 + 4m_u^2\right)^{\frac{3}{2}} - m_L} \qquad D = \gamma \frac{m_L - \sqrt{1 + 4m_u^2}\left(3 + 8m_u^2\right)}{3\left(1 + 4m_u^2\right)^{\frac{3}{2}} - m_L}$$

In order to analyze the stability, we apply a change of variables $z(t) := \mathbf{V}^T\mathbf{L}(t) \in \mathbb{R}^3$, where $\mathbf{V}$ is the eigenvector matrix of the system above, and use the eigenvalue diagonal matrix

$$\mathbf{D} = \begin{bmatrix} -\gamma & 0 & 0 \\ 0 & \lambda_2 & 0 \\ 0 & 0 & -\gamma \end{bmatrix}$$

so that $\lambda_1 = \lambda_3 = -\gamma$, and

$$\lambda_2 = \gamma \frac{m_L - \sqrt{1 + 4m_u^2}(3 + 8m_u^2) + 2(1 + 4m_u^2)^{\frac{3}{2}}}{3(1 + 4m_u^2)^{\frac{3}{2}} - m_L} < 0$$

The oscillatory regime of the DDE in (5.39) is then equivalent to the following one,

$$\dot{\mathbf{z}}(t) = \mathbf{D}\mathbf{z}(t - \tau)$$

In view of the eigen-values in $\mathbf{D}$, the oscillatory and stability limits of the system are:

$$\tau_{oscil} = \frac{1}{e\gamma} \qquad \tau_{stabil} = \frac{\pi}{2\gamma}$$

We have also numerically solved the non-linear system in (5.38) and using a length-dependent delay, different for each segment, that is, $\tau_i(t) = \alpha l_i(t - \delta)$, with $\delta << 1$ the used time-step and $\alpha \in \mathbb{R}_{>0}$. The resulting stability diagram is shown in Fig. 5.6 and resembles the one of the linearized system, but replacing $\tau$ by $\alpha$.

## 5.3.2 Out-of-plane deformation

We consider in this section a strip of tissue where cells are adjacent to each other and a horizontal displacement $\mathbf{u}(t)$ acting on the nodes, as shown in Fig. 5.7.

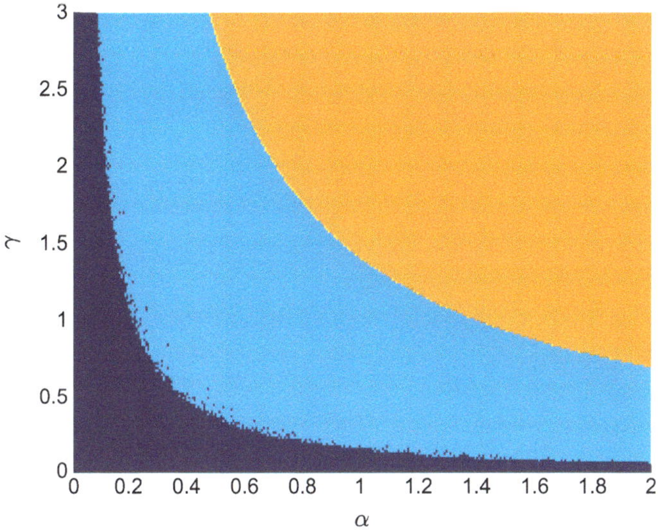


**Figure 5.6** Stability diagram of the cross-section model with on-plane displacements. Dark blue: stable non-oscillatory region. Light blue: unstable oscillatory. Yellow: unstable. Orange: region where parameters lead to negative lengths.

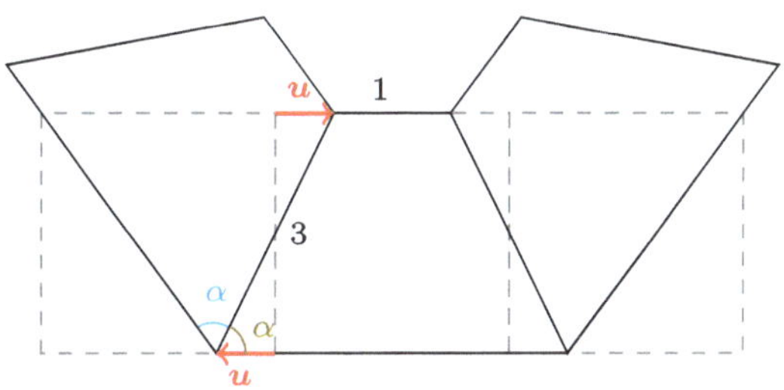


**Figure 5.7** A section of tissue under the displacement $\mathbf{u}(t)$ -in red-. Edges are enumerated.

After assuming a constant height and area, the balance equations for each segment yield the following system of equations:

$$\begin{cases} \dot{L}_1(t) = \gamma(l_0 - 2u(t-\tau) - L_1(t-\tau)) \\ \dot{L}_3(t) = \gamma(\sqrt{l_0^2 + 4u^2(t-\tau)} - L_3(t-\tau)) \\ 0 = l_0^2 + 12u^2(t) + 4u(t)L_1(t) - 4u(t)l_0 - \sqrt{l_0^2 + 4u^2(t)}L_3(t) \end{cases} \tag{5.40}$$

In order to simplify the subsequent equations, let us define the dimensionless parameters $m_u$ and $m_L$, and include a small perturbation of $L_3(0)$ denoted by $\epsilon$ which guarantees the initial equilibrium:

$$u(0) = m_u l_0, \qquad L_1(0) = m_L l_0, \qquad L_3(0) = m_L l_0 + \epsilon,$$

The equilibrium condition at $t = 0$ leads to an expression for $\epsilon$:

$$\epsilon = \epsilon(l_0, m_L, m_u) = l_0\left(-m_L + \frac{1 + 12m_u^2 + 4m_u m_L - 4m_u}{\sqrt{1 + 4m_u^2}}\right) > 0$$

Note that in order to avoid overlapping domains, the parameters must satisfy $m_u \in \left(-\frac{1}{2}, \frac{1}{2}\right)$ and $m_L \in (0, 1)$.

#### 5.3.2.1 Analysis of linearized form

The linearisation of (5.40) yields the following system of DDEs

$$\begin{cases} \dot{L}_1(t) = \gamma(l_0 - 2u(t-\tau) - L_1(t-\tau)) \\ \dot{L}_3(t) = \gamma(\dfrac{l_0^2}{\sqrt{l_0^2 + 4u_0^2}} - L_3(t-\tau) + \dfrac{4u_0}{\sqrt{l_0^2 + 4u_0^2}}u(t-\tau)) \\ 4u_0\left(L_1(t) - L_{10} - 3u_0 - 6u(t) + \dfrac{L_{30}(u_0 - u(t))}{\sqrt{l_0^2 + 4u_0^2}}\right) \\ \qquad = \sqrt{l_0^2 + 4u_0^2}L_3(t) + (4l_0 - 4L_{10})u(t) - l_0^2 \end{cases}$$

Note that from the coefficient of $u(t)$ in the equilibrium equation is zero when $m_L = -12m_u^3 - 5m_u + 1$. Using the last equations, the system of equations above can be reduced to,

$$\begin{Bmatrix} \dot{L}_1(t) \\ \dot{L}_3(t) \end{Bmatrix} = \begin{bmatrix} A & B \\ C & D \end{bmatrix}\begin{Bmatrix} L_1(t-\tau) \\ L_3(t-\tau) \end{Bmatrix} + \gamma\mathbf{b}$$

with

$$\mathbf{b} = \begin{Bmatrix} l_0 - \dfrac{2\left(l_0^2 - 12u_0^2 - 4u_0L_{10} + \dfrac{4u_0^2L_{30}}{\sqrt{l_0^2 + 4u_0^2}}\right)}{-\dfrac{4u_0L_{30}}{\sqrt{l_0^2 + 4u_0^2}} - 4l_0 + 4L_{10} + 24u_0} \\ \dfrac{l_0^2}{\sqrt{l_0^2 + 4u_0^2}} + \dfrac{4u_0\left(l_0^2 - 12u_0^2 - 4u_0L_{10} + \dfrac{4u_0^2L_{30}}{\sqrt{l_0^2 + 4u_0^2}}\right)}{\sqrt{l_0^2 + 4u_0^2}\left(-\dfrac{4u_0L_{30}}{\sqrt{l_0^2 + 4u_0^2}} - 4l_0 + 4L_{10} + 24u_0\right)} \end{Bmatrix}$$

$$A = -\gamma + \frac{8\gamma u_0}{-\frac{4u_0 L_{30}}{\sqrt{l_0^2 + 4u_0^2}} - 4l_0 + 4L_{10} + 24u_0} \qquad B = \frac{-2\gamma\sqrt{l_0^2 + 4u_0^2}}{-\frac{4u_0 L_{30}}{\sqrt{l_0^2 + 4u_0^2}} - 4l_0 + 4L_{10} + 24u_0}$$

$$C = \frac{-\left(\frac{16\gamma u_0^2}{\sqrt{l_0^2 + 4u_0^2}}\right)}{-\frac{4u_0 L_{30}}{\sqrt{l_0^2 + 4u_0^2}} - 4l_0 + 4L_{10} + 24u_0} \qquad D = -\gamma + \frac{4\gamma u_0}{-\frac{4u_0 L_{30}}{\sqrt{l_0^2 + 4u_0^2}} - 4l_0 + 4L_{10} + 24u_0}$$

The eigenvalues of the system matrix are,

$$\lambda_1 = -\gamma < 0, \qquad \lambda_2 = \gamma\left(\frac{-2m_u - m_L + 1}{12m_u^3 + 5m_u + m_L - 1}\right)$$

These expressions allows us to define four different stability and oscillatory regions according to the sign and values of $\lambda_2$, as shown in Fig. 5.8:

- In Regions I, II:

$$\tau_{oscil} = \frac{1}{e\gamma}, \qquad \tau_{stabil} = \frac{\pi}{2\gamma}$$

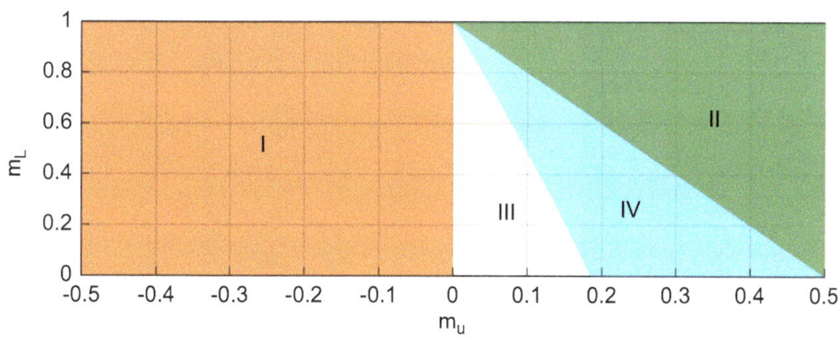


**Figure 5.8** Stability regions according to sign and values of $\lambda_1$ and $\lambda_2$ for out-of-plane deformation of cross-section. $\lambda_1$ defines regions *I* and *II*. $\lambda_2$ defines region *III*. In region *IV*, $\lambda_2 > 0$ and the system is unstable.

- In Region III:

$$\tau_{oscil} = \frac{12m_u^3 + 5m_u + m_L - 1}{e\gamma(2m_u + m_L - 1)}, \qquad \tau_{stabil} = \frac{\pi(12m_u^3 + 5m_u + m_L - 1)}{2\gamma(2m_u + m_L - 1)}$$

- In Region IV, the system is always unstable.

#### 5.3.2.2 Stability of non-linear form

Similarly to Section 5.2.4, we note that the non-linear system in (5.40) admits the solution $\dot{u} = 0$ when $\gamma$ is uniform. This is however not true for the linear system in (5.38). To see this, the equilibrium equation of the linear system is:

$$\sum_i k(l_i - L_i(t))c_i + b_0 = 0, \qquad c_i = \frac{\partial l_i}{\partial u} \text{ and } \dot{l}_i(t) = 0,$$

where $b_0(m_u, m_L) \neq 0$. Differentiation of the first equation leads to contradictory results:

$$\sum_i k\dot{L}_i(t)c_i = 0 \Rightarrow \sum_i k\gamma(l_i - L_i(t))c_i = 0 \quad \forall t \Rightarrow b_0 = 0$$

showing that the linear system does not admit $u = u_0$. When $u = u_0$, the DDE remain unchanged and uncoupled, and thus the stability bounds are those of a single element, that is, $\tau_{stab} < \pi/(2\gamma)$.

## 5.4 Conclusions

We have analyzed the stability and the presence of oscillatory solutions of a simplified planar deformation mode in a regular homogeneous epithelium, represented by a three-bar system, and also a simplified cross-section of the tissue. Resorting to a rheological law with variable rest-length [19,20] and a delayed response [16,17], we have deduced stability and oscillatory bounds, which depend on the remodeling rate $\gamma$ and the delay $\tau$. For increasing values of the remodeling rate, the tissues exhibits a lower viscous response, resulting in a more oscillatory or unstable dynamics of the rest-length.

We have demonstrated that linear stability analysis may restrict the bounds, introducing a dependence on the initial conditions for some initial stressed states. This dependence is in fact absent when considering the non-linear version of the equilibrium equations, as our numerical solutions also illustrate.

Although we have not given general stability conditions for the non-linear system, we have been able to give some results for some general initial conditions. The bounds are in fact identical to those deduced in linear unidimensional analysis [17].

The stability and origin of the oscillatory deformation of cells is now under intensive investigation [3,9]. Although experimentalists have detected and measured delays between the concentration of motor proteins (myosin or cadherin) [2], the relation between these oscillations and eventual drastic tissue deformations remains elusive. Although the work presented here aims at suggesting potential causes behind the observed instabilities, further measurements are needed to corroborate the deduced instability and oscillatory regimes.

## Acknowledgments

J.J.M. and M.A.D. are financially supported by the Spanish Ministry of Science, Innovation and Universities (MICINN) through grant DPI2016−74929-R, while C.B. has been supported by the Maria de Maetzu program of MICINN at Barcelona Graduate School of Mathematics (BSGMath) through grant MdM-2014-0445. J.J.M has been also financed by government of Catalonia through grant 2017 SGR 1278.

## References

[1] A. Munjal, J. Philippe, E. Munro, T. Lecuit, A self-organized biomechanical network drives shape changes during tissue morphogenesis, Nature 524 (7565) (2015) 351−355.

[2] K. Dierkes, A. Sumi, J. Solon, G. Salbreux, Spontaneous oscillations of elastic contractile materials with turnover, Phys. Rev. Lett. 113 (2014) 148102.

[3] J. Solon, A. Kaya-Copur, D. Brunner, Pulsed forces timed by a ratchet-like mechanism drive directed tissue movement during dorsal closure, Cell 58 (137) (2009) 1331−1342.

[4] A. Martin, M. Kaschube, E. Wieschaus, Pulsed contractions of an actin-myosin network drive apical constriction, Nature 457 (2009) 495−499.

[5] K. Kaouri, P.K. Maini, P.A. Skourides, N. Christodoulou, S.J. Chapman, A simple mechanochemical model for calcium signalling in embryonic epithelial cells, J. Math. Biol. 78 (2019) 2059−2092.

[6] H. Oosterwyck, Computational mechanobiology: may the force be with you, J. Math. Biol. 70 (6) (2015) 1323–1326.
[7] J. Notbohm, S. Banerjee, K. Utuje, B. Gweon, Y. Park, HJ, et al., Cellular contraction and polarization drive collective cellular motion, Bioph. J. 110 (12) (2012) 2729–2738.
[8] G. Peyret, R. Mueller, J. d'Alessandro, S. Begnaud, P. Marcq, R. Mège, et al., Sustained oscillations of epithelial cell sheets, Bioph. J. 117 (3) (2019) 464–478.
[9] V. Petrolli, M. Goff, M. Tadrous, K. Martens, C. Allier, O. Mandula, et al., Confinement-induced transition between wave-like collective cell migration modes, Phys. Rev. Letters 122 (16) (2019) 168101.
[10] A. Sumi, P. Hayes, A. D'Angelo, J.C.G. Salbreux, K. Dierkes, J. Solon, Adherens junction length during tissue contraction is controlled by the mechanosensitive activity of actomyosin and junctional recycling, Dev. Cell 47 (4) (2018) 453–463.
[11] L. Roldán, J.J. Muñoz, P. Sáez, Computational modeling of epithelial wound healing: short and long term chemo-mechanical mechanisms, Comp. Meth. Appl. Mech. Engng. 350 (2019) 25–56.
[12] A. Lapytsko, J. Schaber, The role of time delay in adaptive cellular negative feedback systems, J. Theor. Biol. 308 (2016) 64–73.
[13] H. Chi, J. Bell, B. Hassard, Numerical solution of a nonlinear advance-delay-differential equation from nerve conduction theory, J. Math. Biol. 24 (1986) 583–601.
[14] K. Cooke, J. Turi, Stability, instability in delay equations modeling human respiration, J. Math. Biol. 32 (6) (1994) 535–543.
[15] P. Getto, M. Gyllenberg, Y. Nakata, F. Scarabel, Stability analysis of a state-dependent delay differential equation for cell maturation: analytical and numerical methods, J. Math. Biol. 79 (2019) 281–328.
[16] J.J. Muñoz, S. Albo, Physiology-based model of cell viscoelasticity, Phys. Rev. E 88 (1) (2013) 012708.
[17] J.J. Muñoz, M. Dingle, M. Wenzel, Mechanical oscillations in biological tissues as a result of delayed rest-length changes, Phys. Rev. E 98 (1) (2018) 052409.
[18] K. Cavanaugh, M. Staddon, E. Munro, S. Banerjee, M. Gardel, RhoA mediates epithelial cell shape changes via mechanosensitive endocytosis, Dev. Cell 52 (2) (2020). 152-166.e5.
[19] R. Clément, C. Collinet, B. Dehapiot, T. Lecuit, P. Lenne, Viscoelastic dissipation stabilizes cell shape changes during tissue morphogenesis, Curr. Biol. 27 (20) (2017) 3132–3142.
[20] K. Doubrovinski, M. Swan, O. Polyakov, E. Wieschaus, Measurement of cortical elasticity in drosophila melanogaster embryos using ferrofluids, Proc. Natl. Acad. Sci. USA 114 (5) (2017) 1051–1056.
[21] M. Staddon, K. Cavanaugh, E. Munro, M. Gardel, S. Banerjee, Mechanosensitive junction remodelling promotes robust epithelial morphogenesis, Bioph. J. 117 (9) (2019) 1739–1750.
[22] N. Khalilgharibi, J. Fouchard, N. Asadipour, R. Barrientos, M. Duda, A. Bonfanti, et al., Stress relaxation in epithelial monolayers is ocntrolled by actomyosin cortex, Nat. Phys. 15 (2019) 839–847.
[23] T. Erneux, Applied delay differential equations, Surveys and Tutorials in the Applied Mathematical Sciences, vol. 3, Springer, New York, 2009.
[24] R. Corless, G. Gonnet, D. Hare, D. Jeffrey, D. Knuth, On the Lambert W function, Adv. Comp. Math. 5 (1996) 329–359.

[25] H. Shinozaki, T. Mori, Robust stability analysis of linear time-delay systems by Lambert W function: Some extreme point results, Automat 42 (1) (2006) 1791–1799.
[26] F.M. Asl, A.G. Ulsoy, Analysis of a system of linear delay differential equations, J. Dyn. Sys. Meas. Contr 125 (2003) 215–223.
[27] S. Yi, P. Nelson, A. Ulsoy, Delay differential equations via the matrix Lambert W function and bifurcation analysis: application to machine tool chatter, Math. Biosc. Engin. 4 (2) (2007) 255–369.

CHAPTER SIX

# Viscoelastic properties driving collective migration in zebrafish development

**Timo Betz**[1,2]
[1]Third Institute of Physics, University of Göttingen, Göttingen, Germany
[2]Institute of Cell Biology, ZMBE, University of Münster, Münster, Germany

## 6.1 Introduction

The development of a living organism from a single, fertilized cell is at the same time complex but also highly robust [1,2]. From our daily life point of view, complexity and robustness are often perceived as contradictory properties. However, during development, it remains unclear how they can operate hand in hand to result in a perfectly functional entity. This question is one of the fundamental mysteries of developmental biology. Although there is an extensive body of knowledge regarding biochemical signaling, gene expression, and their relation to changes in organism development, many experimental results from the past decades demonstrate that also physical and mechanical changes need to be integrated into these studies to further advance our understanding [3]. This becomes obvious when looking at the different stages of basically any organism development we know of today. Typically, after fertilization, an initial cell duplication phase is followed by a first morphogenic tissue motion that sets already the stage for the further developmental steps [4]. Observing time-lapse movies of developing organisms, it seems that this coordinated and often global collective migration of cells sets in spontaneously, without any obvious external signs. An important example of such collective and coordinated motion is presented by gastrulation where the future digestive tube is formed [5]. From a topological point of view, this is a key moment, since the embryo changes from a simply connected and topologically sphere-like surface to a torus-like, nonsimply connected object. Such topological changes are indeed very fundamental and complex

*Viscoelasticity and Collective Cell Migration.*
DOI: https://doi.org/10.1016/B978-0-12-820310-1.00009-4

processes, which need to be highly regulated to ensure robust outcomes. In the case of gastrulation, this importance is reflected in a famous quote of Lewis Wolper, a pioneer of developments biology, stating, "It is not birth, marriage, or death, but gastrulation which is truly the most important time in your life." [6]

To understand the motion of thousands of cells, each representing a complex entity, inherently requires a careful analysis of forces as well as mechanical properties on both tissue and cell level. Strikingly, the classical notion that mechanics can be seen as the final target of biochemical signaling has been shown to be too simplistic. In this passive view, the biochemistry is at the steering wheel, while mechanics provides simply the final force generation. This view is very intuitive, as it resembles a lot the way we interact with vehicles such as bikes and cars. However, reality for biological tissue has been proven to be much more complicated. It turns out that the signaling itself can depend on forces and mechanical properties of the environment and hence generating a loop between biochemical and mechanical signals. This process is described in the context of mechanotransduction [7]. Hence, to understand how different tissue and cellular motions are triggered and controlled, this feedback requires to not only characterize signaling and gene expression but also to discuss this in the context of cell and tissue mechanics. The fact that our understanding of biomechanics in itself is still limited and many new aspects are discovered continuously pictures how much we have to learn on either side to reach the level where we can quantitatively model and describe the development of complex organism. State-of-the-art descriptions of biological tissue typically use the notion of viscoelastic materials that provide both elastic and fluid properties. The description of tissue and cell mechanics hence requires the integration of different timescales but also add nonlinear regimes and out-of-equilibrium properties to the classical material science point of view [8,9].

Although inherently viscoelastic, the common description of developmental tissue flow typically uses a hydrodynamic approximation. This is due to the rather long timescale of 10 minutes to hours relevant for the morphological changes occurring during development. These long timescales are known to be in the range where shear forces typically relax, thus justifying the hydrodynamic approximation of a fluid. Although seen as a fluid, tissue does apparently differ in many aspects from a simple fluid, for example, in the sense that cells can change size, duplicate, or die. But even more importantly, cells possess the ability to apply active forces on their environment [8]. It is generally thought that especially this activity represents an important factor for the correct morphogenesis. Here, the coordinated generation of mechanical tension leads to flowing cells and collective migration [10–12]. On the other hand, many local changes in viscoelastic material properties diversify such flows, hence also contributing to the required pattern. In this sense, viscoelastic properties might not be the actual origin of the motion but can be seen as a mechanical guidance queue directing flow toward locations with different elastic and viscous properties. Finally, a well-described process called durotaxis turns out to robustly guide cell migration from soft toward stiff substrates, potentially also implicating that the viscoelastic properties of the embryo can tremendously affect the 3D motion of cells and tissue in development [13].

In this chapter, we will focus on the mechanical implications of the early zebrafish development and discuss the different mechanisms and models based on the process of epiboly, doming, gastrulation, and somite formation. The zebrafish is a very attractive model system to study both development and cell and tissue migration because the embryos develop from eggs outside of the maternal body, and hence, no animal has to be scarified. Furthermore, a large number of embryos are accessible, and the organism is transparent for a long time making high-resolution microscopy and modern fluorescence techniques straightforward to be used. Last but not least, both genetically modified embryos are available, and DNA as well as RNA can be injected into the embryo to drive the expression of proteins and fluorescent labels.

## 6.2 Morphogenesis and zebrafish development

Early zebrafish development follows a very well-studied and timed process, illustrated partially in Fig. 6.1. After fertilization, during the zygote stadium, the cytoplasm streams toward the animal pole of the blastodisc, which separates about 45 minutes postfertilization from the yolk cell in the first cell division. This is followed by a series of global cell divisions until the 1000 cell stadium is reached at about 3 hpf. During this time, the embryo is called blastula. In this state, three main cell layers are described [1]. First, an epithelial-like layer that covers the blastula is called enveloping layer (EVL), the interface cells between the blastula and the yolk form the yolk syncytial layer (YSL), and both layers surround the main part of the

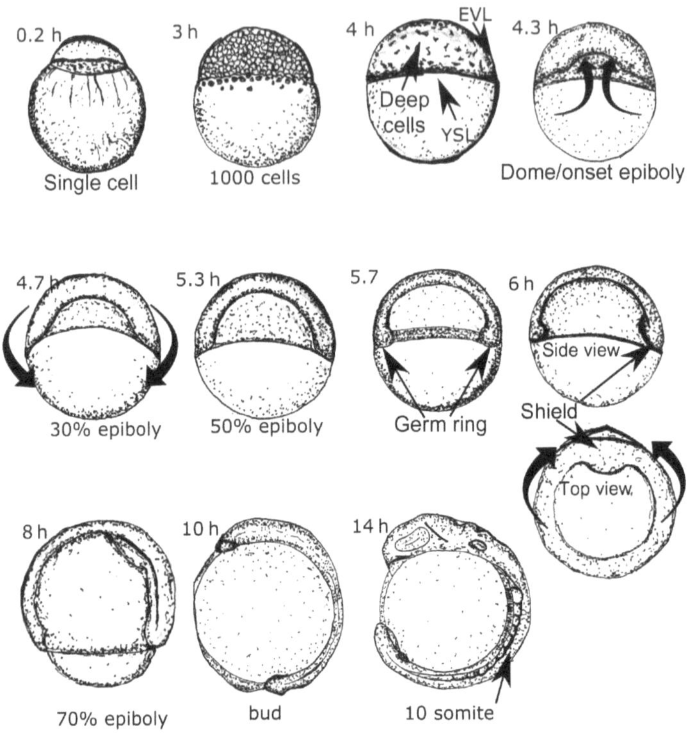


**Figure 6.1** Overview of different stages of zebrafish development where the influence of viscoelastic properties and force generation has been shown to play a key role. Especially the phase of epiboly [4.3–9 hpf (hours post fertilization)], the process of doming, the shield formation and the somite formation have attracted considerable attention. During the early development a series of morphogenic motion is required for the correct development of the organism. The key collective migration processes are epiboly, doming, and gastrulation, which correspond to the germ ring and shield formation. *Images redrawn from C.B. Kimmel, W.W. Ballard, S.R. Kimmel, B. Ullmann, T.F. Schilling, Stages of embryonic development of the zebrafish, Dev. Dyn. 203 (1995) 253–310.*

blastocyte called deep cells. From the deep cells, later the three germ layers (ectoderm, mesoderm, and endoderm) emerge. Looking at the embryo, the side where the blastocyte sits is called animal pole, while the opposite side with the yolk cell is called vegetal pole. Once at the 1000 cell stadium, the first morphological change is observed. The embryo shape changes from a situation where the blastocyte first sits in a cap-like fashion on the yolk to a fully spherical shape (4 hpf), called sphere state. This change already suggests an increase in surface tension, which would readily explain the global shape transformation. After the sphere state the first morphogenetic motion, called epiboly, begins [14]. During this phase, the cells start moving from the animal pole over the yolk cells toward the opposite side, called vegetal pole. Epiboly is the first collective motion of cells during the zebrafish development. This motion is accompanied by an upward motion of the yolk cell under the blastocyte, forming an upward-directed bulge of the yolk cell, called a dome. At the same time the doming of the yolk cells (4.3 hpf) also pushes the deep cells radially away, effectively leading to a homogenization of the blastocyte thickness. During this doming motion the deep cells start migrating toward the vegetal pole. At about 50% coverage of the yolk cells (50% epiboly, 5.25 hpf), the visible germ band forms and the differentiation into the three germ layers starts. Here, the germ band is a thickening zone located at the region where EVL and YSL meet and hence form the advancing front of the blastocyte. At this time a symmetry breaking occurs along the germ band, which is positioned at the equatorial plane of the zebrafish. During this fundamental symmetry breaking, the cells at the germ band start to ingress, which is manifested by changing their direction of motion towards the animal pole of the developing fish. This ingression is the onset for the further differentiation of cells into the three germ layers and corresponds to a gastrulation motion. Morphologically, this upward motion leads to a thickening of the embryo that eventually forms the embryonic shield (6 hpf). During shield formation a global collective motion sets in where a large number of cells start flowing toward a line that later forms the animal. At 75% epiboly (8 hpf) the dorsal side of the embryo can be observed as a distinctly thicker cell mass, while the evacuation zones, from where the cells move toward the future animal body, become visible. At 90% epiboly (9 hpf) the animal axis is clearly visible, and the neuronal systems begin to form. The segmentation of the embryo starts with the formation of the tail bud (10 hpf), which grows while forming distinct somite and leads to the elongation of the tail. In the following the inner organs begin to appear and the larva hatches after 48–72 hpf. Pictures of this process can be found in several locations. A detailed presentation of these processes can be found in Ref. [1].

For the cell migration and mechanics perspective on which we will focus in this chapter, four of these developmental stages, namely, epiboly, doming, gastrulation, and somite formation, are of interest. All these stages have received considerable attention in the past years. To describe the collective motion in these phases, different concepts have been developed such as localized tension generation, wetting, differential friction, yolk plasma flow, and liquid–solid transitions like jamming and unjamming.

## 6.3 Epiboly

Epiboly is among the most striking global motions observed during early zebrafish development, since it shows a transition from rather randomly oriented cell motion to an obviously orchestrated collective migration over the yolk cell. A detailed investigation in the past showed a series of phenotypes that depend on a series of key proteins, suggesting a complex but well-controlled process [14]. The forces which drive epiboly where long thought to originate from a 'purse string' like effect. Here a continous constriction of a circumfencial actomyosin band that spans around the yolk at the equatorial plane effectivly shortens this band. This constriction provoces a gliding of the band towards the vegetal pole of the yolk, hence effectively pulling the blastocyte towards the vegetal pole of the embryo [15]. This view was supported by experiments showing that inhibition of actomyosin contraction leads to a failure of epiboly [16]. More recently, the forces within the actomyosin band have been investigated in more detail and it became apparent that although contractile, the main tension forces are not directed laterally around the embryo, but that there is a high tension along the animal—vegetal axis [17]. This required to abandon the idea of a purse string like mechanism in favor of a direct pulling process. A very convincing experiment in this work was to show that the coverage of the yolk worked robustly even when pushing the embryo in a cylindrical geometry, such as losing the curvature required for the purse string mechanism. In this work the question about force generation and force balance was discussed in detail [18]. It turned out that a viscous coupling of the contractile actomyosin network to the yolk cytoplasm generated the frictional force required to pull the blastocyte over the yolk cell. This model was coined flow-friction motor [4,18]. The importance of the coupling between the yolk cytoplasm and the actin flow is that the pulling forces are not balanced by an elastic interaction, but simply by the contracting network coupled via viscous interaction with the environment (Fig. 6.2A). Such friction motors seem initially somewhat counterintuitive, as they require to constantly generate and destroy the structure that is generating the force. In this example, the actin network is indeed continuously polymerized at the regions of low flow way from the highest force application to the EVL and depolymerized in the region of high contraction. Although both the actin network and the yolk cell are viscoelastic materials, a full viscoelastic model of the

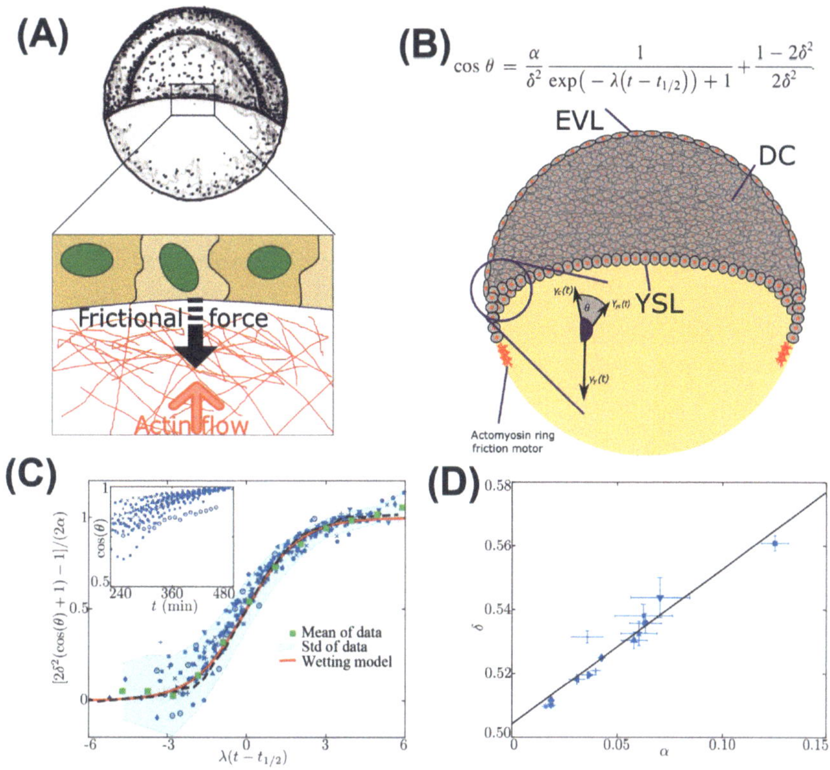


**Figure 6.2** Proposed processes to explain epiboly. (A) The friction motor model uses the finding that epiboly depends on the contraction of an actomyosin network at the interface between the EVL and the yolk. The contraction of the network pulls on the EVL on one side. The counterforce is generated by the friction between the flowing actin network and the yolk cell. Hence, effective pulling is happening due to the friction between the actin network and the underlying yolk cytoplasm. (B) A further model that includes the friction motor is the wetting model. Here, the surface tension at the EVL decreases over time, while the tension at the yolk–medium interface increases. This simple idea leads to a quantitative prediction for the contact angle that was experimentally confirmed. (C) Pooled data of several embryos plotted as time evolution of the contact angle (inset) can be collapsed on a master curve by rescaling the axis using the general wetting equation presented in (B). (D) The wetting model furthermore predicts a nontrivial relation between the different model parameters, which requires not only a linear relation but also a situation where the offset and the slope of the linear dependence are the same. This is precisely confirmed when analyzing the interdependence between $\delta$ and $\alpha$. *EVL*, Enveloping layer.

actomyosin network is not required in this situation because the relaxation time of the actin network (30 seconds) is much shorter than the timescale of the motion involved in epiboly which is 30 minutes. In this limit a simple hydrodynamic approach is enough to explain this early onset of collective migration. The typical relaxation timescales for the actin cytoskeleton in living cells is 30 seconds, which also corresponds to the typical turnover timescale of the dynamic actin network [19]. For the contractile actin network pulling the EVL during epiboly, the relaxation time is even faster and was measured to be in the order of 10 seconds [20]. It should be noted that the stress relaxation timescale of the zebrafish tissue is with 5 minutes much slower than the actin relaxation timescale [21]. However, the stress relaxation of the tissue remains still much faster than the typical time required for epiboly and doming as discussed here.

Another approach to describe epiboly focuses on the role of the yolk cell during this process [20]. Here, the viscoelastic flows of the yolk cytoplasm were measured in detail, and a hydrodynamic regression was used to determine the forces required to explain these flows. This method uses the laws of physics to infer the forces by modeling the tissue flow. Of course, this approach relies on the knowledge of the viscoelastic properties of the sample. The resulting forces acting on the blastocyte and the relevant cells can provide an alternative explanation of epiboly independent of the flow-friction motor. Indeed, several of the experimental results describe that at the onset of epiboly, the previously introduced anisotropic tension in the actomyosin network is absent. Using a viscoelastic approach to infer forces from the measured flow fields nicely shows that taking into consideration the differential flows in the tissue and the surrounding environment (here the yolk cell) gives rise to a complex situation that is far from being understood in terms of one simple process. Especially, the coupling between tissue flow and the viscoelastic properties of both the tissue and the environment is critical for the quantitative description of the underlying processes.

Besides the tendency to diversify the explanation of the processes driving epiboly, another approach was recently published that rather tries to exploit a simplified description of epiboly as a wetting process [22]. The main aim of this work is to develop a reduced model that can still explain epiboly and quantitatively covers the main parameters involved (Fig. 6.2B). In this approach the blastocyte is interpreted as a fluid that generates a classical surface tension with the surrounding interfaces, namely, the yolk cell, and the medium. In this view, epiboly starts in a self-organized manner if the three different surface tensions vary over time until a critical point is reached. Right at that point, the tension in the blastocyte–medium interface becomes smaller than the yolk–medium interface, which energetically favors the spreading of the embryo cells over the yolk. Hydrodynamically, this is equivalent to a wetting of the yolk by the blastocyte. Obviously, the idea to describe the complex onset of collective migration in a developing organism with a simple process such as the spreading of a water droplet over a hydrophilic surface represents a very strong simplification. However, in many situations, it was shown that such fundamental physical pictures allow not only to describe biological tissue surprisingly well [23,24], they also allow the interpretation of the different parameters in a new light, while explaining how these extremely important processes are performed in such a robust matter, allowing to successfully work in a large variety of situation.

To test if the wetting model may also quantitatively describe the process of epiboly, Wallmeyer et al. measured the contact angle between the yolk, the blastocyte cells, and the surrounding medium from 4 to 7 hpf in a light-sheet microscope from four orthogonal views, with a time resolution of up to 30 seconds. Using embryos that express a fluorescently labeled histone protein is allowed to directly visualize the cells' nuclei and follow the motion of all cells. This technique allows obtaining a 3D volume measurement with isotropic resolution after fusion of the images. Tracking of the nuclei then enabled the authors to fit spheres at the blastocyte–medium interface and at the blastocyte–yolk interface. The approximation of a sphere for these interfaces was shown to be excellent during the recording time. Once having the radius and the center of these spheres, it is possible to analytically determine the contact angle. This procedure yields the contact angle at the tissue scale. Another approach was to zoom at the cells at the interface between blastocyte, yolk, and medium and to determine the shape of this cell [25]. This also leads to a contact angle, but on the scale of a single cell, thus neglecting the overall tissue shape. Consequently, the contact angles obtained by these two approaches yield very different results.

As expected, the contact angle during the onset of epiboly is time dependent, with an almost sigmoid shape starting rather constant before the beginning of epiboly, then evolving rapidly during early epiboly and doming and reestablishing at a lower angle after 7 hpf (Fig. 6.2C).

From a theoretical point of view a minimal model was developed covering the whole observed change in contact angle. This model was simply based on the assumption that the surface tensions at the different interfaces are mediated by the concentration of the relevant proteins controlling the tension. Here, the relevant interfacial tensions are at the EVL-to-medium interface ($\gamma_c$), at the YSL ($\gamma_{yc}$), and at the yolk-to-medium interface ($\gamma_y$). To keep the approach simple the analytical expression describing the tension can be developed as function of protein concentration $c(t)$ to first order.

$$\gamma_i(t) = \gamma_{i,0} + \epsilon_i c(t) \tag{6.1}$$

where $i$ represents the interface ($c$, $y$, or $yc$) and $\epsilon_i$ represents an interface specific prefactor. In this approach the biological differences of the three interfaces are formalized by the constant offset tension (indicated by the index 0), and the prefactor $\epsilon_i$ at the particular interface. The time dependence of the protein concentration was then described using again the simple assumption of a protein production–degradation process.

$$\frac{dc}{dt} = -\lambda c + k \tag{6.2}$$

Here, on the one side, proteins are degraded with a certain decay rate $\lambda$, while, on the other, they are produced by translation with a rate $k$. Right after fertilization, basically all proteins are maternal and hence provided by the mother at a given concentration. During development the protein production of the embryo starts, hence leading to a new protein concentration, while the maternal proteins are slowly degraded. With these assumptions, it is possible to generate three equations with two parameters, one for each interface. The solution of the differential equation for the protein concentration also includes two parameters. The force balance for the contact angle finally couples all three equations yielding an expression for the time dependence of the contact angle as a function of a total of eight parameters. To reduce the number of parameters a parameter search was performed that demonstrated that only four independent parameters are required for solutions that fulfilled the requirement that the contact angle in the embryo is smaller than 180 degrees and that it decreases monotonously over time. With these simple assumptions the time-dependent change of the contact angle can be described by a simplified model:

$$\cos(\theta) = \frac{\alpha}{\delta^2} \frac{1}{\exp\left(-\lambda\left(t - t_{1/2}\right)\right) + 1} + \frac{1 - 2\delta^2}{2\delta^2} \tag{6.3}$$

This reduced model yields a sigmoid function that provides an excellent fit for the contact angle and does hence describe the collective motion of >1000 cells over 4 hours (Fig. 6.2C). The relevant parameters turn out to be a time offset $t_{1/2}$ that defines the moment of fertilization, the decay rate constant $\lambda$, and two mechanical parameters $\alpha$ and $\delta$. Here, $\delta = \epsilon_c/\epsilon_y$ is a simple ratio between the impact of the modeled protein concentration of the surface tension at the cell–medium and at the yolk–medium interface. $\alpha = \gamma_{y,0}/\epsilon_y c_s$ describes the ratio between the offset surface tension at the yolk to the final tension on the cell layer toward the end of epiboly (where $c_s$ denotes the final protein concentration at the final, hence stationary time).

An important point of this model is that it remains very general, not making any statement about the proteins in charge; besides that, it is controlling the surface tension. No specific assumptions about this control are made here. Hence, it could be a cell adhesion protein or a motor protein actively leading to the surface tension. All the individual properties of the protein functions are integrated into the two parameters $\alpha$ and $\delta$.

As commonly observed in biology, each embryo behaves in an individual way, although the overall process of epiboly is robust. On the level of measured contact angles, the individuality is reflected in a large spread of angles and time evolution when plotting many experiments in the same graph. As shown in the inset of Fig. 6.2C, such a large spread and individuality is indeed found when plotting 14 experiments together in one graph. This heterogeneity might suggest that epiboly is not highly controlled, which somewhat contradicts the notion that it is one of the most important early processes in zebrafish embryogenesis. Failure of epiboly will lead to the death of the embryo. The mystery of this rather "non-reproducible," but highly relevant process is solved when rescaling the time and contact angle axis according to the theoretical model. Doing so allows to collapse all datasets on a master curve, showing that the model can indeed describe all individual curves (Fig. 6.2C). While this suggests that the model has a certain validity, it can be argued that any sigmoid model will yield the same successful collapse of data. To further test the predictive power of the proposed wetting model, the interdependence of $\alpha$ and $\delta$ was tested. One prediction of the simplified wetting model is that for each embryo a correlation between the mechanical parameters exists that can be directly checked. Mathematically, $\delta$ should linearly depend on $\alpha$ with the condition that the slope of this dependence equals the offset. Strikingly, plotting $\delta$ as function of $\alpha$ does perfectly follow this prediction (Fig. 6.2D). According to the wetting model, the numerical values of the slope correspond to the ratio of pulling forces at the cell–medium interface and the yolk–medium interface. For epiboly to occur, this number must be smaller than 1, which simply means a stronger pull by the yolk compared to the cells. In the experiments a value of 0.5 was found, which suggests that although each embryo is different and independent from the other, the mechanical pulling forces relevant for epiboly are intrinsically well controlled. The wetting model can additionally be implemented in a simulation that shows how, thanks to the change in protein concentration, the blastocyte cells will cover the yolk cell in a process that visually resembles the in vivo observation.

From a biological point of view, these results are quite striking. It suggests that the ratio between the tension at the cell–medium interface and the yolk–medium interface is changing over time in a way that initially the tension $\gamma_c > \gamma_y$ suppresses the spreading of the cells over the yolk. Over time this ratio changes, and once $\gamma_c < \gamma_y$, the cells spread over the yolk cell just as in a simple wetting situation. The final closing at the vegetal pole happens when the ratio of the two tension approaches 1/2, hence the tension at the yolk is simply twice the tension at the cell–medium interface. It is important to note that the model does not predict this value of 0.5. It is a result of the experimentally measured dependency of $\delta$ on $\alpha$. The value of 0.5 is either the result of a feedback mechanism in the zebrafish embryo, or it is a yet unexplained inherent dependence of the two tensions on each other. Unfortunately, it was not yet possible to directly measure the time evolution of these two tensions. Literature values of tension at the yolk show an increase from about 60 to 110 pN/μm, consistent with the proposed model [20].

In this approach, it became evident that it is possible to describe a highly complex and important process by a rather simple physical picture that effectively depends on a minimization of energy. Of course, the zebrafish embryo is much more complex than a drop of water, but this does not mean that fundamental and hence highly robust physical laws could not be used to achieve the required final outcome. In the view of the wetting model, it turned out that a simple change in surface tension can already lead to the observed effects. Interestingly, the parameter search did show that an active force at the yolk is required at the end stage of epiboly to sustain the process. Indeed, the introduced friction motor that was previously demonstrated experimentally has exactly the properties required for the wetting model. In contrast, the friction motor alone is not able to explain the full process of epiboly, as it only starts later, and can hence not explain the initial onset of this collective migration.

## 6.4 Doming

At the onset of epiboly, a striking change in the embryo is represented by the upward bulging of the yolk cell, called doming. Although there exists an inherent link between the process of doming and epiboly, it has inspired many biomechanical models to explain the upward motion of the yolk cell against the epiblast. Since the doming happens at the same time as epiboly, it is suggested that forces and motion observed are tightly related. In fact, the mechanical models explaining doming rely not only on the already mentioned motion of the yolk cytoplasm but also on an active expansion of the EVL, the final cell layer defining the interface between the epiblast and the outside medium. In a series of impressive experiments, it was demonstrated that the active expansion of the EVL is a key process required for the motion of the yolk cell toward the animal pole of the embryo [25]. This effect is surprising, since between the EVL and the region of the upward bulge remains the bulk of the blastocyte, providing a massive mechancial resistance. Still, by transplanting EVL cells from an embryo deficient of the extension capacities on wild type embryos results in doming defects right under the area of the transplantation. The underlying idea of this model is that the expanding EVL reduces the surface tension at the blastocyte—medium interface, similar as in the wetting model. While this tension reduction supports the vegetal motion of the embryo cells, it is not sufficient to explain the upward motion of the yolk and hence doming. To understand this upward motion the intercalation of the deep cells in the blastocyte has been suggested to generate a tensile force that is effectively pulling the yolk upward. The idea of this mechanism is to explain why doming is excluded from regions where EVL extension is inhibited. The thinning of the epiblast during doming is an important process, and the current experimental evidence suggests that the changes in force generation and viscoelastic properties play an important role.

The importance of variation in viscoelastic properties was further demonstrated in another series of elegant experiments. Here the viscoelastic properties of the tissue during doming were directly probed using micropipette aspiration [26]. In these experiments a micropipette with an opening of several cell diameters was inserted in the zebrafish before doming and at the onset of doming at several different positions. Applying a suction pressure allows to aspirate cells of the epiblast into the pipette. The tissue viscosity can be calculated from the aspiration speed and the applied suction pressure. The interesting finding was that during doming the tissue viscosity dropped by a factor of 10 in the area where the yolk cell moves towards the animal pole. At the margin of the embryo, this drop was not observed. In a simple mechanical model that assumes a pushing force either applied by the yolk cell or by the cell layer at the interface between the yolk and the blastocyte, the extensive reduction of viscosity will lead to a much faster indentation in the region of reduced viscosity. Transplanting cells from the margin, which did not undergo the viscosity drop, into the region of doming, it can be seen that doming indeed is perturbed. Here, the cells come from the same organism, and only the viscoelastic properties are argued to be different. Surely, this does not exclude that possible differentiation processes can also play a role. To understand the observed tissue fluidification better the fluid between the cells in the different regions was analyzed. This investigation showed that during the time of doming the cells in the relevant location underwent a cell division, which leads to a detachment from the neighboring cells. Hence, the tissue in these regions becomes loose and less connected, directly explaining the drop of viscosity.

Overall, the process of doming and blastocyte thinning presents another aspect of epiboly, which couples collective cell migration tightly to changes in the viscoelastic properties and well-orchestrated force generation. In the line of development, early epiboly and doming is followed by another large-scale collective motion of the cells representing the embryo, called shield formation or gastrulation.

## 6.5 Gastrulation

In this phase the different germ layers further differentiate, and the first motion of cells against the animal–vegetal directed flow of epiboly arises. In detail, during this process, cells are differentiating in the mesoderm, while the prechordial plate (ppl) ingresses at the dorsal germ ring margin. This motion effectively means that these ingressing cells start to detach from the overall ectoderm layer, attach to the yolk syncycial layer, and begin an active motion toward the animal pole of the embryo. This motion against the overall tissue flow is obviously directly disproving the wetting model for gastrulation. Obviously, the image of a simple fluid that is driven by energy minimization breaks down at this step, and to keep such a rather passive model, many strong assumptions about cell and substrate adhesion would be required. Fluorescence video microscopy marking the actin cytoskeleton of the ingressing cell that starts moving toward the animal pole shows that active, lamelliodia-driven cell migration drives this motion against the general epiboly flow [27]. This is hence in strong contrast to the vegetal pole directed motion of the ectoderm cells. At about the time when the first upward motion of the ppl cells starts, an overall symmetry breaking occurs, which defines the later axis of the developing animal. During this time, the ectoderm cells in the region of the ppl cells are changing their flow pattern, hence leading to a rotational motion that increases the cell density along the forming fish. This motion and the timed occurrence with the cell ingression and the animal pole directed ppl cells were the reason to suggest that the ppl cells of the mesendoderm effectively drag the ectoderm cells along in a friction like process. For such processes to be efficient a viscous coupling needs to be realized between the two cell populations that flow in opposing directions. If in this situation the force transmission of the ppl cells to the yolk cell dominates, the friction force can effectively pull the ectoderm cells along and this symmetry breaking can then have a long-range effect that drives the accumulation of cells along the animal axis. In this view, the ingression of cells might be the key process for gastrulation.

Such a friction-based model requires mechanical interaction between the ppl and ectoderm cells which is mediated by the transmembrane protein E-cadherin. Indeed, E-cadherin is known to accumulating at the interaction zone between the two tissues. It is quite counterintuitive to imagine a sudden change of tissue flow where the cells of the two oppositely migration cell populations show an increase of friction. At first sight, such an increased viscous interaction does rather suggest that the flows will perturb each other and hence slow down. However, looking at it from the organism perspective makes the overall idea that now the cells with the largest tissue integrity and force generation on the relevant substrate will dominate the motion and hence start dragging the softer tissue with it. Once two counterpropagating cell populations are established, a simple hydrodynamic treatment of the tissue is able to explain the large-scale rotating cell motions. The processes described here are eventually forming the neural anlage that will be the basis for the neuronal spline of the fish. But the large-scale motion furthermore drives a large number of cells toward the axial region, thus enhancing the cell density in the location of the future fish, while thinning the cells in the lateral sides. All these processes start at 50% epiboly and go on during the further development of the fish.

Overall it should be mentioned here that the frictional coupling is a striking example where the viscoelastic properties of tissue combined with a controlled and localized force generation have a large-scale effect on the collective migration in the tissue. The beauty of such motion in development is that at the final stage of this process a fully developed organism is generated. Both the coupling of signaling and biochemical protein interactions with changes in viscoelastic properties and force generation lead to highly robust large-scale motion, even when the signaling might be quite local. In this view, the mechanical interaction is automatically providing the large-scale coupling, thanks to fundamental conservation laws, such as mass and momentum conservation, which is at the heart of any hydrodynamic, but also elastic theory.

## 6.6 Somite formation

Up to now, we have focused on collective motion and viscoelastic changes in the early embryo development, with the outcome that not only changes in forces but also changing viscoelastic properties as well as adhesion and surface tension are important drivers of morphogenesis and collective motion. The final section will introduce evidence that a jamming-like transition in the tail formation of the zebrafish provides the mechanical support to direct the tissue growth in the correct region [28]. During tail outgrowth and subsequent somite formations, new cells are provided by a highly proliferative stem cell region, the so-called "tail bud" or mesodermal progenitor (MSP) zone. The growth of this region provides cells to the presomitic mesoderm (PSM). Once the cells get detached from the PSM, their differentiation depends on the state of a biochemical clock that is translating temporal in spatial information [29]. While the biochemistry of this clock is well studied, the simple question about the mechanical changes during this differentiation process has only been recently uncovered. In an elegant study using magnetic oil droplets injected into the zebrafish, it became possible to gain deep insights into both, stress generation, mechanical rigidity, and mobility in the

MSP and the PSM during somite formation [28,30]. The relevant mechanical parameters are tissue prestress, also called supercellular stress, the yield stress of the tissue, the tissue packing, and the cell–cell interaction. All these stresses are in the range of 20–100 Pa and are highly dynamic. In these measurements the anisotropic part of the tissue stress was determined by quantifying the deformation of the injected oil droplet, where the pushing stress could be directly calculated using the known surface tension of the oil. Although very elegant, this method only allows one to measure forces that are acting normal to the oil droplets, as these are liquid, and hence simply relax when shear forces are applied. Any shear component that might still be of high relevance is not captured. However, already these oil droplet–based stress measurement showed that moving from the growing end (the MSP) through the PSM toward the newly formed somite gives rise to an increase in tissue stress, as well as in a stress reorientation from a direction normal to the outgrowth region toward a parallel orientation. The change in orientation and the increase in stress level already suggest that the outgrowth of the MSP is mechanically supported by PSM. A further hint pointing in this direction was given by the measurement of the tissue yield stress, which measures at which force pushing on the tissue, it will not only deform elastically but will also start to permanently flow. Higher yield stress hence corresponds to lower fluidity. The introduced finding that the tissue in the PSM region already sustains larger supracellular tissue stress does suggest that also the yield stress is larger in these regions. This was indeed found and led to the hypothesis that the tissue at the growing MSP region is more fluid than in the differentiating PSM region. Interestingly, using mutants of the cell–cell adhesion protein N-cadherin did not show this spatial difference, and neither the tissue stress nor the yield stress was systematically increased toward the somite. Furthermore, the volume fraction of the fluid in the space between cells was highly variable, leading to an analogy with complex passive material, such as foams or colloids. The volume fraction of the fluid is a key parameter, as it acts like some kind of lubrification factor. If the fluid between the elements generating such complex material exceeds a critical value, the overall material properties change to a fluid behavior. Once in a fluid phase, objects can become quite mobile and exchange places without the need to overcome a large energy barrier. However, if the volume fraction drops below the critical value, the whole material experiences a drastic stiffening, losing the fluid character while a rigid behavior emerges. This process is generally known as a jamming transition. Such classical jamming transition goes hand in hand with a sharp decrease in mobility that can be quantified by the mean-square-displacement and as a drastic reduction of next neighbor exchange rates. A detailed analysis of these quantities could show that all the observed behavior can be explained by a jamming-like transition happening during the differentiation. The classical reason for such jamming or also of similar glass transitions is either found in the mentioned volume fraction of fluid between the cells or in a change of nonequilibrium motion. In glass transitions the solidification is due to a decrease in temperature below a critical value, when the molecules cannot exchange neighbor anymore as the energy barrier for these exchanges is larger than the energy thermally provided. In the case of tissue the agitation is typically not driven by temperature, as this parameter fixes in a narrow physiological regime but by active forces that move the cells. Typically, such forces are driven by actomyosin networks of the cell cytoskeleton. In analogy to the glass transition, these forces are often quantified to be an effective temperature that is entirely driven by biochemical and metabolic energy. In the case of jamming, it was successfully tested that by removing this additional effective temperature, using a drug that inhibits myosin motor activity, the characteristics of the jammed state were already found in the MSP region, which used to be rather fluid when not blocking the active motion with the drug.

These experiments point toward mechanical changes as highly relevant parameters not only required to generate motion but also influencing shapes and the formation of key structures during development. Especially, the notion of a jamming transition is of high interest, as these transitions occur very rapidly and may lead to a spatially close, but mechanically highly different tissue that is inherently similar. In the case of the tail bud growth and the somite formation the rigidification during the jamming might provide the mechanical basis for the continuous outgrowth. In simple words the stiff regime directs the growth along the body axis, hence providing the relevant spatial queues for correct morphogenesis of the fish.

## 6.7 Outlook

In this chapter, we have given an overview of recent experimental evidence that the controlled changes of viscoelastic properties of tissue are not only a relevant parameter to understand tissue flow but also to understand morphogenesis in the case of development. Although this sounds rather obvious from a physics perspective, it only recently became apparent to the field of developmental biology. These new insights were possible thanks to a series of new technical developments that allow now, for the first time, to not only visualize the motion of tissue and the full organism during development but also to determine mechanical parameters. Namely, the spatiotemporal localization of forces and stresses was instrumental not only for a deeper understanding but also for more detailed connections between mechanics in general and the molecular players controlling it. Besides forces, a second new insight was that spatiotemporal changes of viscoelastic material parameters provide a further access to understand collective motion in development. In simple words, to form an object, it is of course possible to push and pull at the right position at the right time; however, another possibility is to switch mechanics from liquid to solid and hence deform predominantly the fluid regions, while applying the same amount of forces. The regulation of such liquid-to-solid transition is only starting to be uncovered, but it is to be expected that active fluctuations in the tissue, similar to the effective temperature, as well as cell—cell adhesion are here the main driving forces.

To gain an even further and deeper understanding the field currently aims at a full description of tissue. For such an analysis, it is necessary to determine tissue flow at the single-cell level, for example, via light-sheet microscopy and single-cell tracking. Although the tools for this already exist, the sheer amount of information is still providing a considerable barrier for many labs to extract the information it contains. Besides the flow fields and density information of the tissue, more insights into the tissue stress are required. In the long run, we need here access to the full stress tensor fields, which might be achievable by recently developed elastic hydrogel spheres. Finally, we need access to the viscoelastic material properties at the tissue scale. This is quite a challenge, as most measurement approaches are invasive and damage or destroy the tissue during the measurement. Here, optical tweezer—based microrheology and the introduced magnetic fluids are prominent candidates to eventually provide the yet missing information. It is to be expected that not only during development but also later the viscoelastic properties of the tissue will become highly relevant for many biological functions.

Once all these parameters are obtained, it is possible to directly apply the modern theoretical models of active fluids to tissue. This will then directly hint if the observed motion can be simply explained by rather physical and hence global properties where the tissue as a whole is shaped, and the single cell is not a part of this global motion. In contrast, violations of the basic assumptions of such hydrodynamic approaches will directly suggest that cells do indeed move as individual objects, each having an intrinsic program steering it at the right location.

It is to be expected that nature uses both hydrodynamic, and hence robust but also simple physical mechanisms, and highly specific cell migration that might be guided by chemical and physical signals. In the end, it is important that the final organism is healthy and can reproduce, and we can very much rely on the fact that evolution has used all possible approaches to ensure this.

## References

[1] C.B. Kimmel, W.W. Ballard, S.R. Kimmel, B. Ullmann, T.F. Schilling, Stages of embryonic development of the zebrafish, Dev. Dyn. 203 (1995) 253–310. Available from: https://doi.org/10.1002/aja.1002030302.

[2] R.M. Warga, C.B. Kimmel, Cell movements during epiboly and gastrulation in zebrafish, Development 108 (1990) 569–580.

[3] N.I. Petridou, Z. Spiró, C.-P. Heisenberg, Multiscale force sensing in development, Nat. Cell Biol. 19 (2017) 581–588. Available from: https://doi.org/10.1038/ncb3524.

[4] C.-P. Heisenberg, Y. Bellaïche, Forces in tissue morphogenesis and patterning, Cell 153 (2013) 948–962. Available from: https://doi.org/10.1016/j.cell.2013.05.008.

[5] S. Weng, J.B. Wallingford, May the force be with you, eLife 7 (2018) e36662. Available from: https://doi.org/10.7554/eLife.36662.

[6] L. Wolpert, The Triumph of the Embryo, Dover Publications Inc, Mineola, NY, 2008.

[7] H. Wolfenson, B. Yang, M.P. Sheetz, Steps in mechanotransduction pathways that control cell morphology, Annu. Rev. Physiol. 81 (2019) 585–605. Available from: https://doi.org/10.1146/annurev-physiol-021317-121245.

[8] W.W. Ahmed, É. Fodor, T. Betz, Active cell mechanics: measurement and theory, Biochim. Biophys. Acta 1853 (2015) 3083–3094. Available from: https://doi.org/10.1016/j.bbamcr.2015.05.022.

[9] H. Turlier, T. Betz, Unveiling the active nature of living-membrane fluctuations and mechanics, Annu. Rev. Condens. Matter Phys. 10 (2019) 213–232. Available from: https://doi.org/10.1146/annurev-conmatphys-031218-013757.

[10] X. Trepat, M.R. Wasserman, T.E. Angelini, E. Millet, D.A. Weitz, J.P. Butler, et al., Physical forces during collective cell migration, Nat. Phys. 5 (2009) 426–430. Available from: https://doi.org/10.1038/nphys1269.

[11] G. Duclos, C. Blanch-Mercader, V. Yashunsky, G. Salbreux, J.-F. Joanny, J. Prost, et al., Spontaneous shear flow in confined cellular nematics, Nat. Phys. 14 (2018) 728–732. Available from: https://doi.org/10.1038/s41567-018-0099-7.

[12] S. Garcia, E. Hannezo, J. Elgeti, J.-F. Joanny, P. Silberzan, N.S. Gov, Physics of active jamming during collective cellular motion in a monolayer, Proc. Natl. Acad. Sci. U.S.A. 112 (2015) 15314–15319. Available from: https://doi.org/10.1073/pnas.1510973112.

[13] A. Haeger, K. Wolf, M.M. Zegers, P. Friedl, Collective cell migration: guidance principles and hierarchies, Trends Cell Biol. 25 (2015) 556–566. Available from: https://doi.org/10.1016/j.tcb.2015.06.003.

[14] S.E. Lepage, A.E.E. Bruce, Zebrafish epiboly: mechanics and mechanisms, Int. J. Dev. Biol. 54 (2010) 1213–1228. Available from: https://doi.org/10.1387/ijdb.093028sl.
[15] D.P. Kiehart, Wound healing: the power of the purse string, Curr. Biol. 9 (1999) R602–R605. Available from: https://doi.org/10.1016/S0960-9822(99)80384-4.
[16] J.C. Cheng, A.L. Miller, S.E. Webb, Organization and function of microfilaments during late epiboly in zebrafish embryos, Dev. Dyn. 231 (2004) 313–323. Available from: https://doi.org/10.1002/dvdy.20144.
[17] M. Köppen, B.G. Fernández, L. Carvalho, A. Jacinto, C.-P. Heisenberg, Coordinated cell-shape changes control epithelial movement in zebrafish and Drosophila, Development 133 (2006) 2671–2681. Available from: https://doi.org/10.1242/dev.02439.
[18] M. Behrndt, G. Salbreux, P. Campinho, R. Hauschild, F. Oswald, J. Roensch, et al., Forces driving epithelial spreading in zebrafish gastrulation, Science 338 (2012) 257–260. Available from: https://doi.org/10.1126/science.1224143.
[19] M. Fritzsche, A. Lewalle, T. Duke, K. Kruse, G. Charras, Analysis of turnover dynamics of the submembranous actin cortex, Mol. Biol. Cell 24 (2013) 757–767. Available from: https://doi.org/10.1091/mbc.E12-06-0485.
[20] A. Hernández-Vega, M. Marsal, P.-A. Pouille, S. Tosi, J. Colombelli, T. Luque, et al., Polarized cortical tension drives zebrafish epiboly movements, EMBO J. 36 (2017) 25–41. Available from: https://doi.org/10.15252/embj.201694264.
[21] E.-M. Schoetz, Dynamics and Mechanics of Zebrafish Embryonic Tissues, VDM Verlag Dr. Müller, 2013, ISBN:978-3-639-00826-5
[22] B. Wallmeyer, S. Trinschek, S. Yigit, U. Thiele, T. Betz, Collective cell migration in embryogenesis follows the laws of wetting, Biophys. J. 114 (2018) 213–222. Available from: https://doi.org/10.1016/j.bpj.2017.11.011.
[23] G. Beaune, T.V. Stirbat, N. Khalifat, O. Cochet-Escartin, S. Garcia, V.V. Gurchenkov, et al., How cells flow in the spreading of cellular aggregates, Proc. Natl. Acad. Sci. U.S.A. 111 (2014) 8055–8060. Available from: https://doi.org/10.1073/pnas.1323788111.
[24] S. Douezan, K. Guevorkian, R. Naouar, S. Dufour, D. Cuvelier, F. Brochard-Wyart, Spreading dynamics and wetting transition of cellular aggregates, Proc. Natl. Acad. Sci. U.S.A. 108 (2011) 7315–7320. Available from: https://doi.org/10.1073/pnas.1018057108.
[25] H. Morita, S. Grigolon, M. Bock, S.F.G. Krens, G. Salbreux, C.-P. Heisenberg, The physical basis of coordinated tissue spreading in zebrafish gastrulation, Dev. Cell 40 (2017) 354–366.e4. Available from: https://doi.org/10.1016/j.devcel.2017.01.010.
[26] N.I. Petridou, S. Grigolon, G. Salbreux, E. Hannezo, C.-P. Heisenberg, Fluidization-mediated tissue spreading by mitotic cell rounding and non-canonical Wnt signalling, Nat. Cell Biol. 21 (2019) 169–178. Available from: https://doi.org/10.1038/s41556-018-0247-4.
[27] M. Smutny, Z. Ákos, S. Grigolon, S. Shamipour, V. Ruprecht, D. Čapek, et al., Friction forces position the neural anlage, Nat. Cell Biol. 19 (2017) 306–317. Available from: https://doi.org/10.1038/ncb3492.
[28] A. Mongera, P. Rowghanian, H.J. Gustafson, E. Shelton, D.A. Kealhofer, E.K. Carn, et al., A fluid-to-solid jamming transition underlies vertebrate body axis elongation, Nature 561 (2018) 401–405. Available from: https://doi.org/10.1038/s41586-018-0479-2.
[29] A.C. Oates, L.G. Morelli, S. Ares, Patterning embryos with oscillations: structure, function and dynamics of the vertebrate segmentation clock, Development 139 (2012) 625–639. Available from: https://doi.org/10.1242/dev.063735.
[30] F. Serwane, A. Mongera, P. Rowghanian, D.A. Kealhofer, A.A. Lucio, Z.M. Hockenbery, et al., In vivo quantification of spatially varying mechanical properties in developing tissues, Nat. Methods 14 (2017) 181–186. Available from: https://doi.org/10.1038/nmeth.4101.

CHAPTER SEVEN

# Oscillations in collective cell migration

**Vanni Petrolli, Thomas Boudou, Martial Balland and Giovanni Cappello**
Laboratoire Interdisciplinaire de Physique, Centre national de la recherche scientifique, Université Grenoble Alpes, Grenoble, France

## 7.1 Introduction

Every multicellular organism begins its life as a single cell, a fertilized egg, and later spontaneously develops into an extremely complex entity, whose existence depends on the precise hierarchy of shapes, structures, and functions of its tissues. All these morphological features, which arise almost identically over thousands of generations, can be traced back to moments where cells spontaneously (i.e., without external guidance) alter their behavior, being its growth, differentiation, or movement. The reproducibility of this phenomenon is, in fact, resting upon its complexity: thousands of processes happen under strict spatiotemporal control, the nature of which is still subject to debate.

The first explanation was provided by Alan Turing in his seminal paper in 1952 [1], where he explained how, based on the interaction between two molecules with different diffusion constants, stable periodic patterns could spontaneously arise from a homogeneous system. If any of these two molecules were to have a biological function, that is, being what Turing refers to as a *morphogen*, this theory could provide an explanation for the patterning of cell behavior [2–4]. Since then, the tools available to measure and observe cells have grown and with them our awareness that cell behavior is inherently correlated to the environment and its mechanical properties. We now know that cells continuously probe the response of the surrounding matrix and that patterns of its mechanical properties can directly affect proliferation [5,6], extrusion [7,8], differentiation [9–11], and migration [12,13]. If, initially, our

*Viscoelasticity and Collective Cell Migration.*
DOI: https://doi.org/10.1016/B978-0-12-820310-1.00004-5

explanations were strictly biochemical, new theories were formulated, where actors of a different nature, mostly mechanical, could also play a role [14–17]. More recently, practical examples of such influence on organ morphogenesis have been found [18–23].

In this regard, the presence of wave-like mechanical patterns in the motion of cells, reported in the past decade, could be significant. These oscillations, which concern cell velocity [24], forces cells applied to the substrate [25], and the stresses within the monolayer [26], have been observed in vitro within cell sheets. When viewed from the perspective of the self-organization of tissues in vivo, the periodic nature of these mechanical phenomena could provide a "reference system," allowing the specification of different subpopulations of cells and, ultimately, leading to supracellular patterning. Moreover, the propagating nature of these mechanical signals could be relevant for cellular communication at larger scales.

In this review, we briefly describe collective cell motion and the emergent properties arising from the coordination of cell behavior. We will then concentrate on the reports of mechanical oscillations in monolayers. The literature on the matter encompasses different cell types and experimental setups. If, on the one hand, each work seems to highlight a specific aspect of these phenomena, on the other, the similarities, both in terms of typical scales and cellular behavior, seem to be painting a common picture. This review is meant to provide a broad overview of these similarities, with a final outlook toward the future challenges to be tackled to assess the biological significance of such wave-like phenomena.

## 7.2 Mechanism of collective cell motion

Before describing the various reports of mechanical oscillations, it is useful to outline the basic concepts behind cell migration and their interaction with their environment.

### 7.2.1 Mechanical machinery

The mechanical machinery of single cells (described in the first chapter of this book and in Refs. [27,28]) allows them to physically interact with their surroundings and exert forces. The cytoskeleton is responsible for this latter function and the forces are relayed to the environment through two main channels. Focal adhesions are protein complexes that span the membrane to bridge the actin network with the external world, in the form of the extracellular matrix [29,30]. Some of the proteins present at focal adhesions are mechanosensitive: responding to the characteristics of the environment, they can recruit different signaling molecules, reinforcing or weakening the cytoskeleton and ultimately affecting gene expression. Similarly, cell–cell junctions, which connect the membranes and cytoskeletons of two adjacent cells, can play a role in regulating many aspects of cell activity [31,32]. This extended mechanical network grants cell layers structural integrity and, together with the bidirectional communication between mechanical and biochemical channels, allows cells to coordinate their behavior at a scale larger than that of the individual. Ultimately, this supracellular communication gives rise to emergent behaviors, which are at the core of many processes typical of morphogenesis [21] and tissue homeostasis [33].

### 7.2.2 Collective cell migration

One such process is collective cell migration (CCM): the coordinated movement of multicellular ensembles, fundamental for wound healing, tissue shape, and cancer progression [34,35]. The most evident advantage of supracellular coordination is the higher persistence of migration. If, on the one hand, single cells have a higher instant velocity, their migration tends to be somehow more erratic. Cell–cell contacts increase the overall persistence, creating a more directional and "efficient" system, which can also generate and sense biochemical and mechanical gradients on a scale bigger than that of single entities [36]. Although CCM refers to many forms of collective behavior in vivo, the vast majority of the research is carried out in vitro, on planar geometry, for its convenience in terms of imaging and force quantification. Moreover, two-dimensional (2D) migration recapitulates many of the general mechanisms characteristic of in vivo conditions [37,38].

### 7.2.3 The role of density

It is well known that, when seeded on 2D substrates, cells proliferate until all available space is filled (confluent state) and will later reach a quiescent state with little to no cell division, a phenomenon also accompanied by a steady decrease in cell motion. These two effects are concomitant and generally referred to as *contact inhibition*. Interestingly, when confluent cells are exposed to empty space, for example, by physically scratching the monolayer, both processes resume, and the resulting invasion of this adhesive area allows for the reliable observation of collective migration. This experiment, historically called wound healing [39], has taught us much of what we know about emerging properties during collective migration (reviewed in Ref. [40]).

Density also plays a fundamental role in controlling the cooperativity of the cell migration. This is better expressed as the velocity-correlation length, which represents the distance over which cells move in a correlated fashion, that is, the size of the cell patch moving in the same direction [41,42] or the radius of a swirl [43,44]. As cells become more dense and slow down their movement, the correlation length first increases to approximately 10 cells lengths or 150 μm [41,45], and later drops to the single cell value for the highest packing. This process is referred to as *jamming* because of its similarity with the behavior of amorphous materials [46]. Interestingly, substrate stiffness has also been shown to affect the extent to which directional cues are relayed [43,47].

A complete picture of the collective phenomena reported in the last decade rests on all these notions: cells feel their environment through focal adhesions, coordinate their behavior thanks to junctions, and this allows collective motion to be influenced by the presence of boundaries and, thus, cell density [48]. Specifically, in this review, we concentrate on mechanical oscillations, a term we use to identify all periodic fluctuations of mechanical parameters, such as cell velocity, the resulting strain, substrate tractions, and stresses. These oscillations are typically present in space and time and can be divided into two major categories: expanding monolayers are characterized by *propagative* waves [26,49–51] (i.e., signals that are traveling through the system like sound waves through a medium), while in fully confined environments (such as circles or rectangles) cell migration is limited to local cell rearrangements and oscillations resemble *standing* waves [8,25,44,52].

## 7.3 Propagative waves

We will first consider the case of propagative mechanical waves, which have been shown for the first time within freely expanding monolayers [26]. This is a stereotypical example not only because, chronologically, it was the first one reporting the presence of mechanical waves, but also because it highlights the existence of both oscillations in velocity and in forces/stresses.

### 7.3.1 Mechanical forces in expanding monolayers

These experiments are the direct evolution of the more classical wound healing [53]. With the help of a polydimethylsiloxane polymeric membrane, cells are first confined to a small space, where they are allowed to proliferate until confluence is reached and motion is almost entirely inhibited. The removal of the membrane exposes cells to a free surface, a stimulus sufficient to trigger the global polarization toward the newly available space and its invasion [54,55]. The major characteristics of the migration of expanding tissues are summarized in the first chapter of this book, while methods for the quantification are in Refs. [40,48].

Overall, the edge of the monolayer advances at a speed of approximately 30 μm/h [26,56], a velocity dependent on the steady increase of density provided by mitosis [26,51]. Cells at the edge are the first ones to react: they polarize toward the wound and consequently begin migrating and exerting forces onto the substrate. Initially limited to a boundary layer near the edge, this behavior is then relayed to the neighboring cell, propagating backward toward the bulk and progressively creating a state of global motion [57]. The study of the forces on the substrate, measured through the deformation of the underlying material, reveals that cell-generate tractions follow a similar trend: initially, the largest forces are localized at the front of the monolayer and perpendicular to its edge and will eventually propagate toward the bulk. Generally, cells far from the leading edge exert smaller forces on the substrate, although important tractions are generated up to several hundreds of micrometers inside the monolayer, where forces are extremely heterogeneous in magnitude and direction [26,58,59].

Force transmission through the substrate, if easily measured, is only half of the physical picture of CCM. The other half is the transmission of tensions at cell—cell junctions, and it lies hidden within the monolayer.

One method to indirectly access this information is monolayer stress microscopy (MSM) [59,60]. The measured local traction exerted by the cells on the substrate is precisely balanced by the internal stress generated within the internal monolayer. By applying Newton's laws, and considering the cell monolayer as a continuous, homogeneous, and isotropic thin sheet, the mechanical stress within the monolayer is calculated as the integral of the traction forces over the tissue area. The internal stresses $\sigma_{ij}$ are those required to balance the measured traction forces, irrespective of whether monolayer material is active or passive, elastic or viscoelastic, linear or nonlinear. With the additional hypothesis that the cell monolayer is elastic and uncompressible, one can indirectly compute the planar strain tensor $\epsilon_{ij}$. This mechanical stress accumulates from the edge toward the bulk of the tissue [58], a trend also validated with the use Förster resonance energy transfer sensors [61]. Like substrate tractions, stresses are heterogeneous, with areas under compressive and areas under tensile stress.

These results describe a situation where single cells engage in a global tug-of-war: a given cell within the monolayer feels the pulling force exerted at the front through its neighbors and, in turn, pulls on the substrate along the same direction, but opposite orientation. The balance of forces is ultimately satisfied when, through the junctions, the cell pulls on the neighbor immediately behind it, describing a mechanical signal which, traveling backward through the colony, generates global forward migration. This integration of local forces along the tissue has the net result of a global state of tensile stress [26,58,59,62,63], believed to be an important ingredient to achieve global polarization in migrating tissues. The assumptions mentioned above to apply MSM are much more restrictive than they seem. In fact, multicellular systems are not homogeneous, because they are made up of cells, extracellular matrix, and interstitial fluid, they are anisotropic, because cells possess a natural polarization, and are constantly kept out of equilibrium by an internal energy source. As a consequence, it remains difficult to measure in a reliable manner the stress within a 2D monolayer. To circumvent this limit, Serra-Picamal et al. measured the strain rate $\dot{\epsilon}_{ij}$ in an independent manner, by particle imaging velocimetry [26]. In the following of this chapter, we will mainly focus on experiments based on the direct measurements of the deformation field, which does not require assumptions on the constitutive model of the cell monolayer.

### 7.3.2 Oscillations in expanding monolayers

Superimposed to this general landscape is the presence of marked oscillations. The trends described velocity and tractions propagating backward through the tissue are not only typical of the initial moment but also continue in a fluctuating manner throughout the expansion of the monolayer. By averaging their measurables over $y$ (i.e., to obtain a measurement as a function of the distance from the edge, $x$), Serra-Picamal et al. obtained the kymographs in Fig. 7.1, which demonstrate in a clear way the presence of periodic mechanical waves of velocity, stress, and tissue strain rate ($\dot{\varepsilon}_{xx}$) [26]. Cells at the edges of the monolayer exert forces pulling inward (i.e., away from the edge), but as we move further within the tissue we notice nearly periodic oscillations in the direction of the forces (Fig. 7.1F). The presence of traveling perturbations is more transparent in the graph of the strain rate, calculated from the velocity ($\dot{\varepsilon}_{xx} = \partial v_x / \partial x$): wave-like excitations originate from the edge, propagate backward into the tissue, past the midline, and toward the opposite edge at a speed of

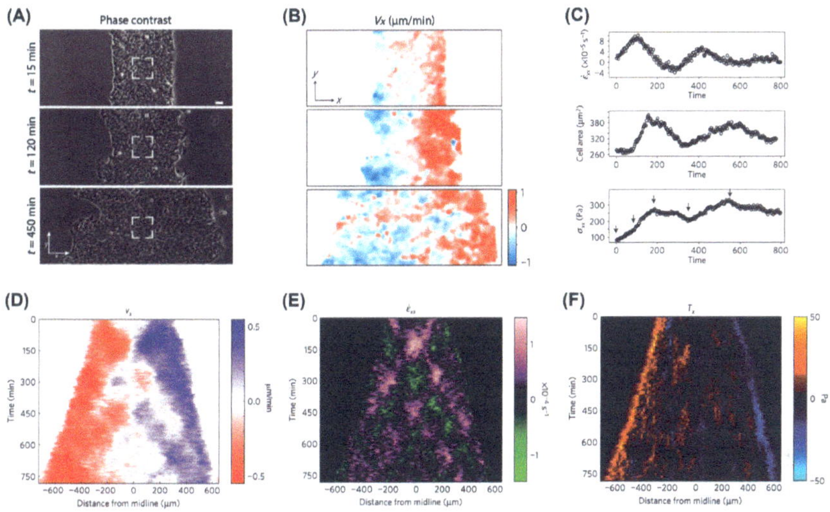


**Figure 7.1** Mechanical waves in expanding monolayers and at repulsive interfaces. A confluent tissue is grown within a rectangular hole in a PDMS membrane, which is in turn resting on top of a collagen substrate. The membrane is lifted and the tissue starts invading the available space (A), moving outward with a velocity measured with PIV (B). The mechanical behavior of the tissue is described by averaging the components in the *x*-direction over y (the tissue edge) to generate kymographs of *x*-velocity ($V_x$, D), strain rate ($\dot{\varepsilon}_{xx}$, E) and *x*-tractions ($T_x$, F). The average values of $\dot{\varepsilon}_{xx}$, cell area, and monolayer stress component ($\sigma_{xx}$) at the monolayer midline [averaged over the *white square* in (A)] are reported in (C). *PDMS*, polydiméthylsiloxane; *PIV*, particle image velocimetry. *Adapted from X. Serra-Picamal, V. Conte, R. Vincent, E. Anon, D.T. Tambe, E. Bazellieres, et al., Mechanical waves during tissue expansion, Nat. Phys. 8 (8) (2012) 628–634. http://www.nature.com/nphys/journal/v8/n8/full/nphys2355.html%5Cn; http://www.nature.com/nphys/journal/v8/n8/pdf/nphys2355.pdf.*

roughly 60 μm/h, describing what the authors call "X-waves" [26]. If, normally, one would expect stress to rise monotonically from the edges toward the midline, X-waves modulate this behavior so that monolayer stress in the center of the tissue oscillates with a periodicity of ~5 hours (see Fig. 7.1C). Interestingly, these fluctuations are in phase with those of cell area in a trend reminiscent of purely elastic behavior (see Section 7.5.3).

The importance of cell–cell junctions for the propagation of waves is confirmed by the authors by disrupting adherens junction with calcium chelation. In these conditions the speed of the advancing monolayer is unaffected, but no waves are present. Similarly, treatment with blebbistatin, which prevents myosin II activity, demonstrates the importance of contractility for the accumulation and fluctuation of stress [26,59]. Interestingly, similar velocity oscillations, with comparable periodicity (~4 hours, ~700 μm traveling at ~170 μm/h), are also observed at the edges of cancerous cell monolayers, albeit with circular symmetry [64].

### 7.3.3 Colliding tissues

Surprisingly, mechanical waves are not only distinctive of expanding monolayers, but they have also been observed in situations where a migrating tissue meets an obstacle, being it another epithelium or a wall [49,65]. When two repulsive tissues meet, they create an interface characterized by parallel cell alignment and strong tractions. As time proceeds, cell density increases and the tissue start jamming, a condition necessary for the appearance of waves at boundaries. Once more, cells at the interface exert forces pulling away from the edge, followed by periodic oscillations in their direction toward the bulk. Rather than homogenizing, this pattern seems to be stable in time, in what is a long-lived supracellular cooperative behavior. The periodicity is approximately 50 μm, a value comparable with those obtained for traveling oscillations in expanding monolayers [26,58]. If these characteristics might have been expected based on previous works, a peculiarity of colliding epithelia is the presence of deformation/velocity waves propagating backward across the monolayer at a velocity of ~120 μm/h, without attenuation. Interestingly, similar waves were also observed at the boundary with a repulsive wall and identical tissues, making velocity waves a generic feature of repulsive interfaces [49].

### 7.3.4 Orientation at external boundaries

CCM has also been studied in the presence of externally imposed boundaries, defining regions of space that cells cannot invade. In general, cells show a tendency to align tangentially to a boundary between adherent and nonadherent substrates [7,66–68]. Drawing general conclusions is not trivial because the specifics of this alignment depend strongly on cell type and density.

Most of the works on cellular alignment have been carried out on spindle-shaped fibroblasts, which develop weaker cell–cell adhesions and whose interactions rely mainly on steric hindrance [66,67,69]. These properties make quantification easier, but mean that the knowledge obtained on these cell types is harder to transfer on epithelial cells, whose strong cell–cell junctions and propensity to assume hexagonal shape make for a slightly different behavior. Madin-Darby Canine Kidney (MDCK) monolayers growing on 230 μm wide rings, all cells, including the ones in the center, align within 15 degrees of the tangential direction [68]. This is unsurprising because we know that cells coordinate their migration velocity over a well-defined correlation distance, ~150 μm for MDCKs, so we expect any forced orientation to propagate within the tissue for approximately the same length. If cells at the edges can only migrate parallelly to the border, then an overall orientation is to be expected. Similar results are obtained for MDCKs confined on strips of different widths, confirming that cell orientation increases with decreasing width and becomes global below ~100 μm [50] (see Fig. 7.2A).

### 7.3.5 Collective migration with confining borders

The presence of boundaries, together with the preferential orientation they cause, enabled scientists to isolate and reproduce some of these behaviors.

The tendency of cells to correlate their velocity over a specific distance produces, during CCM on planar geometry, swirling motion. Patches of cells spontaneously begin rotating, a behavior that can be replicated on round patterns (i.e., disks) of different sizes. The most rudimentary form is two cells confined on small circular islands, situation in which the angular symmetry is spontaneously broken and cells are observed to rotate in a coordinated manner around the center of the pattern [70,71]. On a larger scale the presence of boundaries and the preferential orientation they cause enhance this phenomenon [44], the persistence of which increases

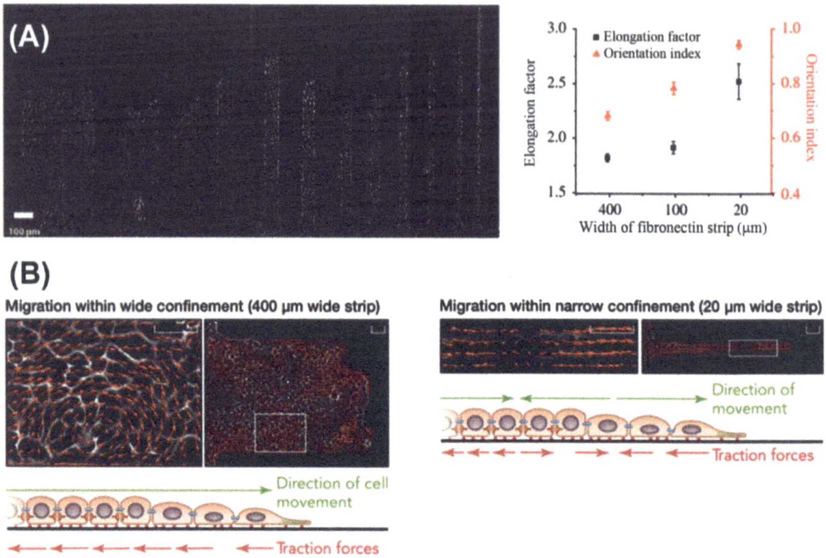


**Figure 7.2** Cellular orientation in presence of external boundaries. (A) MDCK monolayer migrating on adhesive strips of different widths (*left*) and the quantification of elongation ratio and orientation for 3 widths (*right*). (B) PIV measurements of MDCK cells migrating on wide channels (*top*) showing the presence of vortices. On more narrow channels (*bottom*) cells migrate in a contraction–relaxation manner. The schemes summarize the distributions of traction forces for both behaviors. *MDCK*, Madin-Darby Canine Kidney; *PIV*, particle image velocimetry. *Adapted from (A) S.R.K. Vedula, M.C. Leong, T.L. Lai, P. Hersen, A.J. Kabla, C.T. Lim, et al., Emerging modes of collective cell migration induced by geometrical constraints, Proc. Natl. Acad. Sci. U.S.A. 109 (32) (2012) 12974–12979. https://doi.org/10.1073/pnas.1119313109. http://www.pnas.org/content/109/32/12974.abstract and (B) R.K. Vedula, A. Ravasio, C.T. Lim, B. Ladoux, Collective cell migration: a mechanistic perspective, Physiology 28 (6) (2013) 370–379. https://doi.org/10.1152/physiol.00033.2013.*

in a density-dependent manner [72]. Before reaching confluence, cell clusters move mostly in a disorganized manner, occasionally displaying collective rotation, although with extremely short persistence. Once confluence is reached, the tissues begin rotating in a "solid disk-like" synchronized and collective manner, with stochastic inversions of the rotation direction [8]. As expected, the increase of density associated to cell division produces a slow decrease of the rotational speed, with average velocity values going from ~25 to ~8 µm/h, and an eventual jamming of the tissue [44]. There seems to be a critical diameter (~250 µm) of the circular patterns above which global rotation is substituted by smaller vortices and transient coordinated flows. This is reasonable when viewed with the velocity-correlation length (~150 μm) in mind. For small circles the whole area falls within this distance, and cells can coordinate their motion globally. When the diameter approaches twice the correlation length, cells at opposing sides of the colony behave "independently," and multiple swirls are formed. Since supracellular coordination depends on cell–cell junctions, it is unsurprising that MDCKs where cadherins are downregulated, as well as cancerous cell lines, cannot give rise to collective rotations.

These experiments tell us that increasing the confinement length above the typical velocity-correlation distance drives a transition between migration modes, and this is also verified on adhesive strips [50]. By creating a cell reservoir, and then releasing the tissue onto a prepatterned substrate, Vedula and colleagues confined the wound-healing migration of MDCK tissues to strips of different widths (see Fig. 7.2B). As one might expect, wide channels allow the formation of vortices (diameter ~300 μm), a behavior reminiscent of that of unconfined wounds. If under these circumstances the velocity of the advancing edge is approximately constant (~20 μm/h) [73], the vortices become less prominent when the migration of cells is confined to strips whose width is smaller than the velocity-correlation length (100–200 μm). In this situation, cell velocity becomes, on average, oriented in the direction of the strip, and this causes an increase in the speed of the advancing edge for thinner channels (up to ~40 μm/h) [50]. In this regime the velocity of cells follows a plug-flow-like distribution perpendicularly to the channel, and the overall flow is well described by a constant drift component and a diffusion-like contribution dependent on local cell density [73]. Vortices disappear altogether under the strongest confinement. In very narrow strips (~20 μm), cells become more elongated, the velocity becomes strongly aligned in the direction of the strip, and the front of the tissue advances more rapidly. Moreover, particle image velocimetry (PIV) analysis shows that in highly confined geometries, MDCKs switch to a contraction–relaxation or "caterpillar-like" migration mode, the cycle of which spans more or less 100 μm [50] (see Fig. 7.2B). Interestingly, the advance of monolayers in wide strips is associated with a monotonic increase in the stress component, in line with CCM in wound-healing experiments. Cells migrating by contraction–relaxation, on the other hand, exert on the substrate an alternation of positive and negative forces. Even though not specifically reported, the periodicity of these oscillations is approximately 4 hours and 250 μm, values in agreement with the mechanical phenomena presented earlier.

It is also worth noting that the motion of cells on strips has been compared with that on external surface of glass wires, where the relationship between radius and the velocity of the advancing tissue-front is somehow conserved, a fact suggesting that the boundary conditions control this effect rather than curvature [74]. On the other hand, under tubular confinement (i.e., cells migrating on the inner surface of a tube), the velocity increases with the radius [75]. Whereas these results are in disagreement, the transition between classical migration and a contraction–relaxation mode is observed for smaller radii in both conditions [74,75] and is attributed to the stronger relative effect of cell division in smaller spaces by Yevick et al. [74]. Interestingly, the experiments on glass wires report an additional phenotype: for radii smaller than ~40 μm, cells at the advancing edge escape the monolayer and move back and forth with velocities in the order of 100 μm/h, albeit no periodicity is reported by the authors [74].

Together, these results hint at the necessity of considering the role of curvature on the appearance of mechanical waves.

### 7.3.6 Velocity oscillations in confined channels

In a recent article [51], Tlili et al. repeated wound-healing experiments while forcing MDCK tissues to migrate along functionalized adhesive strips (Fig. 7.3A). Even though they do not report a strong effect of the confinement width, when compared to Ref. [50], the lack of swirling motion and globally directed quasilaminar flow place their experiments in the central region, where migration is strongly guided, albeit without the presence of the contraction–relaxation phenotype. The authors inhibit cell division to disentangle the contribution of migration and proliferation, and they report faster advance of the leading edge (~40 μm/h) compared to similar examples where mitosis was unaffected. The first observation in such conditions is that, as the tissue spreads and cells in the back progressively join the forward migration, the average velocity decreases from the front toward the bulk. Interestingly, the velocity is inversely proportional to local cell density, in a trend that does not depend on the distance from the leading edge.

On top of this average velocity profile, the authors observe multiple periods of backward propagating velocity waves (see Fig. 7.3B), essentially describing cells that slow down and accelerate with a period of ~2 hours and a wavelength of ~1 mm. The absence of cell division guarantees a better contrast and, thanks to the lack of jamming, longer observations (~20 hours). If the nature of the waves is similar to those reported before, their phase

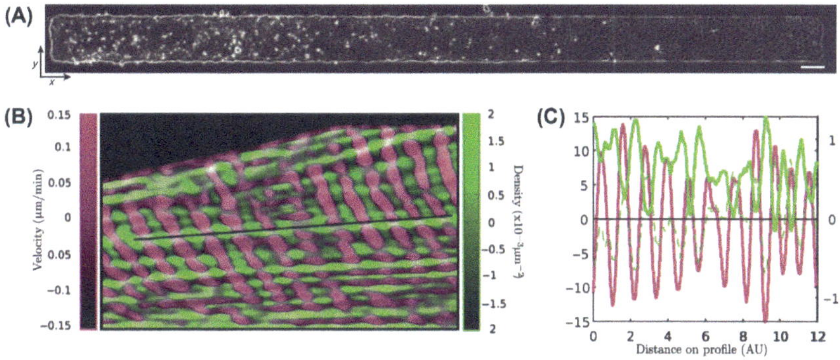


**Figure 7.3** Mechanical waves in confined expanding monolayers. (A) Brightfield image of MDCK cells expanding in confined channels without proliferation (scale bar = 100 μm). (B) Superimposed kymographs of small-scale variations in local cell density (*green*) and velocity (*purple*). The space along *x* is reported on the vertical axis (from 0 at the bottom to 3 mm at the top) and time on the horizontal axis (from 0 on the left to 25 h on the right). The dark space in the top-left corner represents the area not yet invaded by the tissue. (C) Plots of density (*green*) and velocity (*purple*) along the black line in (B). The phase opposition between the quantities plotted is highlighted by inverting the sign of the density plot (*dashed green line*). *MDCK*, Madin-Darby Canine Kidney. *Adapted from S. Tlili, E. Gauquelin, B. Li, O. Cardoso, B. Ladoux, H. Delanoë-Ayari, et al., Collective cell migration without proliferation: density determines cell velocity and wave velocity, R. Soc. Open Sci. 5 (5) (2018) 172421. http://rsos.royalsocietypublishing.org/lookup/doi/10.1098/rsos.172421.*

velocity is much higher, around 600 μm/h. Moreover, the wave crests in Fig. 7.3B are curved, evidence that the wave slows down near the bulk, where density is higher. There is, in fact, a strong link between velocity oscillations and inhomogeneities in cell numbers. The authors report steady fluctuations of density (period ~200 μm) parallel to the confining edge (vertical direction in Fig. 7.3B), which seem to deform with the stretching of the tissue. More interestingly, density oscillations are present in the perpendicular direction, in a fashion that follows velocity waves in phase opposition (Fig. 7.3C). The authors also demonstrate that lamellipodia inhibition slows or hinders the formation of waves altogether, confirming once more the fundamental role played by actin contractility in these phenomena.

Overall, these evidences describe backward propagating mechanical waves as a common feature of all the boundaries of migrating monolayers, being them with free space or repulsive walls. Even though the values vary between reports, there is agreement on the general magnitude for the periodicity of these phenomena, in the order of hours and hundreds of micrometers. An interesting correlation is observed between mechanical waves and waves of extracellular signal-regulated kinases (ERK) activation, which have been reproduced in three of the situations mentioned in this section: free expansion, swirling motion, and contraction–relaxation migration [76]. Strikingly, the periodicity of ERK waves (~4 hours and 250 μm) agrees with the typical time and space of mechanical oscillations, a similarity which begs further experimental work.

## 7.4 Standing waves in fully confined monolayers

Mechanical waves have also been reported in fully confined environments, where collective motion only happens through neighbor exchange rather than directed expansion. In this situation the available space is finite, which means that continuous proliferation inevitably leads to contact inhibition and the absence of coordinated movement. The presence of oscillatory phenomena is therefore restricted to a narrow time window between confluence and the jammed state [8,24,77].

Even though each account concentrates on a different geometry, from thin lines [chosen to go as close to a one-dimensional (1D) system as possible] to wider 2D monolayers (such as rectangles or circles), they all report the presence of periodic motion resembling that of a standing wave, with specific periodicity in time and space.

### 7.4.1 Breathing oscillations

Oscillatory movements have been observed in epithelia confined to round patterns of radii smaller than ~300 μm. Contrary to the swirling motion presented earlier, these oscillations happen along the radial direction, displaying standing waves of highly coordinated outward and inward cell movement (Fig. 7.4) [8,25,52]. The two phenomena, collective rotation and these "breathing" oscillations, are not mutually exclusive but are rather superimposed and observed together. With respect to the tangential velocity, radial oscillations are characterized by a lower velocity amplitude, which depends on the account, from ~15 [52] to less than 1 μm/h [25]. Oscillations are present as fluctuations of this radial velocity, which is generally averaged over the angular coordinate for better visualization (Fig. 7.4B-*right*). The cells in the center of the disk and those at the far edge tend to be still, whereas the

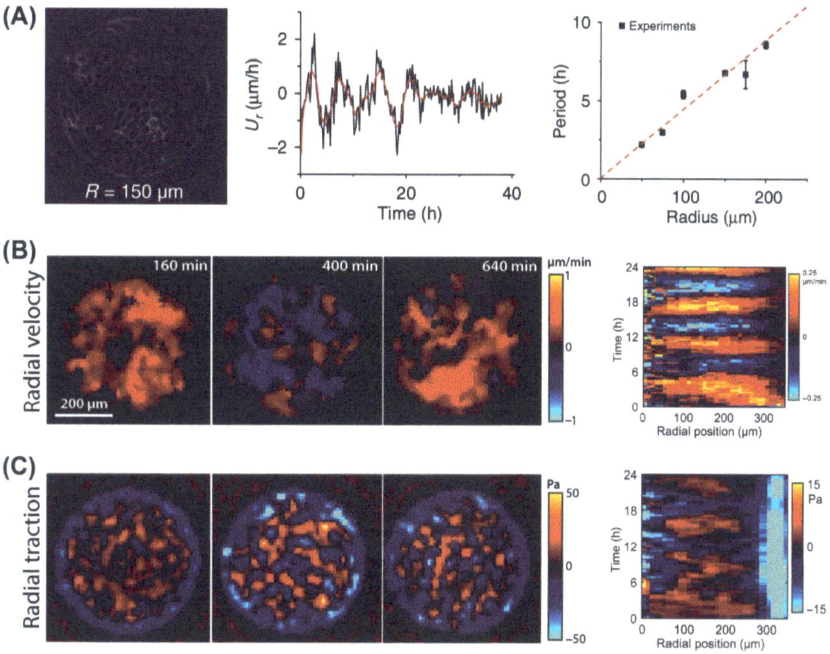


**Figure 7.4** Breathing oscillations in epithelial tissues confined to round adhesive patterns. (A) A phase-contrast image of MDCK cells confined on a disk-like pattern (*left*) and the evolution of the average radial velocity in time (*center*). The typical time period for these oscillations can be quantified as a function of the radius of the tissue (*right*). (B) Examples of the distribution of radial velocity over time for MDCK cells on patterns of radius equal to 350 μm. The velocity is averaged over the angular direction to yield a kymograph (*right*). (C) Cell-generated tractions on the substrate (*left*) for the examples in (B) are reported together with a kymograph of the average over the angular direction (*right*). *MDCK*, Madin-Darby Canine Kidney. *Adapted from (A) M. Deforet, V. Hakim, H. Yevick, G. Duclos, P. Silberzan, Emergence of collective modes and tri-dimensional structures from epithelial confinement, Nat. Commun. 5 (1) (2014) 3747. https://doi.org/10.1038/ncomms4747. http://eutils.ncbi.nlm.nih.gov/entrez/eutils/elink.fcgi?dbfrom = pubmed&id = 24796352&retmode = ref&cmd = prlinks&5Cnpapers2://publication/doi/10.1038/ncomms4747; and (C) J. Notbohm, S. Banerjee, K.J.C. Utuje, B. Gweon, H. Jang, Y. Park, et al., Cellular contraction and polarization drive collective cellular motion, Biophys. J. 110 (12) (2016) 2729–2738. https://doi.org/10.1016/j.bpj.2016.05.019.*

maximum of the radial velocity is at the mid-point between the two positions [8]. The amplitude of the cell displacements is the order of micrometers, and the period a few hours. Although this latter quantity is in general agreement with those for expanding monolayers, the period of breathing oscillations depends strongly on the radius of the round patch (Fig. 7.4A-*right*) [8].

Surprisingly, even though radial motion is very limited, oscillations in cell velocity are accompanied by similar standing waves in the radial component of the tractions on the substrate, with patterns of inward and outward pointing forces over time (Fig. 7.4C) [25]. These oscillations only interest the central part of the tissue, as cells at the edges constantly pull themselves toward the exterior, in a manner coherent with kenotaxis [78]. One might be tempted to assume that tractions and velocities are interdependent, an assumption carefully disproved by the authors, who demonstrate that no correlation between traction and velocities, nor their relative angles, exists [25]. The study of stresses within the monolayer led to an interesting conclusion: tensions and cell area (i.e., strain) are in phase, implying that the tissue behaves elastically with no viscous contribution (see Section 7.5.3).

Given the correlation between the presence of waves and ERK activation advanced recently [76], it is interesting to point out that breathing oscillation was suppressed upon ERK inhibition [25]. This is loosely in contrast with what is reported, admittedly in a very different situation, in [78], indicating that ERK activity has no significant role in kenotaxis.

These breathing oscillations share many similarities with those mechanical waves present in expanding monolayers, like the strong dependence on cell–cell junctions and contractility, and might be explained by the same basic principles of force transmission and mechanotransducive coupling with polarization [25]. One big difference lies in that, whereas velocity waves in wound healing are propagating through the tissue, on circular patterns the phase difference between radial positions is insignificant.

## 7.4.2 Multinodal standing waves

The discrepancies in the reports presented so far, namely, the different spatiotemporal coherence and dependence on confinement, can be seen as evidences that traveling waves and breathing oscillations are independent. On the other hand, the similarities between these phenomena suggest that they might be two faces of the same coin, each emerging for specific geometries and length scales. Discriminating between these interpretations requires more suitable experiments, specifically tailored to understand whether and how the presence of boundaries play a role in the emergence of oscillations, and if it can affect the typical periodicity.

Two simultaneous and independent papers recently tackled this question, both reporting the appearance of extended standing-wave-like velocity oscillations in fully confined environments [24,77]. Contrary to breathing oscillations, these works extend their observations to larger systems to uncover the presence of nodes, that is, positions of zero net movement, and study the effect of the confinement on the characteristics of cell motion, albeit from different points of view.

#### *7.4.2.1 Waves on quasi-one-dimensional lines*

The first work, published by our research group [24], was carried out trying to understand whether the environment alone dictates the way cells oscillate, or if their behavior is a consequence of some innate feature. This was tackled by searching for the most simplified experimental conditions to reproduce wave-like phenomena, a bottom-up approach resulting in the creation of thin 40 μm wide strips of different lengths ($L_X$). In accordance with previous reports [50], this geometry forces cell migration along the longitudinal direction, allowing the authors to average over the transverse direction and only study the motion along one dimension, thus creating what is referred to as *quasi-1D* migration.

Wave-like oscillations are reported for MDCK cells on all strip lengths, although two clear phenotypes can be distinguished: *global oscillations*, where all cells move alternatively leftward and rightward in phase (see schema in Fig. 7.6A), and *multinodal oscillations*, characterized by multiple iterations of these global movements, in phase opposition and separated by nodes (Fig. 7.5A). The relative incidence of the two behaviors depends strongly on the length of the line, with a clear transition between modes around a critical length $L_X^c \simeq 400\ \mu m$ (Fig. 7.5B). We distinguish three situations: (1) on lines shorter than $L_X^c$, global oscillations are overwhelmingly prevalent; (2) when LX is comparable with $L_X^c$, tissues display both phenotypes; (3) when LX is larger than $L_X^c$, multinodal oscillations statically dominate (Fig. 7.6B). Moreover, a quantification of the typical wavelength and time period confirms the presence of two regimes: for relatively short strips, the time period scales linearly with the tissue size, resembling the trend reported for breathing oscillations (the wavelength, imposed by the confinement, cannot be calculated), but when multinodal waves dominate, both time period and wavelength are approximately independent of $L_X$ and can be estimated as $\lambda \sim 380\ \mu m$ and $T \sim 5$ hours. An effective phase velocity is calculated from these values as $u = \lambda / T \sim 80\ \mu m/h$, which is approximately independent of the tissue length, and is coherent with previous observations.

These results suggest that, when possible, cells spontaneously establish multinodal oscillatory modes with predetermined features. When confined to spaces smaller than those characteristic of a full wavelength (i.e., 400 μm),

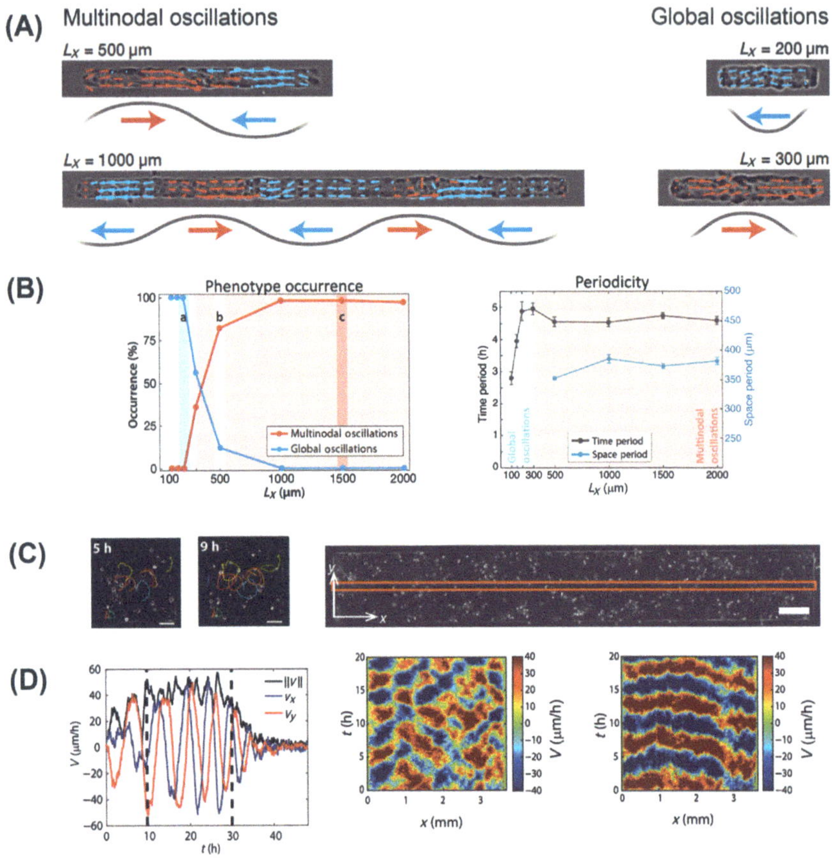


**Figure 7.5** Wave-like oscillations in epithelial layers. (A) Velocity field measured by PIV. Velocities pointing in the positive (negative) *x*-axis direction are shown in red (blue), in agreement with the arrows reported under the image. Lines longer than 500 μm display multinodal oscillations (*left*). Stripes shorter than 300 μm exhibit global oscillations (*right*). (B) Frequency of multinodal and global oscillations (*left*), and the characteristic time and space periodicity (*right*), for different stripes length. (C) Snapshots of a confluent HaCaT layer on square (*left*) and rectangular patterns (*right*). (D) Temporal evolution of the velocity field on the square pattern (*left*), and the kymograph of the velocity field on the rectangular pattern (*right*), respectively, along the long (*x*) and the short (*y*) axis. *PIV*, particle image velocimetry. *Adapted from (B) V. Petrolli, M. Le Goff, M. Tadrous, K. Martens, C. Allier, O. Mandula, et al., Confinement-induced transition between wave-like collective cell migration modes, Phys. Rev. Lett. 122 (16) (2019) 168101. https://link.aps.org/doi/10.1103/PhysRevLett.122.168101 and (D) G. Peyret, R. Mueller, J. D'Alessandro, S. Begnaud, P. Marcq, R.-M. Mège, et al., Sustained oscillations of epithelial cell sheets, Biophys. J. 117 (3) (2019) 464–478. https://linkinghub.elsevier.com/retrieve/pii/S0006349519305028.*

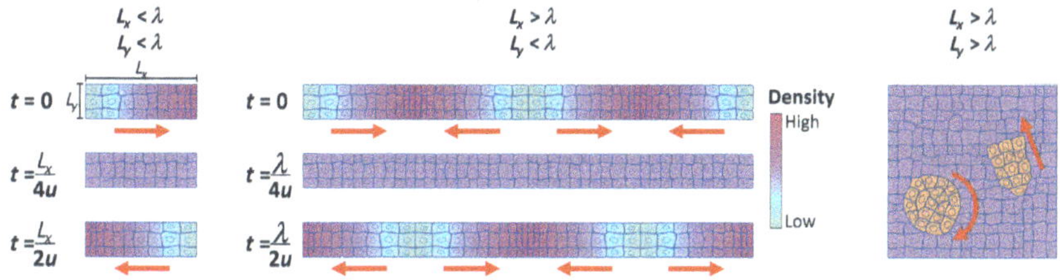


**Figure 7.6** Schematic view of velocity waves in different geometries. (A) Global oscillations appear when cells are confined in both directions below the coherence length $\lambda$, with a time period scaling linearly with the confinement size $L_x$. $u = \lambda/T$ ($u \simeq 80$ μm/h) is the effective phase velocity of the wave. (B) Multinodal oscillations are established in quasi-one-dimensional geometry, along the direction where the confinement length is larger than $\lambda$. For any $L_x > \lambda$, the time period of the oscillations is $T$. (C) Swirling motion and collective migration appear in unconfined geometries (both $L_x$ and $L_y$ are both larger than the coherence length $\lambda$).

nodes disappear and the overall behavior changes. The presence of this transition, together with the fact that time and space periods do not adapt to the confinement, leads to the conclusion that velocity oscillations are intrinsically encoded in the biology of the cell.

Interestingly, the presence of these phenomena in confined environments fits well with the physical interpretation of standing waves in as the sum of two identical waves traveling in opposite directions. For the sake of simplicity, this can be compared to what happens in a laser cavity where photons are reflected at the edges and interfere with the following ones. Although a simple reflection does not capture the complexity of the underlying phenomenon, something similar might be happening on thin lines, where standing velocity waves are the superimposition of waves reflected at the shorter edges. Moreover, this interpretation is somehow supported by the observation that traveling velocity waves are typical of all repulsive boundaries [49]. Since biological systems are noisy, cells might be able to oscillate with a wider range of wavelengths, to adapt their motion to the particular length of the line in a resilient manner.

### 7.4.2.2 Standing waves on extended two-dimensional systems

If the quasi-1D geometry only allows for "binary" cell polarity and, thus, simplified motion, the presence of standing waves has also been reported on larger systems where cell migration happens along two directions in a correlated manned [77]. Human keratinocytes (HaCaTs) confined on square patterns (side length 200–1000 μm) display globally coordinated movement along elliptical trajectories (see Fig. 7.5C), coherent with a polarity vector that rotates in time. When separated into its $x$ and $y$ components, the velocity field displays standing-wave-like oscillations in both directions, with a $\pi/2$-phase shift, the typical period (in time) and amplitude (in μm) of which increase linearly with the confinement size (2–10 hours, 20–100 μm) (see Fig. 7.5D) [77]. This behavior is reminiscent of the *global oscillations* on thin lines, albeit on a much larger scale. The discrepancy between reports can be explained considering the velocity-correlation length of the cell types employed: global oscillations happen for line lengths smaller than the correlation of MDCKs (~150 μm), a condition which is also met for HaCaT (correlation length ~800 μm) on the squares tested by Peyret and colleagues.

Interestingly, this length scale is also associated to the amplitude of the oscillations in velocity, which increases linearly as long as the square edge is smaller than the cell velocity-correlation length, after which it becomes approximately constant. This behavior hints at the interplay between pattern size and velocity correlation as an important regulator of 2D oscillatory motion. The role of the environment becomes that of prompting cell polarity to rotate in time, a feat that is best achieved when the two length scales match. On the contrary, when confinement is too large or too small, motion is hindered and oscillations are less pronounced. Moreover, the same behavior is reported for HaCaTs on circular patterns.

From this perspective the symmetry of the confining geometry directly causes a symmetry in oscillatory motion in $x$ and $y$. To investigate how the two directions are coupled, the authors confine cells on long rectangles (width 200–500 μm, length 2000–3000 μm), where the rotatory nature of velocity and cell polarity is confirmed. In this case, they report the presence of global oscillations along the short direction and a standing wave with multiple nodes along $x$, describing a phenomenon reminiscent of multinodal oscillations. Interestingly, the phase coherence length (i.e., 1/2 the wavelength) along both directions approximately matches the tissue width, leading to the conclusion that the characteristics of the oscillations only depend on the smallest confinement size [77]. The authors also confirm this observation on expanding monolayers (i.e., rectangles with a free edge), where the scale of the velocity oscillation along the long axis is determined by the size of the shorter axis.

This might, at first, appear that this statement is in contrast with the previous reports of the wavelength being independent on the boundary conditions [24]. However, the authors report a linear behavior, albeit with an intercept that is far from zero. This might represent the intrinsic value of the phase coherence length, which is then tailored by the properties of the environment on large systems. Interestingly, a quick estimation of such value is in the range of 100–200 μm, surprisingly coherent with previous reports on very different cell types.

To assess the possible downstream effect of these oscillations the authors calculate the divergence of the velocity field to prove that these oscillations are accompanied by fluctuations in density and, thus, phases of compression and extensions, although a direct measurement of the correlation between nuclear density and waves is lacking. Further experiments also revealed that areas of low cell density are characterized by high YAP nuclear localization and thicker cellular layers, hinting at the possible role of these oscillations in controlling cell behavior from a biochemical point of view. On a side note, related to the possible alignment between cell velocity and traction [25], the authors state that, only on a local scale, the velocity of each cell aligns with the traction forces with a delay of ~30 minutes.

## 7.5 Mechanical considerations

All the phenomena described represent different aspects of collective migration and, as such, are strongly dependent on cell communication and, more generally, force production [48], a consideration common to all the works presented. On top of actin contractility and the formation of lamellipodia, other processes are responsible for mechanical stresses, such as cell division [79,80], extrusion [7,52], and volume changes [81]. The effect of some of these factors on mechanical oscillations has already been tackled, the effect of others remains to be assessed in the future.

### 7.5.1 Affecting the correlation length

Inhibition of myosin II activity, generally carried out with blebbistatin, leads to the strong decrease of tractions and monolayer tensions [78] and, even though the integrity of cell–cell junctions is unaffected, a drop in correlation length [47]. If, on the one hand, during wound healing, this has no effect on the migration velocity and the front of the monolayer advances at the same pace [25,26]; on the other, inhibition of contractility causes a loss of all emerging properties. Monolayers treated with blebbistatin lose the dependence of velocity on confinement geometry [50,74,75], the contraction–relaxation phenotype, swirling motion, and mechanical oscillations in general [8,25,26,49]. These reports agree that cell contractility is essential for oscillations, the only exception being the paper of Peyret et al. [77]. After treatment with a high dosage of blebbistatin, the authors observe a sharp drop in cell velocity and the persistence of oscillatory motion, albeit with a larger period and smaller amplitude. These observations lead to the conclusion that cytoskeletal activity only promotes cell motion and their coordination, without having a fundamental role in the presence of oscillations [77]. The discrepancy might be attributable to the different cell type used.

Disruption of adherens junctions has a similar effect, preventing coordinated motion [47,50] and stress buildup [49,59,82], ultimately hindering the propagation of mechanical waves [26,51] and oscillations as a whole [8,77].

One conclusion to be drawn from these evidences is that any treatment affecting mechanical communication between cells and, thus, the correlation length of their motion should have an effect on the typical periodicity of mechanical waves. It has been previously reported that communication through the substrate, dependent on its elastic modulus, can affect the transmission of directional cues [47] and the extent of the velocity correlation [43]. In the future, it might be interesting to verify whether this parameter has an effect on mechanical waves. Dissipation processes, such as friction within the monolayer or with the substrate, could also introduce a viscous screening length [67]. Both internal and external frictions should be modified, possibly by changing the concentration of adhesive proteins on the surface and by tuning the strength of cell–cell junctions concomitantly. A precise characterization of dissipation processes, and a better understanding of their role in cell migration, would probably help us to explain the formation and disruption of collective patterns [83].

A different approach is to utilize biochemical treatments to change cell behavior. Good candidates are RAB5, shown to restore highly coordinated collective motion even after jamming [84], and various growth factors. EGF, for example, is involved in the inhibition of cell motion during jamming [85], and it has been associated to a change in the elastic modulus of monolayers and an increase in the period of breathing oscillations [25].

### 7.5.2 The role of mitosis

As discussed earlier, cell density is also a very important determinant for CCM (reviewed in Ref. [48]). It plays a fundamental role in contact inhibition and jamming and contributes to the invasion of space during wound healing [26] and the presence of mechanical oscillations [49,51]. Interestingly, although a transient arrest of mitosis does not alter oscillations in expanding tissues, where cell division is regarded as a perturbation in cell flow and a disturbance for the appearance of waves [51], proliferation has been associated with the contraction–relaxation motility on thin strips [74], two phenomena which share many similarities. Even though there have been contrasting reports, intuitively, mitosis should play a role in controlling CCM, and we already know it to be a fundamental factor for the growth of all living organisms. If periodic velocity waves do play a role in controlling cell behavior in vivo at the supracellular scale, possibly the most attractive perspective for this field, the role of mitosis should be specifically regarded in more detail in future experiments.

### 7.5.3 Elastic behavior of expanding tissues

Cell layers are generally considered as viscoelastic materials (see Chapter 2, The Basic Concept of Viscoelasticity). Over relatively short timescales the response is elastic, and its origin is thought to be related to the passive spring-like behavior of the cytoskeleton and intercellular adhesions [86]. On longer timescales (~hours), tissues behave as active materials, accommodating deformations by remodeling cell–cell junctions, cell–cell rearrangements [87] or divisions [88], processes that dissipate stress in a manner reminiscent of viscous materials.

A peculiarity of the mechanical waves observed by Serra-Picamal et al. is that stress and strain are in phase, describing what is effectively an elastic behavior, on timescales in the order of hours [26]. If the viscoelastic description captures the monolayer expansion, it fails to account for the generation of propagative waves in an overdamped environment. The authors suggest that the introduction of two additional assumptions explains this effective long-term elasticity. First, cell motility increases when the surrounding cells create space. Second, above a given threshold of strain, the cytoskeleton first reinforces then fluidizes over different timescales [25,26,82].

Even though the assumption of an effective elastic behavior is far from being universally accepted, other works seem to be pushing the idea forward. Similar mechanical oscillations also emerge in round disks, where Notbohm et al. observed, once again, that cell velocity and monolayer tension are in phase. Moreover, phases of outward motion were associated with an average cell stretching and, contrarily, inward motion with compression, further implying an elastic relationship with a modulus of 113 ± 28 Pa. An inhibition of actin contractility leads to a decrease in this value, demonstrating that elasticity is of an active nature [25]. Another measurement of this quantity was obtained by stretching a tissue between flexible poles, technique that enabled Harris et al. to estimate the elastic modulus at 20 ± 2 kPa for an MDCK cell monolayer [86]. A linear relationship between strain rate and stress was also observed in expanding tissues [82]. Interestingly, the stress buildup is independent of the substrate rigidity, therefore highlighting the role of cell–cell junctions. Without confinement, the traction forces exerted by cell colonies scale with their size, a linear relationship that agrees with the interpretation of the tissue as a (contractile) elastic layer coupled to the substrate [89].

In specific situations, such as the migration of monolayers over thin strips, tissues can extend over nonadherent areas, called *epithelial bridges* [90]. As cell–substrate interactions are inhibited outside of the strips, the structural integrity of such bridges is ensured thanks to adherens junctions and actin contractility, which self-assemble into a supracellular actin-cable under strong tension. The behavior of epithelial bridges in response to geometrical and mechanical perturbations is coherent with that of an elastic material. Interestingly, this is true for primary human keratinocytes, but not for MDCK cells, a discrepancy attributed to the more fluid nature of MDCK tissues [90].

Another peculiar instance of the mechanical response of tissues was reported by Latorre and colleagues. While studying the formation of 3D epithelial domes, they noticed that epithelial sheets could sustain massive deformations under constant tension thanks to the coexistence of super-stretched cells with barely stretched ones, in a behavior reminiscent of the superelasticity of some alloys [91].

### 7.5.4 Effective inertia at the cellular level

The propagation of mechanical waves is, from a physical point of view, not of any particular complexity. It can generally be explained as a continuous interconversion between potential and kinetic energy, damped by friction forces. In this scenario the presence of oscillations critically depends on the ratio between friction forces ("fighting" the oscillation) and the inertia (i.e., the mass, defining a timescale for the oscillation). When friction is negligible, the system is underdamped and waves can propagate indefinitely. Within tissues, on the other hand, being the mass of the cell extremely small, inertia is negligible compared to friction [63], an overdamped situation where wave propagation is, in principle, impossible. In this case, to explain the presence of the mechanical waves we have just reviewed, an additional phenomenon must be present to play the role of "effective inertia" [49] and restore oscillatory behaviors (see Chapter 5 of this book). Even though the biological nature of this feedback is not clear yet, merlin seems like the most obvious candidate, theoretical works have taken this factor into account by introducing an additional term. Even though a full review of the models utilized to explain oscillations in tissues is beyond the scope of this work (see Refs. [92–94]), it is nonetheless interesting to quickly glance at the different solutions embraced in literature.

One example is that of Vicsek models [95], where the velocity of each cell is assumed to align to that of its neighbors [8]. This alignment happens on a typical timescale, a fundamental quantity for the appearance of coordinated motion [96], and is affected by rotational noise. The group of Marchetti introduces a feedback between the local strain and the rate of change in contractile tension. Simply put, larger cells will exert stronger contractile forces. This, together with the assumption of tissue elastic behavior, explains the presence of oscillations [25,97]. Other possibilities are a dynamic response from the cytoskeleton, first reinforcing and then fluidizing in response to stretch [26], a slow realignment of cell polarity to the local strain [51] or to the resultant forces on the cell surface [77].

## 7.6 Discussion

Since the first reports, mechanical waves have been observed in a variety of situations, both in confined and unconfined environments. Even though the most studied model is MDCK cells, other types have been employed, and despite the intrinsic variability between them, all reports seem to agree on the typical times and space scale: mechanical oscillations happen with a periodicity of ~0.5–1 mm and ~5 hours. Moreover, the effective velocity of transmission of mechanical signals, being them traveling or standing waves, is between 60 and 120 μm/h.

In all situations, the characteristics of the phenomena observed are associated to the degree of cooperativity of cell motion, quantified as the velocity-correlation length. On freely expanding monolayers, this cooperation manifests itself as the presence of swirls [44], patches of cells moving parallelly [40] and finger-like protrusions [98]. When contractility or cell–cell junctions are disrupted, the expansion of the monolayer is driven solely by a density gradient, and these peculiar motion patterns are not present. Similarly, the presence of mechanical waves depends on the active communication between cells. Peculiar motion patterns, such as the contraction–relaxation phenotype, only happen when the width of the channel is below the correlation length [50] and the spacing between velocity waves launched the front of the monolayer is approximately twice this quantity [26]. This interpretation is also coherent with the presence of a transition between oscillatory regimes for values approximately equal to double the velocity-correlation length (Fig. 7.6A and B) [24]. As a gross simplification, having cells moving in opposite directions requires two groups of coordinated cells, which, in a quasi-1D geometry, requires twice the correlation length. Interestingly, this change in cell behavior coincides with the transition between global coordinated rotation and the presence of multiple swirls for disks of diameter ~300 μm [44].

There seem to be two contrasting points of view on the role of the confinement: the presence of waves in expanding monolayers [26,51,64], together with the observation of steady time and space periods on lines of different length [24], suggests that oscillations might stem from a quantity intrinsic to the cell, which does not depend on the properties of the surroundings. On other hand, the period of breathing oscillations [8,25] and collective elliptical movements [77] depends directly on the size of the colony (Fig. 7.6C). Viewed with the correlation length in mind, these two conclusions might be coherent. Peyret and colleagues report that the period and amplitude of oscillations scale linearly with tissue size, a trend similar to that followed in a regime of global oscillations on quasi-1D lines [24]. The discrepancy stems from the fact that, whereas thin lines are designed to span a wide range of lengths, both considerably smaller and larger than the correlation length, the squares in Ref. [77] remain in the former regime, a fact that could hide the presence of a transition for larger system. From this point of view, the experiments in Ref. [77] remain in the "2D version" of the global oscillation regime, controlled by the smaller of two confining lengths. The presence of a transition in both directions might be assessed in the future. Moreover, the dependence of the space periodicity on confinement size (see Fig. 7.6A and B) describes a linear behavior with a threshold (i.e., the intercept is around 200 μm).

It seems plausible that some intrinsic property of cells is at the root of the periodic motion, which manifests itself whenever cell density is at levels preceding jamming. Above a certain threshold, and when the size of the confinement is of the same order as the correlation length, the cues relative to the presence of edges can be transmitted through the colony and can actively tune the periodicity. The edges might act as cues for the synchronization of cell motion in time, as after the release of a border, or in space.

Furthermore, breathing oscillations fit in this be interpreted as global oscillations, with an additional cylindrical symmetry. By only considering the radial direction, the system can be reduced to one dimension, where the radius plays the role of the length. The two phenotypes might thus represent similar phenomena, an assumption confirmed by the fact that global oscillations are measured for lengths smaller than 400 μm [24], and all works reporting breathing oscillation are carried out on disks whose radius is smaller than 350 μm [8,25,52]. Whether different migration modes are characteristic of larger disks, is not clear.

In this perspective, all reports of mechanical oscillation represent different manifestations of the same phenomenon, guided by an intrinsic quantity and tuned by the symmetry and characteristic size of the available space, which highlight specific aspects of the underlying oscillations. Nonetheless, if breathing oscillations and standing waves describe periodic inversions of velocity and, thus, cell polarity, a fundamental difference with elliptical movement remains to be assessed. When seeded onto round patterns in fact, HeCaT cells do not display swirling motion [77], a discrepancy that sets them apart from other reports.

## 7.7 Conclusion and perspective

The presence of mechanical waves acquires new importance when viewed in light of the intricate relationship between biochemical signaling and mechanosensitive pathways, and their role in understanding the process of morphogenesis. All events leading to complex structures and shapes require cells in different positions to behave differently, in terms of movement, proliferation, differentiation, and death [21]. For this to happen, those same cells need to somehow "be aware" of the state of the tissue and their particular location. The concept is summarized as *positional sensing* [62] and, as we now understand, happens through a bidirectional relationship between genetic and epigenetic factors. Mechanical signals are a major player in this latter category [23], their effects on cell behavior are well documented in 2D in vitro systems [5,7–9] and are being understood on more complex environments [13,20,22]. In many of these cases, the specific setup prompts the emergence of a long-range mechanical landscape, used as a reference system. An example is that of cells plated on square patters, where the edges are under stronger stress compared to the interior [5]. Every single cell then feels the local state and acts accordingly, in this particular case enhancing or reducing proliferation rate. The coordination offered by the mechanical reference system leads to the emergence of a global phenomenon, at a scale much larger than that of the single cell sphere of influence.

In this context, the presence of traveling waves makes sense as a mean to tune the local guidance and exchange information over distances bigger than the immediate surroundings. In fact, the typical velocity of mechanical waves is in the order of ~100 μm/h, a value approximately 10-fold larger than the average velocity of cells within a monolayer (between 4 and 12 μm/h, depending on cell density [41,45] and the properties of the environment [57,99]). Even though the two phenomena might be somehow reconciled considering reflection at the edges and a "laser cavity" interference, there is a fundamental difference between traveling and standing waves.

The mechanical stimulus of a traveling wave is only transient, and affects, at different times, all cells equally. Considering the velocity of propagation and the size of a cell, the timescale of the stimulus (~minutes) is in the range of cytoskeletal rearrangements, but it is much shorter than that needed for a change in gene expression (~hours) and, thus, a downstream biochemical response. If it is true that some degree of periodicity is generally present [26,49,51], this phenomenon is observed in expanding monolayers, associated with a deformation of the system. This makes the assessment of an eventual long timescale effect much more challenging.

Standing waves, on the other hand, are intrinsically periodic and are observed in fully confined systems, where cells at a certain location are exposed to a regular mechanical stimulus over a longer timescales (~days [24,25,77]). These factors make standing waves a promising candidate for the exploration of possible downstream effects, a hypothesis which seems to be common to most of these reports. This idea stems from two basic considerations: standing waves are characterized by nodes and antinodes in cell velocity, the spacing of which seems to be controlled by an intrinsic quantity; and cells located at the antinodes appear to be alternatively stretched and compressed. This selective mechanical stimulus could somehow distinguish those cells, creating two subpopulations which undergo a different fate and ultimately lead to supracellular organization. The report of a transition between migratory modes [24] suggests that standing waves might also provide a "time-zero" for the synchronization of cell behavior. If we consider a growing system, for example, an embryo, below a certain size-threshold, in a regime of global oscillations, no node is present. As the embryo grows, such threshold is eventually reached, defining a precise moment (dependent on the system size) for the shift in cell behavior. Even though this is nothing but a hypothesis, it is interesting to notice that this intrinsic metric (~400 μm) matches the spacing between hair follicles in the mouse embryo [2] and feather buds during chicken embryogenesis [22], both organs whose correct development depends on coordinated cell migration.

This interpretation, even though extremely enticing, is still suffering from a general lack of documented biological consequences. If mechanical oscillations have been characterized on a number of different systems, their effects on the behavior of cells are still unclear. The obvious candidates for such role are those molecules involved in mechanotransduction, chief among which is YAP/TAZ [100]. This possibility has been recently explored [77], although a direct causal link between cell velocity and YAP translocation is still absent. Other promising molecules are β-catenin and merlin, for their mechanotransductive role in morphogenesis [101,102] and in collective orientation during CCM [103], respectively. The activity of ERK, whose wave-like activation has been shown to be concomitant with mechanical waves [76], might also be significant.

If the presence of a downstream biological effector could shed light on the possible function of these oscillations, understanding their origin remains an important perspective. Specifically, two issues seem to have the biggest relevance: the source of the active elasticity which seems to characterize the response of monolayers to stresses and the nature of the feedback mechanism allowing for the presence of sustained oscillations in biological (overdamped) systems. Regarding this second matter, most models rely on an interaction between cell polarity and another quantity, generally of a mechanical nature, making a direct measurement of stresses at cell–cell junctions [83] and a univocal way to assess cell polarity compelling future directions. Moreover, neither of these factors can be regarded at without considering the underlying architecture of the cytoskelton, which remains an extremely important player for supracellular signal propagation and behavioral control in general.

In conclusion, mechanical oscillations have been characterized by a variety of geometries and cell types, and their presence is now widely accepted. Even if differences between reports remain to be assessed, the many similarities seem to paint a common picture, each observation representing a single point of view, selected by the experimental perspective, rather than being a phenomenon in its own right. In the future, a full understanding of waves requires a better characterization of the underlying biological processes, both in terms of the components that make oscillations possible and a potential downstream signaling.

## References

[1] A.M. Turing, The chemical basis of morphogenesis, Philos. Trans. R. Soc. London, Ser. B, Biol. Sci. 237 (641) (1952) 37–72. Available from: http://www.royalsocietypublishing.org/doi/10.1098/rstb.1952.0012.

[2] J.D. Glover, K.L. Wells, F. Matthäus, K.J. Painter, W. Ho, J. Riddell, et al., Hierarchical patterning modes orchestrate hair follicle morphogenesis, PLoS Biol. 15 (7) (2017) e2002117. Available from: https://dx.plos.org/10.1371/journal.pbio.2002117.

[3] K.J. Painter, G.S. Hunt, K.L. Wells, J.A. Johansson, D.J. Headon, Towards an integrated experimental–theoretical approach for assessing the mechanistic basis of hair and feather morphogenesis, Interface Focus 2 (4) (2012) 433–450. Available from: https://doi.org/10.1098/rsfs.2011.0122.

[4] D. Dalle Nogare, A.B. Chitnis, Self-organizing spots get under your skin, PLoS Biol. 15 (12) (2017) e2004412. Available from: https://doi.org/10.1371/journal.pbio.2004412.

[5] C.M. Nelson, R.P. Jean, J.L. Tan, W.F. Liu, N.J. Sniadecki, A.A. Spector, et al., Emergent patterns of growth controlled by multicellular form and mechanics, Proc. Natl. Acad. Sci. U.S.A. 102 (33) (2005) 11594–11599. Available from: https://doi.org/10.1073/pnas.0502575102.

[6] S.J. Streichan, C.R. Hoerner, T. Schneidt, D. Holzer, L. Hufnagel, Spatial constraints control cell proliferation in tissues, Proc. Natl. Acad. Sci. U.S.A. 111 (15) (2014) 5586–5591. Available from: http://www.ncbi.nlm.nih.gov/pubmed/24706777.

[7] T.B. Saw, A. Doostmohammadi, V. Nier, L. Kocgozlu, S. Thampi, Y. Toyama, et al., Topological defects in epithelia govern cell death and extrusion, Nature 544 (7649) (2017) 212–216. Available from: http://www.nature.com/doifinder/10.1038/nature21718.

[8] M. Deforet, V. Hakim, H. Yevick, G. Duclos, P. Silberzan, Emergence of collective modes and tri-dimensional structures from epithelial confinement, Nat. Commun. 5 (1) (2014) 3747. Available from: https://www.nature.com/articles/ncomms4747.

[9] A.J. Engler, S. Sen, H.L. Sweeney, D.E. Discher, Matrix elasticity directs stem cell lineage specification, Cell 126 (4) (2006) 677–689. Available from: https://doi.org/10.1016/j.cell.2006.06.044.

[10] S.A. Ruiz, C.S. Chen, Emergence of patterned stem cell differentiation within multicellular structures, Stem Cell 26 (11) (2008) 2921–2927. Available from: http://doi.wiley.com/10.1634/stemcells.2008-0432.

[11] X. Xue, Y. Sun, A.M. Resto-Irizarry, Y. Yuan, K.M. Aw Yong, Y. Zheng, et al., Mechanics-guided embryonic patterning of neuroectoderm tissue from human pluripotent stem cells, Nat. Mater. 17 (7) (2018) 633–641. Available from: http://doi.org/10.1038/s41563-018-0082-9.

[12] R.K. Vedula, A. Ravasio, C.T. Lim, B. Ladoux, Collective cell migration: a mechanistic perspective, Physiology 28 (6) (2013) 370–379. Available from: https://doi.org/10.1152/physiol.00033.2013.

[13] E.H. Barriga, K. Franze, G. Charras, R. Mayor, Tissue stiffening coordinates morphogenesis by triggering collective cell migration in vivo, Nature 554 (7693) (2018) 523–527. Available from: http://doi.org/10.1038/nature25742.
[14] G.F. Oster, J.D. Murray, A.K. Harris, Mechanical aspects of mesenchymal morphogenesis, J. Embryol. Exp. Morphol. 78 (1983) 83–125. Available from: http://www.ncbi.nlm.nih.gov/pubmed/6663234.
[15] J.D. Murray, On the mechanochemical theory of biological pattern formation with application to vasculogenesis, C.R. Biol. 326 (2) (2003) 239–252. Available from: https://doi.org/10.1016/S1631-0691(03)00065-9.
[16] N. Gjorevski, C.M. Nelson, The mechanics of development: models and methods for tissue morphogenesis, Birth Defects Res. C Embryo Today 90 (3) (2010) 193–202. Available from: https://doi.org/10.1002/bdrc.20185.
[17] K.J. Painter, W. Ho, D.J. Headon, A chemotaxis model of feather primordia pattern formation during avian development, J. Theor. Biol. 437 (2018) 225–238. Available from: https://doi.org/10.1016/j.jtbi.2017.10.026.
[18] N. Desprat, W. Supatto, P.A. Pouille, E. Beaurepaire, E. Farge, Tissue deformation modulates twist expression to determine anterior midgut differentiation in drosophila embryos, Dev. Cell 15 (3) (2008) 470–477. Available from: https://doi.org/10.1016/j.devcel.2008.07.009.
[19] N. Gjorevski, C.M. Nelson, Endogenous patterns of mechanical stress are required for branching morphogenesis, Integr. Biol. (Camb.) 2 (9) (2010) 424–434. Available from: http://www.pubmedcentral.nih.gov/articlerender.fcgi?artid = 3074564&tool = pmcentrez&rendertype = abstract.
[20] A.E. Shyer, A.R. Rodrigues, G.G. Schroeder, E. Kassianidou, S. Kumar, R.M. Harland, Emergent cellular self-organization and mechanosensation initiate follicle pattern in the avian skin, Science 357 (6353) (2017) 811–815. Available from: http://www.sciencemag.org/lookup/doi/10.1126/science.aai7868.
[21] D. Gilmour, M. Rembold, M. Leptin, From morphogen to morphogenesis and back, Nature 541 (7637) (2017) 311–320. Available from: https://doi.org/10.1038/nature21348.
[22] W.K.W. Ho, L. Freem, D. Zhao, K.J. Painter, T.E. Woolley, E.A. Gaffney, et al., Feather arrays are patterned by interacting signalling and cell density waves, PLoS Biol. 17 (2) (2019) e3000132. Available from: http://dx.plos.org/10.1371/journal.pbio.3000132.
[23] C.P. Heisenberg, Y. Bellaïche, Forces in tissue morphogenesis and patterning, Cell 153 (5) (2013) 948–962. Available from: https://doi.org/10.1016/j.cell.2013.05.008.
[24] V. Petrolli, M. Le Goff, M. Tadrous, K. Martens, C. Allier, O. Mandula, et al., Confinement-induced transition between wave-like collective cell migration modes, Phys. Rev. Lett. 122 (16) (2019) 168101. Available from: https://link.aps.org/doi/10.1103/PhysRevLett.122.168101.
[25] J. Notbohm, S. Banerjee, K.J.C. Utuje, B. Gweon, H. Jang, Y. Park, et al., Cellular contraction and polarization drive collective cellular motion, Biophys. J. 110 (12) (2016) 2729–2738. Available from: http://doi.org/10.1016/j.bpj.2016.05.019.
[26] X. Serra-Picamal, V. Conte, R. Vincent, E. Anon, D.T. Tambe, E. Bazellieres, et al., Mechanical waves during tissue expansion, Nat. Phys. 8 (8) (2012) 628–634. Available from: http://www.nature.com/nphys/journal/v8/n8/pdf/nphys2355.pdf.
[27] D.A. Fletcher, R.D. Mullins, Cell mechanics and the cytoskeleton, Nature 463 (7280) (2010) 485–492. Available from: http://www.nature.com/articles/nature08908.
[28] C.C. DuFort, M.J. Paszek, V.M. Weaver, Balancing forces: architectural control of mechanotransduction, Nat. Rev. Mol. Cell Biol. 12 (5) (2011) 308–319. Available from: http://www.pubmedcentral.nih.gov/articlerender.fcgi?artid = 3564968&tool = pmcentrez&rendertype = abstract.

[29] B. Geiger, J.P. Spatz, A.D. Bershadsky, Environmental sensing through focal adhesions, Nat. Rev. Mol. Cell Biol. 10 (1) (2009) 21–33. Available from: http://www.nature.com/articles/nrm2593.
[30] T. Iskratsch, H. Wolfenson, M.P. Sheetz, Appreciating force and shape—the rise of mechanotransduction in cell biology, Nat. Rev. Mol. Cell Biol. 15 (12) (2014) 825–833. Available from: http://doi.org/10.1038/nrm3903.
[31] T.J.C. Harris, U. Tepass, Adherens junctions: from molecules to morphogenesis, Nat. Rev. Mol. Cell Biol. 11 (7) (2010) 502–514. Available from: http://www.nature.com/articles/nrm2927.
[32] M.A. Garcia, W.J. Nelson, N. Chavez, Cell-cell junctions organize structural and signaling networks, Cold Spring Harb. Perspect. Biol. 10 (4) (2018) a029181. Available from: http://cshperspectives.cshlp.org/lookup/doi/10.1101/cshperspect.a029181.
[33] F. Martino, A.R. Perestrelo, V. Vinarský, S. Pagliari, G. Forte, Cellular mechanotransduction: from tension to function, Front. Physiol. 9 (2018) 1–21. Available from: https://www.frontiersin.org/article/10.3389/fphys.2018.00824/full.
[34] P. Friedl, D. Gilmour, Collective cell migration in morphogenesis, regeneration and cancer, Nat. Rev. Mol. Cell Biol. 10 (7) (2009) 445–457. Available from: http://www.ncbi.nlm.nih.gov/pubmed/19546857.
[35] P. Friedl, S. Alexander, Cancer invasion and the microenvironment: plasticity and reciprocity, Cell 147 (5) (2011) 992–1009. Available from: http://doi.org/10.1016/j.cell.2011.11.016.
[36] R. Mayor, S. Etienne-Manneville, The front and rear of collective cell migration, Nat. Rev. Mol. Cell Biol. 17 (2) (2016) 97–109. Available from: http://www.nature.com/doifinder/10.1038/nrm.2015.14.
[37] P. Martin, Parallels between tissue repair and embryo morphogenesis, Development 131 (13) (2004) 3021–3034. Available from: http://dev.biologists.org/cgi/doi/10.1242/dev.01253.
[38] P. Hayes, J. Solon, Drosophila dorsal closure: an orchestra of forces to zip shut the embryo, Mech. Dev. 144 (2017) 2–10. Available from: https://linkinghub.elsevier.com/retrieve/pii/S0925477316300983.
[39] J.C. Yarrow, Z.E. Perlman, N.J. Westwood, T.J. Mitchison, A high-throughput cell migration assay using scratch wound healing, a comparison of image-based readout methods, BMC Biotechnol. 4 (1) (2004) 21.
[40] B. Ladoux, R.-M. Mège, Mechanobiology of collective cell behaviours, Nat. Rev. Mol. Cell Biol. 18 (12) (2017) 743–757. Available from: https://doi.org/10.1038/nrm.2017.98.
[41] A. Puliafito, L. Hufnagel, P. Neveu, S. Streichan, A. Sigal, D.K. Fygenson, et al., Collective and single cell behavior in epithelial contact inhibition, Proc. Natl. Acad. Sci. U.S.A. 109 (3) (2012) 739–744. Available from: http://www.pnas.org/cgi/doi/10.1073/pnas.1007809109.
[42] S. Garcia, E. Hannezo, J. Elgeti, J.-F. Joanny, P. Silberzan, N.S. Gov, Physics of active jamming during collective cellular motion in a monolayer, Proc. Natl. Acad. Sci. U.S.A. 112 (50) (2015) 15314–15319. Available from: http://www.pnas.org/lookup/doi/10.1073/pnas.1510973112.
[43] T.E. Angelini, E. Hannezo, X. Trepat, J.J. Fredberg, D.A. Weitz, Cell migration driven by cooperative substrate deformation patterns, Phys. Rev. Lett. 104 (16) (2010) 1–4. Available from: https://doi.org/10.1103/PhysRevLett.104.168104.
[44] K. Doxzen, S.R.K. Vedula, M.C. Leong, H. Hirata, N.S. Gov, A.J. Kabla, et al., Guidance of collective cell migration by substrate geometry, Integr. Biol. 5 (8) (2013) 1026. Available from: http://xlink.rsc.org/?DOI = c3ib40054a.
[45] L. Petitjean, M. Reffay, E. Grasland-Mongrain, M. Poujade, B. Ladoux, A. Buguin, et al., Velocity fields in a collectively migrating epithelium, Biophys. J. 98 (9) (2010) 1790–1800. Available from: http://doi.org/10.1016/j.bpj.2010.01.030.

[46] M. Sadati, N. Taheri Qazvini, R. Krishnan, C.Y. Park, J.J. Fredberg, Collective migration and cell jamming, Differentiation 86 (3) (2013) 121–125. Available from: https://linkinghub.elsevier.com/retrieve/pii/S0301468113000170.
[47] M.R. Ng, A. Besser, G. Danuser, J.S. Brugge, Substrate stiffness regulates cadherin-dependent collective migration through myosin-II contractility, J. Cell Biol. 199 (3) (2012) 545–563. Available from: https://doi.org/10.1083/jcb.201207148.
[48] T. Chen, T.B. Saw, R.-M. Mège, B. Ladoux, Mechanical forces in cell monolayers, J. Cell Sci. 131 (24) (2018) jcs218156. Available from: https://doi.org/10.1242/jcs.218156.
[49] P. Rodríguez-Franco, A. Brugués, A. Marín-Llauradó, V. Conte, G. Solanas, E. Batlle, et al., Long-lived force patterns and deformation waves at repulsive epithelial boundaries, Nat. Mater. 16 (10) (2017) 1029–1037. Available from: http://www.nature.com/articles/nmat4972.
[50] S.R.K. Vedula, M.C. Leong, T.L. Lai, P. Hersen, A.J. Kabla, C.T. Lim, et al., Emerging modes of collective cell migration induced by geometrical constraints, Proc. Natl. Acad. Sci. U.S.A. 109 (32) (2012) 12974–12979. Available from: https://doi.org/10.1073/pnas.1119313109.
[51] S. Tlili, E. Gauquelin, B. Li, O. Cardoso, B. Ladoux, H. Delanoë-Ayari, et al., Collective cell migration without proliferation: density determines cell velocity and wave velocity, R. Soc. Open Sci. 5 (5) (2018) 172421. Available from: http://rsos.royalsocietypublishing.org/lookup/doi/10.1098/rsos.172421.
[52] L. Kocgozlu, T.B. Saw, A.P. Le, I. Yow, M. Shagirov, E. Wong, et al., Epithelial cell packing induces distinct modes of cell extrusions, Curr. Biol. 26 (21) (2016) 2942–2950. Available from: https://doi.org/10.1016/j.cub.2016.08.057.
[53] D.L. Nikolić, A.N. Boettiger, D. Bar-Sagi, J.D. Carbeck, S.Y. Shvartsman, Role of boundary conditions in an experimental model of epithelial wound healing, Am. J. Physiol. Cell Physiol. 291 (1) (2006). Available from: https://doi.org/10.1152/ajpcell.00411.2005.
[54] E. Ostuni, R. Kane, C.S. Chen, D.E. Ingber, G.M. Whitesides, Patterning mammalian cells using elastomeric membranes, Langmuir 16 (20) (2000) 7811–7819.
[55] M. Poujade, E. Grasland-Mongrain, A. Hertzog, J. Jouanneau, P. Chavrier, B. Ladoux, et al., Collective migration of an epithelial monolayer in response to a model wound, Proc. Natl. Acad. Sci. U.S.A. 104 (41) (2007) 15988–15993. Available from: http://www.pubmedcentral.nih.gov/articlerender.fcgi?artid = 2042149&tool = pmcentrez&rendertype = abstract.
[56] M. Bindschadler, J.L. McGrath, Sheet migration by wounded monolayers as an emergent property of single-cell dynamics, J. Cell Sci. 120 (5) (2007) 876–884. Available from: https://doi.org/10.1242/jcs.03395.
[57] C. De Pascalis, S. Etienne-Manneville, Single and collective cell migration: the mechanics of adhesions, Mol. Biol. Cell 28 (14) (2017) 1833–1846. Available from: https://doi.org/10.1091/mbc.e17-03-0134.
[58] X. Trepat, M.R. Wasserman, T.E. Angelini, E. Millet, D.A. Weitz, J.P. Butler, et al., Physical forces during collective cell migration, Nat. Phys. 5 (6) (2009) 426–430. Available from: http://www.scopus.com/inward/record.url?eid = 2-s2.0-67650227472&partnerID = tZOtx3y1.
[59] D.T. Tambe, C.C. Hardin, T.E. Angelini, K. Rajendran, C.Y. Park, X. Serra-Picamal, et al., Collective cell guidance by cooperative intercellular forces, Nat. Mater. 10 (6) (2011) 469–475. Available from: http://doi.org/10.1038/nmat3025.
[60] D.T. Tambe, U. Croutelle, X. Trepat, C.Y. Park, J.H. Kim, E. Millet, et al., Monolayer stress microscopy: limitations, artifacts, and accuracy of recovered intercellular stresses, PLoS One 8 (2) (2013) 12. Available from: https://doi.org/10.1371/journal.pone.0055172.

[61] C. Gayrard, C. Bernaudin, T. Déjardin, C. Seiler, N. Borghi, Src- and confinement-dependent FAK activation causes E-cadherin relaxation and β-catenin activity, J. Cell Biol. 217 (3) (2018) 1063–1077. Available from: https://doi.org/10.1083/jcb.201706013.
[62] X. Trepat, J.J. Fredberg, Plithotaxis and emergent dynamics in collective cellular migration, Trends Cell Biol. 21 (11) (2011) 638–646. Available from: http://doi.org/10.1016/j.tcb.2011.06.006.
[63] M. Théry, Wave of migration, Nat. Phys. 8 (8) (2012) 583–584. Available from: http://www.nature.com/articles/nphys2374.
[64] T.D. Yang, H. Kim, C. Yoon, S.K. Baek, K.J. Lee, Collective pulsatile expansion and swirls in proliferating tumor tissue, N. J. Phys. 18 (10) (2016). Available from: https://doi.org/10.1088/1367-2630/18/10/103032.
[65] A. Zaritsky, D. Kaplan, I. Hecht, S. Natan, L. Wolf, N.S. Gov, et al., Propagating waves of directionality and coordination orchestrate collective cell migration, PLoS Comput. Biol. 10 (7) (2014) 15–19. Available from: https://doi.org/10.1371/journal.pcbi.1003747.
[66] G. Duclos, S. Garcia, H.G. Yevick, P. Silberzan, Perfect nematic order in confined monolayers of spindle-shaped cells, Soft Matter 10 (14) (2014) 2346–2353. Available from: https://doi.org/10.1039/c3sm52323c.
[67] G. Duclos, C. Blanch Mercader, V. Yashunsky, G. Salbreux, J. Prost, J.-F. Joanny, et al., Spontaneous shear flow in confined cellular nematics, Nat. Phys. 14 (2018). Available from: http://doi.org/10.1038/s41567-018-0099-7.
[68] M.J. Raymond, P. Ray, G. Kaur, A.V. Singh, L.Q. Wan, Cellular and nuclear alignment analysis for determining epithelial cell chirality, Ann. Biomed. Eng. 44 (5) (2016) 1475–1486. Available from: https://doi.org/10.1007/s10439-015-1431-3.
[69] N.D. Bade, R.D. Kamien, R.K. Assoian, K.J. Stebe, Edges impose planar alignment in nematic monolayers by directing cell elongation and enhancing migration, Soft Matter 14 (33) (2018) 6867–6874. Available from: https://doi.org/10.1039/c8sm00612a.
[70] C. Brangwynne, S. Huang, K.K. Parker, D.E. Ingber, Symmetry breaking in cultured mammalian cells, In Vitro Cell. Dev. Biol.—Anim. 36 (9) (2000) 563–565.
[71] S. Huang, C.P. Brangwynne, K.K. Parker, D.E. Ingber, Symmetry-breaking in mammalian cell cohort migration during tissue pattern formation: role of random-walk persistence, Cell Motil. Cytoskeleton 61 (4) (2005) 201–213. Available from: https://doi.org/10.1002/cm.20077.
[72] F.J. Segerer, F. Thüroff, A. Piera Alberola, E. Frey, J.O. Rädler, Emergence and persistence of collective cell migration on small circular micropatterns, Phys. Rev. Lett. 114 (22) (2015) 228102. Available from: https://link.aps.org/doi/10.1103/PhysRevLett.114.228102.
[73] A.K. Marel, M. Zorn, C. Klingner, R. Wedlich-Söldner, E. Frey, J.O. Rädler, Flow and diffusion in channel-guided cell migration, Biophys. J. 107 (5) (2014) 1054–1064. Available from: https://doi.org/10.1016/j.bpj.2014.07.017.
[74] H.G. Yevick, G. Duclos, I. Bonnet, P. Silberzan, Architecture and migration of an epithelium on a cylindrical wire, Proc. Natl. Acad. Sci. U.S.A. 112 (19) (2015) 5944–5949. Available from: https://doi.org/10.1073/pnas.1418857112.
[75] W. Xi, S. Sonam, T.B. Saw, B. Ladoux, C.T. Lim, Emergent patterns of collective cell migration under tubular confinement, Nat. Commun. 8 (1) (2017) 1517. Available from: http://www.nature.com/articles/s41467-017-01390-x.
[76] K. Aoki, Y. Kondo, H. Naoki, T. Hiratsuka, R.E. Itoh, M. Matsuda, Propagating wave of ERK activation orients collective cell migration, Dev. Cell 43 (3) (2017) 305–317. Available from: https://doi.org/10.1016/j.devcel.2017.10.016.
[77] G. Peyret, R. Mueller, J. D'Alessandro, S. Begnaud, P. Marcq, R.-M. Mège, et al., Sustained oscillations of epithelial cell sheets, Biophys. J. 117 (3) (2019) 464–478. Available from: https://linkinghub.elsevier.com/retrieve/pii/S0006349519305028.

[78] J.H. Kim, X. Serra-Picamal, D.T. Tambe, E.H. Zhou, C.Y. Park, M. Sadati, et al., Propulsion and navigation within the advancing monolayer sheet, Nat. Mater. 12 (9) (2013) 856–863. Available from: http://doi.org/10.1038/nmat3689.
[79] N.S. Rossen, J.M. Tarp, J. Mathiesen, M.H. Jensen, L.B. Oddershede, Long-range ordered vorticity patterns in living tissue induced by cell division, Nat. Commun. 5 (2014) 1–7. Available from: http://doi.org/10.1038/ncomms6720.
[80] A. Doostmohammadi, S.P. Thampi, T.B. Saw, C.T. Lim, B. Ladoux, J.M. Yeomans, Celebrating Soft Matter's 10th anniversary: cell division: a source of active stress in cellular monolayers, Soft Matter 11 (2015) 7328–7336. Available from: https://doi.org/10.1039/c5sm01382h.
[81] S.M. Zehnder, M. Suaris, M.M. Bellaire, T.E. Angelini, Cell volume fluctuations in MDCK monolayers, Biophys. J. 108 (2) (2015) 247–250. Available from: http://doi.org/10.1016/j.bpj.2014.11.1856.
[82] R. Vincent, E. Bazellières, C. Pérez-González, M. Uroz, X. Serra-Picamal, X. Trepat, Active tensile modulus of an epithelial monolayer, Phys. Rev. Lett. 115 (24) (2015) 248103. Available from: https://link.aps.org/doi/10.1103/PhysRevLett.115.248103.
[83] L. Kurzawa, M. Balland, Lost in mechanobiology, what's next?: missing tools related to the physics of the system, Biol. Cell 111 (8) (2019) 213–215. Available from: https://doi.org/10.1111/boc.201900001.
[84] C. Malinverno, S. Corallino, F. Giavazzi, M. Bergert, Q. Li, M. Leoni, et al., Endocytic reawakening of motility in jammed epithelia, Nat. Mater. 16 (5) (2017) 587–596. Available from: https://doi.org/10.1038/nmat4848.
[85] J.H. Kim, K. Kushiro, N.A. Graham, A.R. Asthagiri, Tunable interplay between epidermal growth factor and cell-cell contact governs the spatial dynamics of epithelial growth, Proc. Natl. Acad. Sci. U.S.A. 106 (27) (2009) 11149–11153. Available from: https://doi.org/10.1073/pnas.0812651106.
[86] A.R. Harris, L. Peter, J. Bellis, B. Baum, A.J. Kabla, G.T. Charras, Characterizing the mechanics of cultured cell monolayers, Proc. Natl. Acad. Sci. U.S.A. 109 (41) (2012) 16449–16454. Available from: https://doi.org/10.1073/pnas.1213301109.
[87] C. Bertet, L. Sulak, T. Lecuit, Myosin-dependent junction remodelling controls planar cell intercalation and axis elongation, Nature 429 (2004) 667–671. Available from: https://doi.org/10.1038/nature02590.
[88] T.P.J. Wyatt, A.R. Harris, M. Lam, Q. Cheng, J. Bellis, A. Dimitracopoulos, et al., Emergence of homeostatic epithelial packing and stress dissipation through divisions oriented along the long cell axis, Proc. Natl. Acad. Sci. U.S.A. 112 (18) (2015) 5726–5731. Available from: http://www.pnas.org/content/112/18/5726.short.
[89] A.F. Mertz, S. Banerjee, Y. Che, G.K. German, Y. Xu, C. Hyland, et al., Scaling of traction forces with the size of cohesive cell colonies, Phys. Rev. Lett. 108 (19) (2012) 198101. Available from: https://link.aps.org/doi/10.1103/PhysRevLett.108.198101.
[90] S.R.K. Vedula, H. Hirata, M.H. Nai, A. Brugués, Y. Toyama, X. Trepat, et al., Epithelial bridges maintain tissue integrity during collective cell migration, Nat. Mater. 13 (1) (2014) 87–96. Available from: https://doi.org/10.1038/nmat3814.
[91] E. Latorre, S. Kale, L. Casares, M. Gómez-González, M. Uroz, L. Valon, et al., Active superelasticity in three-dimensional epithelia of controlled shape, Nature 563 (7730) (2018) 203–208. Available from: http://www.nature.com/articles/s41586-018-0671-4.
[92] V. Hakim, P. Silberzan, Collective cell migration: a physics perspective, Rep. Prog. Phys. 80 (7) (2017) 076601. Available from: http://stacks.iop.org/0034-4885/80/i = 7/a = 076601?key = crossref.b018e931a71e958cf98caf654df4d881.
[93] S. Banerjee, M.C. Marchetti, Continuum models of collective cell migration, Adv. Exp. Med. Biol. 1146 (2019) 45–66. Available from: http://doi.org/10.1007/978-3-030-17593-1_4.

[94] R. Alert, X. Trepat, Physical models of collective cell migration, Annu. Rev. Condens. Matter Phys. 11 (1) (2020) 77–101. Available from: https://doi.org/10.1146/annurev-conmatphys-031218-013516.
[95] T. Vicsek, A. Czirk, E. Ben-Jacob, I. Cohen, O. Shochet, Novel type of phase transition in a system of self-driven particles, Phys. Rev. Lett. 75 (6–7) (1995). Available from: https://doi.org/10.1103/PhysRevLett.75.1226.
[96] F. Giavazzi, M. Paoluzzi, M. Macchi, D. Bi, G. Scita, M.L. Manning, et al., Flocking transitions in confluent tissues, Soft Matter 14 (18) (2018) 3471–3477. Available from: https://doi.org/10.1039/c8sm00126j.
[97] S. Banerjee, K.J.C. Utuje, M.C. Marchetti, Propagating stress waves during epithelial expansion, Phys. Rev. Lett. 114 (22) (2015). Available from: https://doi.org/10.1103/PhysRevLett.114.228101.
[98] M. Vishwakarma, J. Di Russo, D. Probst, U.S. Schwarz, T. Das, J.P. Spatz, Mechanical interactions among followers determine the emergence of leaders in migrating epithelial cell collectives, Nat. Commun. 9 (1) (2018) 3469. Available from: 10.1038/s41467-018-05927-6.
[99] R. Sunyer, V. Conte, J. Escribano, A. Elosegui-Artola, A. Labernadie, L. Valon, et al., Collective cell durotaxis emerges from long-range intercellular force transmission, Science 353 (6304) (2016) 1157–1161. Available from: http://www.sciencemag.org/cgi/doi/10.1126/science.aaf9754. <http://www.sciencemag.org/cgi/doi/10.1126/science.aaf7119>.
[100] T. Panciera, L. Azzolin, M. Cordenonsi, S. Piccolo, Mechanobiology of YAP and TAZ in physiology and disease, Nat. Rev. Mol. Cell Biol. 18 (12) (2017) 758–770. Available from: http://doi.org/10.1038/nrm.2017.87.
[101] S.G. Pai, B.A. Carneiro, J.M. Mota, R. Costa, C.A. Leite, R. Barroso-Sousa, et al., Wnt/beta-catenin pathway: modulating anticancer immune response, J. Hematol. Oncol. 10 (1) (2017) 1–12. Available from: https://doi.org/10.1186/s13045-017-0471-6.
[102] D. Pinheiro, Y. Bellaïche, Mechanical force-driven adherens junction remodeling and epithelial dynamics, Dev. Cell 47 (1) (2018) 3–19. Available from: https://doi.org/10.1016/j.devcel.2018.09.014.
[103] T. Das, K. Safferling, S. Rausch, N. Grabe, H. Boehm, J.P. Spatz, A molecular mechanotransduction pathway regulates collective migration of epithelial cells, Nat. Cell Biol. 17 (3) (2015) 276–287. Available from: http://www.nature.com/doifinder/10.1038/ncb3115.

CHAPTER EIGHT

# Flow dynamics of 3D multicellular systems into capillaries

**Karine Guevorkian[1], Françoise Brochard-Wyart[1] and David Gonzalez-Rodriguez[2]**
[1]Laboratoire Physico-Chimie Curie, Institut Curie, PSL Research University, Sorbonne Université, CNRS, Paris, France
[2]Université de Lorraine, LCP-A2MC, F-57000 Metz, France

## 8.1 Introduction

The life of a multicellular organism starts from a single cell, which grows, divides, differentiates, and forms various organs. Throughout this journey, cells experience mechanical cues such as internal or external forces, and geometrical constraints, affecting their motion and function. Hence, understanding how multicellular structures respond to mechanical signals is crucial to decipher mechanisms governing embryonic development or pathological events such as tumor formation and metastasis [1−3]. Mechanical signals induce cell shape changes, which in turn trigger biochemical responses, generating feedback loops [2,4]. During embryonic morphogenesis, forces at the tissue scale lead to cellular rearrangements, giving rise to collective responses, as observed in early stages of *Xenopus* [5,6] and *Drosophilla* [7,8] development, during embryonic elongation in *C. elegans* [9,10], as well as in chicken and zebrafish embryos [11−13].

The response of a material to an applied force is governed by its constitutive material properties. Multicellular living matter exhibits complex rheological behaviors [14,15], such as liquid-like [16] or solid-like [5,17,18] viscoelasticity, which determine their function [19]. In a seminal work, Steinberg and coworkers proposed an analogy between tissues and liquids to attribute a surface tension to soft embryonic tissues [16], which explained their mutual enveloping capacity [20]. In this picture, the surface tension is a direct consequence of cellular cohesion, which may be

*Viscoelasticity and Collective Cell Migration.*
DOI: https://doi.org/10.1016/B978-0-12-820310-1.00008-2

modified by individual cellular cortical tension [21]. Similarly, viscosity can be defined for liquid-like tissues as a measure of the resistance to cellular rearrangement in a stressed tissue [22,23]. In vivo tissues actively adjust their rheological properties, in space and time, to properly execute morphogenetic events [14,24–27], making the cross-talk between signaling and mechanics very rich to explore.

The experimental techniques to study tissue mechanics can be divided into three major groups. The first group consists of techniques that provide information about local material properties of the tissue and hence can be used to study rheological variations over the cell scale. Such techniques include atomic force microscopy [28] and active or passive microrheology [29–32]. The second group of techniques probe the intermediate scale between the cell and the tissue. They consist of cell-size deformable probes, inserted into a cellular aggregate of embryo to characterize local mechanical properties of the tissue [33–35]. Lastly, the third group of techniques measures global tissue mechanical properties by applying a controlled pull [36,37], by squeezing them between two plates at an imposed strain [16,23,38], or by aspirating them inside capillary tubes at a controlled stress [39,40].

In this chapter, we will focus on the technique of micropipette aspiration, which is a versatile tool [41–43] to study the mechanics of soft and biological systems over a large range of length scales spanning cell nucleus [44], single cell [45,46], developing mouse oocyte [47], cellular aggregates [40], tissues explants [39], and in vivo organs [48]. We will first detail key practical considerations to setup the technique, and then present a case study where micropipette aspiration serves to characterize the mechanical properties of spherical cellular aggregates. We will discuss the manifestations of cell activity in response to the applied stress, which modify the aggregate's rheological behavior. Then, we will describe how to characterize multicellular aggregate permeability by using a microfluidic system. We will close the chapter by discussing application perspectives of the aspiration technique.

## 8.2 Micropipette aspiration technique: a practical guide

We will open this chapter by giving a hands-on step-by-step guide for setting up a micropipette aspiration system, to study the mechanical properties of multicellular systems such as cellular aggregates or tissue explants. The technique is rather simple and inexpensive, within the reach of biological laboratories. Moreover, as we will show later, the data analysis does not depend on the precise shape of the sample (cell aggregate, tissue explant, etc.), as long as the sample is large enough compared to the micropipette diameter, which facilitates the application of the technique to ex vivo tissue explants (Fig. 8.1D), organoids, and potentially to in vivo tissues.

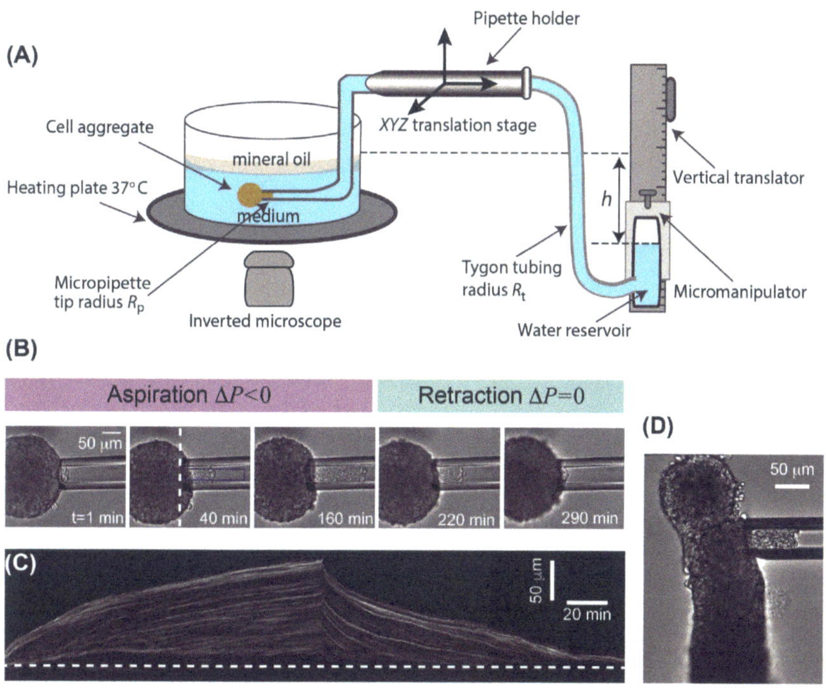


**Figure 8.1** Micropipette aspiration technique. (A) The experimental setup. (B) Time series of the aspiration of an aggregate into a pipette at a pressure 1470 Pa (15 cm $H_2O$) for almost 3 hours, and retraction of the tongue as the aspiration pressure is set to zero. (C) Kymograph corresponding to the rectangular box in (B). The dashed line presents the entrance of the pipette, (D) Aspiration of the presomitic mesodermal tissue from a chicken embryo. *Images (B–C) are reproduced from K. Guevorkian, J.-L. Maître, Micropipette aspiration: a unique tool for exploring cell and tissue mechanics in vivo, in: Cell Polarity and Morphogenesis, Methods in Cell Biology, vol. 139, Academic Press, 2017, pp. 187–201 with permission.*

## 8.2.1 Preparation and mounting of glass micropipettes

The central element of this system is the micropipette. Therefore great care should be taken to obtain micropipettes with straight tips and long tapers to ensure the uniformity of the applied stress to the tissue. Conic tapers change the dynamics of aspiration and complicate the interpretation of the data. Preparation of the micropipettes takes place in two steps. First, the glass capillaries are extruded using a laser puller (Sutter Instrument, USA), and then the tip is scored at the desired diameter, either manually or by using a microforge instrument (Narishige, Japan). For multicellular samples, such as tissue explants or cell aggregates, the diameter of the micropipette tip should be at least 30–40 $\mu$m (several cell diameter) and can go up to few hundred micrometers if the sample is large enough. A microforge can be used to obtain diameters up to 50 $\mu$m, but for larger diameters, the tip should be scored manually, either using a ceramic tile (Sutter Instrument) or a second micropipette. Once the tip is scored, it should be fire-polished by gently approaching the tip to the heated wire of the microforge. Further information on the laser-puller parameters and scoring techniques can be found in The Pipette Cookbook [49]. We have found that borosilicate glass capillaries with 0.5 mm/1 mm (inner/outer) diameter yield the best tips for our experiments.

The shape of the micropipettes can be adapted to fit into the experimental chamber by simply heating the capillary with a flame or heated wire. For example, to introduce the pipette into a Petri dish, it should be bent in a step shape, as shown schematically in Fig. 8.1A. Simply hold the pipette with the desired bending location above a flame and let it bent under the action of its weight. This step should be performed before the passivation step.

Once the micropipettes are pulled and scored, they need to be passivated to prevent cells from adhering to the surface of the glass. To this end, we freshly prepare 0.1 mg/mL PolyEthyleneGlycol-PolyLysin (PLL (20)-g[3.5]-PEG(2), SuSos AG, Switzerland) solution in 7.3 pH HEPES from a 1 mg/mL stock solution. Note that this diluted solution can be kept at 4°C for up to 1 week. The coating is considerably enhanced if the pipette surface is activated in a plasma cleaner before the PEG-PLL treatment. For large tips such as the ones used here, spontaneous capillary suction suffices to fill them with PEG-PLL. A piece of Parafilm is placed in a large Petri dish, and 30–50 $\mu$L PEG-PLL drops are deposited on the

Parafilm. With the help of a spacer, which can be made of modeling paste, the pipettes are positioned slightly tilted, with one pipette tip in each drop. One can verify that liquid runs up the pipette tip within few seconds. To preserve the humidity, small pieces of wet paper tissues are placed in the Petri dish, which is then closed with the cover. For the passivation to take effect, this setup should be kept for 1 hour at room temperature. For an experiment the following day, micropipettes can be left in the drops overnight at 4°C, but care should be taken to prevent the drops from drying up. Before use, the pipettes are rinsed with PBS solution.

To connect the micropipettes to the pressure system, a micropipette holder (Narishige, Japan) is used, which in turn is mounted on an XYZ translation stage, as shown in Fig. 8.1A. The other end of the micropipette holder is connected to the water reservoir via Tygon tubing. The three-dimensional micromanipulator is used to navigate the pipette inside the experimental chamber. Once the sample to be aspirated has been located, a very small pressure is applied by lowering the pressure system just enough to hold the sample by sucking it slightly. Then we raise the pipette a few hundred micrometers above the bottom of the chamber to prevent contact between the sample and the bottom of the chamber during the experiment.

### 8.2.2 Experimental chamber

The choice of the experimental chamber depends on the volume of the medium available for experimentation. For small volume, a sheet of parafilm can be folded several times to give a U-shaped band of 5 mm × 5 mm × 5 cm. The band is then sandwiched in between the coverslip and the slide by gently heating the slide on a heating plate and compressing the system from the top. To assure the sealing, VALAP (Vaseline, Lanolin, Paraffin 1:1:1) can be used at the contours. The aggregates and the micropipette are then introduced from the open side, which is later sealed by injection of mineral oil (M5904, Sigma-Aldrich) at the interface to prevent evaporation.

If the volume of the solution is not a limiting factor, we suggest using Petri dishes as experimental chambers, because the aggregates (or tissue explants) survive longer in larger volumes of culture medium. In this geometry, the micropipette needs to be bent to enter the dish, as discussed above. Once the pipette is positioned, the top of the dish is covered with a thin layer of mineral oil.

### 8.2.3 Pressure system

In our experiments we use a water reservoir system to create a suction pressure, as shown in Fig. 8.1A. A 50 mL home-made plastic or plexiglass reservoir, connected with a Tygon tubing to the micropipette holder, is mounted on a micromanipulator, which in turn is mounted on a vertical linear translator. The micromanipulator allows precise positioning of the water reservoir, of particular importance for locating the null pressure. The vertical translator with a large step size (UniSlide Rapid advance, Velmex Inc., USA) is used to change the pressure quickly.

Before each experiment, it is essential to level the height of the water reservoir with the liquid in the chamber, such that there is no pressure difference. The position of the *zero* pressure is found by locating the exact position of the water reservoir where a trapped small debris in the micropipette does not get sucked in or pushed out. A pressure difference is applied by simply lowering the reservoir by a distance $h$, which yields a suction pressure: $\Delta P = \rho g h$, where $\rho$ is the water (medium) density $\approx 1000\ kg/m^3$, and $g = 9.81\ m/s^2$ is the gravitational acceleration. We note that the volume of the thin layer of oil is negligible and does not change the density.

In this system the accuracy of the applied pressure is around 20 Pa, which is much smaller than the suction pressures used for soft tissues (ranging from 100 Pa to 5 kPa). An alternative to the water reservoir system is a microfluidic pump, as has been described elsewhere [42].

### 8.2.4 Aggregate preparation

In the experiments described in this chapter, murin sarcoma (S180) cell lines transfected to express E-Cadherin molecules [50] were used. This technique to produce spherical aggregates can be applied to other cell lines as well. The cells were cultured under 5% $CO_2$ atmosphere in DMEM enriched with 10% calf serum. Confluent cells were detached by incubation in Trypsin/EDTA for 5 minutes, diluted in $CO_2$ equilibrated culture medium, and centrifuged for 3 minutes at 1300 rpm. The pellets were resuspended in 5 mL of $CO_2$ equilibrated medium, at a concentration of 400,000 cells/mL, in 25 mL Erlenmeyer flasks that are placed in an orbital shaker at 75 rpm at 37°C for 18–24 hours, similar to procedures described in Ryan et al. [51]. With this technique, spherical aggregates of sizes varying from 100 to 500 $\mu m$ in diameter were obtained. Before each experiment, aggregates were resuspended in

$CO_2$-independent medium enriched with glucose and 10% calf serum and transferred into the experimental chamber. Spherical cellular aggregates can also be prepared using the traditional hanging-drop technique [52] or microwell systems [53].

### 8.2.5 Video microscopy and image analysis

The video imaging is performed using an inverted microscope, functioning in bright field, equipped with $\times 10$ or $\times 20$ air objectives and a CCD camera. To observe cell rearrangements, confocal microscopy can be used as well. A heated microscope stage is used to keep the sample at 37°C. The image acquisition frequency varies between one frame per second to one frame per 30 seconds depending on the speed of aspiration and retraction. Fig. 8.1B shows a typical image sequence of an aspiration and retraction cycle, whereas Fig. 8.1D demonstrates the applicability of the technique to a mesoderm explant dissected from a 2-day chicken embryo.

To analyze the dynamics of the advancing tongue inside the pipette, manual tracking plugins available in Fiji (Image J) or custom-made automated routines can be used. A rapid and efficient way of observing and analyzing the dynamics is to use the kymograph modules available in Fiji (Image J) software as is shown in Fig. 8.1C for a typical aspiration/retraction cycle. A rectangular region is selected representing the whole aspirated tongue. Then, the Reslice function is used to produce a kymograph for each horizontal line in the form of a stack. To obtain the well-contrasted image presented in Fig. 8.1C, the Standard Deviation function in Fiji is applied to the stacks. Next, an edge-detection algorithm readily gives the aspiration and retraction profiles as a function of time. We have used this methodology to analyze the shivering profiles, as will later be discussed (Fig. 8.3A).

### 8.2.6 Microfluidic channels, alternatives to micropipettes

Besides their extensive use in single-cell studies, microfluidic channels can be used to study the flow properties of multicellular systems in confined geometries. Microfluidics have the advantage of providing channels of reproducible, precise dimensions as well as more complex geometries, like the one discussed in Section 8.6 to measure the permeability of cellular aggregates. However, in using these channels, few corrections should be brought to the analysis represented in the next section, to take into account the noncircular cross-section of the channel, and the asymmetry imposed on the flow of the aggregate due to the flattening of the side of the sample facing the coverslip.

## 8.3 Viscoelastic behavior of cellular aggregates

Here we summarize how the micropipette aspiration technique reveals the viscoelastic characteristic of some embryonic tissues and tumors. An elastic material builds up stress upon deformation, and it regains its initial shape when the applied force is removed. The ratio of the applied stress (force per area), $\sigma$, to strain (deformation), $\varepsilon$, gives the elastic modulus, $E = \sigma/\varepsilon$. Thus a higher $E$ denotes a material that is less deformable. In contrast, a viscous material deforms as long as the force is applied and does not regain its original shape once the force is removed. The dynamic viscosity coefficient is given by $\eta = \sigma/\dot{\varepsilon}$, where $\dot{\varepsilon}$ denotes the strain rate. Complex fluids, like living tissues, exhibit viscoelastic properties [23,55], meaning that unlike purely elastic or viscous materials, the nature of their response to an applied stress is time-dependent. The simplest type of viscoelastic behavior exhibits a characteristic time, $\tau \approx \eta/E$, which describes the transition from a short-time elastic regime into a long-time viscous regime.

The elastic and viscous properties of a tissue depend on several factors. The elasticity mainly depends on the structure of the extracellular matrix, the cell–cell adhesion, and the cellular cytoskeleton. The viscosity of a tissue originates from the dissipation due to cellular rearrangements. Moreover, when embryonic tissue explants are excised from the embryo, they round up to a final spherical form, like liquid drops, to minimize their surface energy. This is a manifestation of an effective surface tension [16,23], which has been related to the strength of cell–cell adhesion, for example, by modulating the expression of cadherin molecules, and to cortical tension [20,21].

To quantify viscoelastic parameters, spherical cellular aggregates are used as in vitro model systems for epithelial tissue. The advantages of these systems are that one can obtain aggregates of controlled and variable sizes, and modulate the strength of cell–cell adhesion by using cell lines expressing different levels of cadherin molecules.

### 8.3.1 Aspiration force and surface tension

Liquids tend to minimize the area of their free surface under the action of their surface tension. This is why an oil droplet in water will take a spherical shape. Cellular aggregates display a similar behavior, characterized by an effective surface tension arising from intercellular adhesion and cellular cortical tension [16,23]. The aspiration experiment shown in Fig. 8.1B–D demonstrates that once the aspiration pressure is released, the aspirated tongue is slowly reabsorbed inside the aggregate. Here, the restoring force is the effective surface tension, which is also the driving force in rounding experiments [56].

In the following we will derive the equations of a viscoelastic liquid model describing the aspiration of a cellular aggregate into a micropipette. We start by quantifying the net aspiration force deforming the aggregate, which, as we will see, is modified by the effect of its surface tension.

Let us consider an aggregate of resting radius $R_0$. We apply an aspiration pressure, $\Delta P$, with a pipette of radius $R_p$, which corresponds to a total aspiration force $f_a = \pi R_p^2 \Delta P$. For large pressures, the aspirated length, $L$, inside the pipette will be larger than $R_p$, as shown schematically in Fig. 8.2A. In this case, the free energy of the system, $F$, is given by:

$$F = (4\pi R^2 + 2\pi R_p L)\gamma - \Delta P \pi R_p^2 L, \tag{8.1}$$

where $\gamma$ is the effective surface tension of the aggregate, and $R$ is the radius of the aggregate. Considering that volume is conserved during aspiration, we obtain:

$$\frac{4}{3}\pi R_0^3 = \frac{4}{3}\pi R^3 + \pi R_p^2 L \Rightarrow 4\pi R^2 \delta R + \pi R_p^2 \delta L = 0. \tag{8.2}$$

Therefore the effective aspiration force that contributes to aggregate deformation, $f = -\frac{\delta F}{\delta L}|_V$, is given by:

$$f = \pi R_p^2 (\Delta P - \Delta P_c), \tag{8.3}$$

where $\Delta P_c = 2\gamma\left(\frac{1}{R_p} - \frac{1}{R}\right)$ corresponds to Laplace's law and denotes a critical pressure above which the aggregate will flow inside the pipette. For $R \gg R_p$, the radius $R$ of the aggregate can be considered constant, $R \approx R_0$, $\Delta P_c$ approximated by:

$$\Delta P_c = 2\gamma\left(\frac{1}{R_p} - \frac{1}{R_0}\right). \tag{8.4}$$

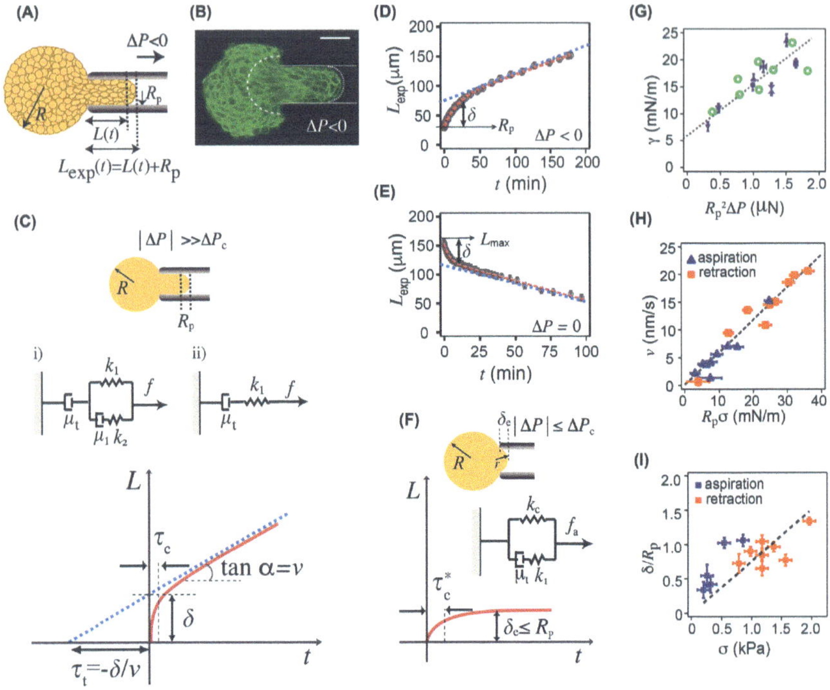

**Figure 8.2** Rheology of cellular aggregates. (A) Schematic representation of an aspirated aggregate. (B) Confocal image of an aspirated aggregate, where the E-cadherins are tagged with GFP. The stretched cells can be observed inside the pipette and in the deformation zone within a radius $R_p$ into the aggregate body depicted with the white dashed line. The pointed line shows the hemispherical cap, whose formation indicates the full development of the capillary resistance. $\Delta P = 500$ Pa, scale bar is 50 $\mu$m. (C) Schematic diagram showing the experimentally measurable parameters, and the rheological models for $\Delta P \gg \Delta P_c$. (i) the complete model with tissue and cell viscoelasticity (modified Kelvin–Voigt model), and (ii) a simplified model that neglects cell viscoelasticity (Maxwell liquid model). (D) Aspiration and (E) Retraction curves for an aggregate of radius $R_0 = 180$ $\mu$m and pipette radius $R_p = 35$ $\mu$m at $\Delta P = 1176$ Pa. Dashed lines are fits to the data using Eq. (8.12). (F) Schematic diagram showing the experimentally measurable parameters, and the rheological model for $\Delta P \leq \Delta P_c$. (G) Effective surface tension as a function of applied force. Open symbols are obtained from Eq. (8.17), filled symbols are from Eq. (8.16), and using the measured value of $\eta$. (H) Aggregate flow velocity as a function of stress. (I) Normalized deformation as a function of stress. *Image (B) is reproduced from K. Guevorkian, D. Gonzalez-Rodriguez, C. Carlier, S. Dufour, F. Brochard-Wyart, Mechanosensitive shivering of model tissues under controlled aspiration. Proc. Natl. Acad. Sci. U.S.A. 108 (33) (2011) 13387–13392 and images (D, E, G, H) are reproduced from K. Guevorkian, M.J. Colbert, M. Durth, S. Dufour, F. Brochard-Wyart. Aspiration of biological viscoelastic drops. Phys. Rev. Lett. 104 (2020) 218101 with permission.*

We note that for the case where $\Delta P \leq \Delta P_c$, the aggregate will respond with a small deformation $\delta_e \leq R_p$, as shown schematically in Fig. 8.2F, and the aggregate does not flow at long time. In this regime, $\Delta P = 2\gamma\left(\frac{1}{r} - \frac{1}{R_0}\right)$, where $r$ is the radius of curvature of the aspirated protrusion. $r$ is related to the deformation $\delta_e$ through

$$\frac{1}{r} = \frac{2\delta_e}{R_p^2 + \delta_e^2}. \tag{8.5}$$

Because in this regime $r > R_p$, the amount of pressure resisted by the surface tension is smaller than $\Delta P_c$. This regime ends when the applied aspiration pressure becomes $\Delta P = \Delta P_c$, which corresponds to $\delta_e = r = R_p$. At this threshold of pressure, the capillary response of the aggregate is fully developed, as demonstrated by the formation of a hemispherical cap at the end of the tongue, as seen in Fig. 8.2B. Beyond this critical aspiration pressure (corresponding to a maximal capillary force $f_{c\ \max} = \pi R_p^2 \Delta P_c$), the aggregate flows over a long time. In this second regime of larger deformation, the effective aggregate rheology can be represented by Fig. 8.2C(i), with a net force for aggregate deformation $f$ given by Eq. (8.3).

We can use Laplace's law to measure the surface tension of aggregates as follows. We start by applying a weak aspiration pressure, which we then increase in small steps while monitoring the length of the tongue $L$. When $L = R_p$, $\Delta P$ equals $\Delta P_c$, from which $\gamma$ can be estimated. While measuring precisely the pressure at which $L = R_p$ remains tedious due to the slow dynamics of the system, this method provides a rough estimate for $\gamma$, which in the case of aggregates of S180 murin sarcoma cell lines used in our studies is 5 mN/m $< \gamma < 7$ mN/m. Note that the surface tension measured with this technique denotes the surface tension of the aggregate at rest, which we refer to as $\gamma_0$ in Section 8.3.5. We will later discuss a second approach to measure the surface tension, which can be obtained from the dynamics of aspiration.

## 8.3.2 Elastic deformation

Now let us look at the aspiration dynamics inside the pipette when the aggregate is subjected to a large constant strain $\sigma = f/A_0$, where $A_0 = \pi R_p^2$, leading to $\sigma = \Delta P - \Delta P_c$. Fig. 8.2B shows a typical creep curve with two distinct regimes; a rapid deformation defined by $\delta$, followed by a constant flow at a rate $v$. Similar behavior has also been observed for Chinese hamster ovary cell aggregates, using the parallel plate compression technique [57].

Taking the strain $\varepsilon = \delta/R_p - 1$, we obtain

$$E = \frac{\sigma}{\varepsilon} = C_e \frac{R_p(\Delta P - \Delta P_c)}{\delta - R_p}. \tag{8.6}$$

The constant $C_e$ has been calculated numerically and shown to depend on geometric parameters such as the ratio between sample size and the pipette radius as well as to the wall thickness of the pipette tip [39]. For our experimental conditions $C_e \simeq 1$.

Similarly, over a long time, the aggregate behaves as a viscous fluid flowing at a constant velocity $v$, as we discuss below. The strain rate is $\dot{\varepsilon} = v/R_p$, leading to the viscosity:

$$\eta = \frac{\sigma}{\dot{\varepsilon}} = \frac{1}{C_v} \frac{R_p(\Delta P - \Delta P_c)}{v}, \tag{8.7}$$

where the coefficient $C_v$ depends on the geometry. Also note that in order to measure the aggregate viscosity, intercellular friction must be the dominant dissipation mechanism, as we discuss in the next section.

### 8.3.3 Viscous dissipation

Based on experimental observations, aggregate flow into the pipette is assumed to exhibit a plug flow profile. Moreover, in the following discussion we will assume aggregate permeability to be negligible, meaning that the aggregate does not behave as a porous material. In Section 8.6 we will revisit this assumption, and confirm that porosity is negligible for the aggregates used in this study. We will discuss the modifications to the equations needed if tissue porosity becomes important.

The dissipative forces have several origins: (1) the viscous deformation of the aggregate as it penetrates into the pipette, $f_v$, (2) the friction of the aggregate sliding over the micropipette wall, $f_w$, and (3) the viscous dissipation due to the flow of liquid along the tube connecting the pipette and the water reservoir, $f_l$. Therefore the total dissipative force, $f_{\text{diss}}$ is

$$f_{\text{diss}} = f_v + f_w + f_l. \tag{8.8}$$

$f_v = C_v \pi \eta R_p v$, where $\eta$ is the viscosity of the cell aggregate, results from the rearrangement of the cells with respect to one another at the entrance of the micropipette. The numerical factor $C_v$ has been calculated analytically for the entrance of a viscous liquid inside an orifice and shown to be: $C_v = 3\pi$ [58]. The validity of this coefficient for a cylindrical capillary has been verified experimentally by studying the dynamics of extraction of a viscous oil through a nonadhesive capillary [59].

Now let us consider the dissipation caused by the sliding of the aspirated tongue on the micropipette with a friction coefficient $k$. The friction force is given by $f_w = 2\pi R_p kLv$, which depends on the length $L$ of the aspirated tongue. For $L$ much larger than the critical length $L_c = \frac{3\pi\eta}{2k}$, $f_w$ dominates over $f_v$, leading to $L \sim t^2$, whereas when $L \ll L_c$, $L \sim t$. We have estimated $k$ and $L_c$ by aspirating an aggregate entirely inside a pipette at a constant suction pressure $\Delta P$ and measuring its advancement speed $v$. Once the aggregate is completely inside the pipette, $\Delta P_c = 0$ and $f_v = 0$, therefore from 3 we obtain

$$\pi R_p^2 \Delta P = 2\pi R_p k L^* v, \tag{8.9}$$

where $L^*$ is the length of the aggregate slug inside the micropipette. For the case where the micropipette surface is rendered nonadhesive, for example, through PEG-PLL coating, we obtain $k \approx 10^8$ Pa s/m. Taking the value of $\eta \approx 10^5$ Pa s, $L_c \sim 5\ \text{mm}^{-1}$ cm. Because the length of the tongue in our experiments does not exceed a few hundreds of micrometers, $f_w$ is negligible compared to $f_v$. This is confirmed by the observation of a constant $v$ $(L \sim t)$ over the duration of our experiments, as mentioned above.

In the absence of aggregate permeability, the dissipation due to viscous liquid flow inside the tube induced by the aggregate motion is given by $f_l = 8\pi\eta_l l_t v \times \left(\frac{R_p}{R_t}\right)^4$, where $l_t$ and $R_t$ are the length and the radius of the tube, respectively. The critical tube length above which the effect of $f_l$ becomes important is $l_t^* = \frac{3\pi}{8}\frac{\eta}{\eta_l}\left(\frac{R_t}{R_p}\right)^4 R_p$. In our experiments, $R_p/R_t \approx 0.1$, $20 < R_p < 50\ \mu$m, and $\eta_l = 10^{-3}$ Pa s. Taking $\eta \approx 10^5$ Pa s [60], $l_t^* \approx 10^7$ m, meaning that for cellular aggregates where $\eta \gg \eta_l$, $f_l$ is negligible.

Therefore the relevant dissipation only occurs at the entrance of the micropipette, yielding the following relationship for the viscosity $\eta$:

$$\eta = \frac{R_p(\Delta P - \Delta P_c)}{3\pi v}. \tag{8.10}$$

## 8.3.4 Rheological models for aggregate flow into the capillary

As mentioned above, the cell aggregates behave like a viscoelastic solid at short times, and like a viscous liquid at long times, as can be seen from the creep curve shown in Fig. 8.2C. To combine these two regimes, we use a modified Kelvin–Voigt model as shown in Fig. 8.2C(i) consisting of a Kelvin–Voigt body built by two springs with constants $k_1$ and $k_2$ and a dashpot with constant $\mu_1$, connected in series to a dashpot with constant $\mu_t$. This four-element system accounts for three regimes: a very fast, purely elastic response seen as a jump at the beginning of the experiment, a viscoelastic solid-like behavior, and a long-term fluid flow. The displacement $L(t)$ of the tongue is given by

$$L(t) = \frac{f}{k_1}\left(1 - \frac{k_2}{k_1 + k_2} e^{-t/\tau_c}\right) + \frac{f}{\mu_t} t. \tag{8.11}$$

The first term represents the viscoelastic solid behavior. At $t \to 0$, there is a fast elastic jump given by $f/k_1 + k_2$, then after a characteristic time $\tau_c = \frac{\mu_1(k_1 + k_2)}{k_1 k_2}$ the deformation reaches a value $\delta = f/k_1$. The second term describes the viscous flow characterized by a constant velocity $v = f/\mu_t$. We can express these mechanical parameters in term of $E$ and $\eta$, using Eq. (8.3) with Eqs. (8.6) and (8.10), leading to $k_1 = \pi R_p E$ and $\mu_t = 3\pi^2 R_p \eta$. Similarly, $\mu_1 = \pi R_p \eta_c$ and $\tau_c$ describe the cell-scale viscoelasticity. Another relevant timescale is the one describing the transition of the tissue from the elastic regime to the viscous regime, which is given by $\tau = \mu_t/k_1 = 3\pi\eta/E$.

As discussed in Section 8.3.1, Eq. (8.11) is only valid for $\Delta P > \Delta P_c$ in which the length $L$ refers to the aspirated tongue, excluding the hemispheriacal cap (see Fig. 8.2A). However, the experimentally measured length, $L_{exp}$, includes the hemospherical cap of radius $R_p$ as shown schematically in Fig. 8.2A. Therefore, in order to employ Eq. (8.11) to adjust the data, we should apply the correction $L = L_{exp} - R_p$, as shown for the data in Fig. 8.2D. The aspiration dynamics for $\Delta P \leq \Delta P_c$ will be discussed below. It will be shown that in this regime, we reach a maximum deformation $\delta = R_p$ for $\Delta P = \Delta P_c$. On the other hand, we see that when $\Delta P = \Delta P_c$ from Eq. (8.11) we get $L_{exp} = R_p$, satisfying the continuity condition between the two regimes.

For simplicity, Eq. (8.11) can be written as the following:

$$L(t) = \delta\left(1 - \kappa e^{-t/\tau_c}\right) + \nu t, \tag{8.12}$$

where $\kappa = \frac{k_2}{k_1 + k_2}$. The viscoelastic parameters are obtained at each aspiration pressure by adjusting Eq. (8.12) to the aspiration and retraction curves with the following corrections:

$$
\begin{aligned}
L_{asp} &= L_{\exp} - R_p, \\
L_{ret} &= L_{\max} - L_{\exp},
\end{aligned}
\tag{8.13}
$$

where $L_{\max}$ is the maximum length of the aspirated tongue, just before the pressure is set back to zero, as shown in Fig. 8.2E.

If the cellular relaxation parameters, $\tau_c$ and $\mu_1$, are not sought, the first rheological model in Fig. 8.2C(i) can be simplified to a Maxwell liquid model Fig. 8.2C(ii), consisting of a dashpot with constant $\mu_t$ in series with a spring, with spring constant $k_1$. This amounts to fitting a line to the linear viscous regime as shown with the dashed blue line in Fig. 8.2C–E. From the slope $v$, and intercept $\delta$, the viscosity and the elasticity can be deduced, as explained in Section 8.3.5.

### *Small $\Delta P$ regime*

Let us now investigate the response of the aggregate when the aspiration pressure is smaller than the critical pressure $\Delta P_c$. In this case, the resisting capillary force is smaller than that in the large pressure regime, $f_c < \pi R_p^2 \Delta P_c$. By applying Laplace's law to an aggregate aspirated by a length $L = \delta < R_p$, we obtain

$$
f_c = \pi R_p^2 \times 2\gamma \left( \frac{2\delta}{\delta^2 + R_p^2} - \frac{1}{R_0} \right). \tag{8.14}
$$

Assuming $R_0 \gg R_p$, Eq. (8.14) can be linearized in two limiting cases: (1) $\delta \ll R_p$, leading to $f_c \approx 4\pi\gamma\delta$, and (2) $\delta \approx R_p$, leading to $f_c \approx 2\pi\gamma R_p$. In both limits, the capillary force $f_c$ is proportional to the deformation $\delta$, suggesting that it can be represented by a (nonlinear) spring. To investigate the aspiration dynamics in this regime, we thus adopt the simple rheological model shown in Fig. 8.2F. Here, we add a spring of constant $k_c$, accounting for the restoring nature of the capillary effect, to our simplified Maxwell model for the tissue. Note also that, because in this regime the capillary resistance is included in the rheological model via the additional spring $k_c$, the force acting on the system is the total aspiration force, $f_a = \pi R_p^2 \Delta P$. The solution for this model is

$$
\delta(t) = \frac{f_a}{k_c} \left( 1 - \frac{k_1}{k_c + k_1} e^{-t/\tau_c^*} \right), \tag{8.15}
$$

where $\tau_c^* = \frac{\mu_t(k_1 + k_c)}{k_1 k_c}$ is the characteristic time for the relaxation of elastic response. This analysis shows that for small aspiration pressures, the system will reach a finite deformation $\delta_e = f_a/k_c \sim \Delta P/\gamma$, up to a maximum value $\delta_e = R_p$ for $\Delta P = \Delta P_c$, beyond which the aggregate will flow at long time.

### 8.3.5 Rheological parameters for murin sarcoma cell aggregates

Here we will show the application of the above procedure to characterize the mechanical properties of S180 murin sarcoma cellular aggregates. The aspiration pressures used for these aggregates were between 1 and 3 kPa. In Fig. 8.2D–E we show an example of an aspiration and retraction cycle for $\Delta P = 1176$ Pa. The red dashed lines correspond to the adjustment of Eq. (8.12), with the corrections given by Eq. (8.13) accounted for. The blue dashed lines correspond to the simplified Maxwell model. We use the values of $v^a$ and $v^r$, the flow rates during aspiration and retraction, respectively, to estimate the surface tension as follows. From Eq. (8.10), we have

$$\begin{aligned} v^a &= \frac{R_p(\Delta P - \Delta P_c)}{3\pi\eta} \\ v^r &= \frac{R_p \Delta P_c}{3\pi\eta} \end{aligned} . \tag{8.16}$$

From this set of equations we can obtain $\Delta P_c$ as

$$\Delta P_c = \Delta P \times \frac{v^r}{v^a + v^r}. \tag{8.17}$$

For the specific set of experiments presented in Fig. 8.2D–E, we obtained $\Delta P_c = 690$ Pa. This pressure is significantly larger than the critical pressure that we would obtain from Laplace's equation, Eq. (8.4), with $\gamma_0 = 6$ mN/m ($\Delta P_{c0} = 300$ Pa). This suggests that $\gamma$ has increased as a function of the applied pressure, $\Delta P$. Indeed, Fig. 8.2G shows that the measured $\gamma$ systematically increases as the applied force increases. We can estimate the surface tension of an aggregate at rest by extrapolation of the linear regression to the data, which gives $\gamma_0 = 6$ mN/m. This result is in agreement with the surface tension estimate that we obtain from measuring the critical Laplace pressure, as well as with the values reported for compression plate experiments [38]. We assume that the change in the surface tension could be due to active responses of cells to the applied force. We will see in the next section that in certain aspiration conditions, this activity becomes directly observable. For comparison, in the case of an inert liquid, like an oil droplet, $\gamma$ would not depend on the aspiration pressure.

Next, we investigate the dependence of the flow rate on the applied stress, $\sigma$. Note that during aspiration, $\sigma = (\Delta P - \Delta P_c)$, and during retraction $\sigma = \Delta P_c$. As can be seen in Fig. 8.2H, $v$ has a linear relationship with $R\sigma$, in agreement with Eq. (8.10), where the slope yields $1/(3\pi\eta)$. The average value of the viscosity is $\eta = (1.9 \pm 0.3) \times 10^5$ Pa s.

An alternative way of deriving the viscosity, $\eta$, is by eliminating $\Delta P_c$ in Eq. (8.16), which gives the following relationship between viscosity and the flow rates:

$$\eta = \frac{R_p \Delta P}{3\pi(v^a + v^r)}. \tag{8.18}$$

For example, for the experiments in Fig. 8.2D–E, $\eta = 2.4 \times 10^5$ Pa s. We emphasize that in both of the proposed approaches to estimate $\eta$, $\Delta P_c$ is assumed not to change between the linear regimes of the aspiration and retraction curves.

The elastic modulus, $E$, can be obtained from the deformation $\delta$, defined as the intercept of the linear flow regime and the vertical axis (see Fig. 8.2C–E). $\delta$ is related to the applied elastic stress by $\sigma_e = E\delta/R_p$. The stress during aspiration is $\sigma = \sigma_e = \Delta P - \Delta P_c$, whereas during the retraction, the net elastic stress is $\sigma_e = \Delta P$. Indeed, upon release of the applied aspiration pressure, the spring of constant $k_1$ is prestressed. Therefore the net change in elastic stress when the aspiration pressure is removed is $\sigma_e = \Delta P_c - (\Delta P - \Delta P_c) = \Delta P$. Therefore we have

$$\begin{aligned} \delta^a &= \frac{R_p(\Delta P - \Delta P_c)}{E}, \\ \delta^r &= \frac{R_p \Delta P}{E}, \end{aligned} \tag{8.19}$$

where $\delta^a$ and $\delta^r$ are the magnitudes of the deformation for aspiration and retraction curves, respectively. Fig. 8.2I shows the normalized $\delta$, as a function of stress, where the inverse of the slope gives the average value $E = 1300 \pm 150$ Pa.

As shown in Fig. 8.2B, $\tau_t = \delta/v$ is obtained by the $x$-axis intercept of the linear fit to the experimental curve. Our estimated values are $\tau_t^a = 2900 \pm 1100$ and $\tau_t^r = 1700 \pm 450$ seconds, obtained from the aspiration and the retraction curves, respectively. Using Eqs. (8.16) and (8.19), we find $\tau = 3\pi\eta/E = \tau_t^a = 2900 \pm 1100$ and $\tau = \tau_t^r \times (\Delta P_c/\Delta P) = 1200 \pm 200$ seconds. We thus notice a discrepancy between the $\tau$ obtained from the aspiration and retraction curves. However, on average we can estimate that the tissue relaxation time is of the order of few tens of minutes.

In summary, in this section we showed that cellular aggregates flow inside a cylindrical capillary like a viscoelastic liquid droplet when the applied pressure is in the range of 1–3 kPa. We then developed a mechanical model with two characteristic time scales; one at the cellular level, $\tau_c = \mu_1/k_1 \approx \eta_c/E \sim 100 - 200$ seconds, and one at the tissue level, $\tau = 3\pi\eta/E = 25 \pm 10$ minutes. The active response of the cells to the applied stress manifests itself in an increase of the effective surface tension. In the next section, we will further explore active cellular responses to an external stress.

## 8.4 Active response of cellular aggregates to mechanical stimuli

As discussed in the previous section, the active response of an aggregate to an external stress globally manifests itself with an increase of the effective surface tension. In this section, we explore the behavior of aggregates of smaller sizes in response to aspiration pressures that are weaker than previously explored. As will be demonstrated, in these specific conditions, we observe concrete signs of cellular mechanosensitivity, leading to pulsation of the aggregates. We present an analytical model that shows how these pulsations can be attributed to the active response of cells to stretching.

### 8.4.1 Aggregate pulsation or shivering

For pressures ranging between 500 Pa $\leq \Delta P < 1500$ Pa, applied to aggregates measuring between 70– and 150 $\mu$m in radius, we observe unexpected pulsatile behavior of the aggregates that we call *shivering* [54]. This pulsatile behavior is suppressed when the aggregates are treated with 10 $\mu$M blebbistatin, which inhibits myosin II activity responsible for cell contraction, reinforcing the assumption that shivering originates from cellular activity. The kymograph in Fig. 8.3A shows that, as the aggregate is aspirated inside the pipette, random contractions are detected inside the aspirated tongue. A careful observation of the kymograph reveals that, in

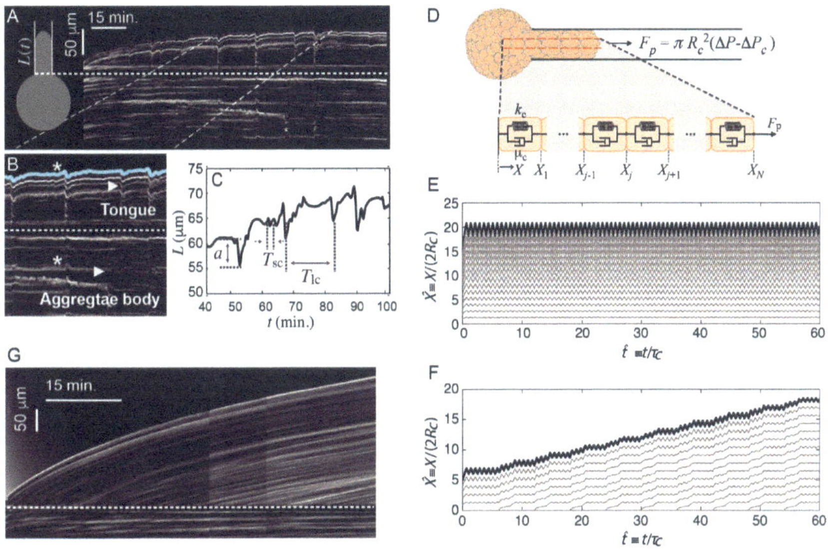


**Figure 8.3** Shivering of a cellular aggregate under aspiration. (A, B) Kymograph showing the advancement $L(t)$ of the aggregate inside the pipette as a function of time, for an aggregate of $R_0 = 70\ \mu$m and $R_p = 37\ \mu$m, at a pressure of $\Delta P = 500$ Pa. (C) Tracking of the leading edge of (B), where $a$ is the amplitude of the large contractions, and $T_{sc}$ and $T_{lc}$ are the periods of the small and large contractions, respectively. (D) One-dimensional model of aggregate shivering. Each cell is represented as a Kelvin–Voigt element to model the viscoelastic response of the cell. The viscous flow of the tissue is modeled by periodically adding one cell to the left end of the chain. (E) Aspiration curves predicted by the one-dimensional model for parameter values corresponding to a typical experiment ($\hat{F}_p = 0.5$, $N = 15$). The active cell response is characterized by $\hat{F}_m = 0.5$, $\epsilon_{crit} = 0.5$, $h = 100$, and $\hat{\tau}_s = 0.25$. The black curve corresponds to the position of the rightmost cell edge, $X_N$, and the gray curves correspond to the other cell edges along the chain. (F) Aspiration curves for the same parameter values but periodically increasing the number of cells to account for the viscous aggregate flow. The computation is started with $N = 5$ and one cell is added at the left end every 6 unit times, corresponding to the experimental observation that $\tau_{flow} \approx 6\tau_c$. The final number of cells is $N = 15$, as in the previous plot. (G) For similar aspiration conditions as in (A), with $\Delta P = 785$ Pa, the inhibition of Myosin II activity by blebbistatin treatment suppresses the shivering behavior and makes the aggregate more fluid. *Images (A–F) are reproduced from K. Guevorkian, D. Gonzalez-Rodriguez, C. Carlier, S. Dufour, F. Brochard-Wyart, Mechanosensitive shivering of model tissues under controlled aspiration. Proc. Natl. Acad. Sci. U.S.A. 108 (33) (2011) 13387–13392 with permission.*

some cases (shown with asters in the figure), these contractions are *global*, meaning that they are concomitant with a contraction in the aggregate body outside of the pipette. Other contractions are *local*, detected either in the tongue or in the body (shown with triangles).

To characterize these contractions, we develop an edge detection routine applied to the kymograph to determine the position of the advancing tongue inside the pipette as a function of time, as shown in Fig. 8.3B and C. We categorize two groups of contractions: (1) large contraction when the amplitudes is larger than 3 $\mu$m, (2) small contraction with amplitude $\approx$ 1 $\mu$m. For large contractions, we find a monotonic decrease of the amplitude from $a \approx 5\ \mu\text{m}$ for $\Delta P = 500$ Pa, down to $a \approx 1 - 2\ \mu\text{m}$ as the aspiration pressure increases to $\Delta P = 1200$ Pa. For each aspirated aggregate, we observe a few large contractions with period 500 seconds $< T_{lc} <$ 1000 seconds, and many small contractions with a period of $T_{sc} \approx 200$ seconds. We note that these aggregates have a $\Delta P_c \approx 200 - 300$ Pa, corresponding to $\gamma_0 \approx 5$ mN/m. For low aspiration pressures too close to $\Delta P_c$, contractions are difficult to observe, because the aggregate barely enters the pipette, and for high $\Delta P > 1200$ Pa the amplitude of contraction becomes very small. Indeed, we did not see shivering for the experiments described in the previous section, where the aspiration pressures were larger than this threshold. In the following, we present a model to interpret these observations.

### 8.4.2 Model of delayed cell contraction

Our proposed model to interpret aggregate pulsation combines the viscoelastic behavior of a group of cells with a time-delayed active response. The latter hypothesis was inspired by a theoretical model previously developed to explain collective cellular oscillations during the dorsal closure in *Drosophilla* embryo [8]. Conveniently, in our system, the cells are confined in the cylindrical geometry of the micropipette tip, and subject to a unidirectional force, along the axis of the micropipette, which allows us to simplify the analysis to one dimension. Therefore to build our active mechanical model, we will simply analyze the response of a row of cells to a force applied to an imaginary pipette of radius $R_c$, equivalent to a cell radius, leading to $F_p = \pi R_c^2(\Delta P - \Delta P_c)$, as shown schematically in Fig. 8.3D. Here, we represent each cell as a Kelvin–Voigt element, with a spring constant $k_c = \pi R_c E$, and a dashpot of constant $\mu_c = \pi R_c \eta_c$. The applied force $F_p$ stretches the cells along the array. If cell deformation surpasses a critical strain $\varepsilon_{\text{shiv}}$, the cell will switch on an active contraction force $F_m$. Activation of this contraction force is not instantaneous, but rather it will occur after a time delay $\tau_s$ from the time when $\varepsilon_{\text{shiv}}$ is surpassed. This strain-dependent response can be expressed as

$$F_s(t) = \begin{cases} 0 & \text{if } \varepsilon(t-\tau_s) \leq \varepsilon_{\text{shiv}} \\ F_m & \text{if } \varepsilon(t-\tau_s) > \varepsilon_{\text{shiv}}, \end{cases} \tag{8.20}$$

where $F_s(t)$ is the contractile force that a cell exerts at time $t$ when it has been strained by $\varepsilon$ at time $(t-\tau_s)$. The corresponding continuum description is

$$\tau_c \frac{\partial \varepsilon}{\partial t} + \varepsilon = \frac{F_p - F_s}{2k_c R_c}. \tag{8.21}$$

where $\tau_c = \eta_c / E$ is the viscoelastic relaxation time of a cell. Integration of Eq. (8.21) yields

$$\varepsilon(\hat{t}) = \hat{F}_p\left(1 - e^{-\hat{t}}\right) + \varepsilon_c e^{-\hat{t}} - e^{-\hat{t}} \int_0^{\hat{t}} \hat{F}_m H(\varepsilon(\hat{\tau} - \hat{\tau}_s)) e^{\hat{\tau}} \mathrm{d}\hat{\tau}, \tag{8.22}$$

where $H$ is the Heaviside function and the hat symbol, ^, indicates quantities that are nondimensionalized using $\tau_c$ as the unit time and $2k_c R_c$ as the unit force. To simplify the writing, the hat symbols are omitted in what follows.

In this model, the shivering oscillations arise because of the active contractility switching alternatively on and off and making the cell deformation oscillate around $\varepsilon_{\text{shiv}}$. Because the constitutive equation is linear, the oscillation period $T_{sc}$ is divided into a half period over which the active force is on and the cell contracts, and a half period over which it is off and the cell expands. To obtain the equation for $\varepsilon(t)$, we evaluate the integral in Eq. (8.21) taking as an initial condition $\varepsilon(t=0) = \varepsilon_{\text{shiv}}$. For $0 < t < \tau_s$, that is, before the active contraction force is switched on, Eq. (8.22) reduces to

$$\varepsilon(t) = F_p(1 - e^{-t}) + \varepsilon_c e^{-t}. \tag{8.23}$$

For $t > \tau_s$, the contractile force is switched on. As a result, if contraction is sufficiently strong, cell deformation decreases, and eventually becomes again smaller than $\varepsilon_{\text{shiv}}$. This occurs at $t = T_{sc}/2$. Therefore at the time $T_{sc}/2 + \tau_s$, the contraction force eventually switches back off. Integration of Eq. (8.22) for $\tau_s < t < T/2 + \tau_s$ yields

$$\varepsilon(t) = F_p(1 - e^{-t}) + \varepsilon_c e^{-t} - F_m(1 - e^{-(t-\tau_s)}). \tag{8.24}$$

Finally, we consider the time $T_{sc}/2+\tau_s<t<T_{sc}$, when the active contraction is switched off. Integration of Eq. (8.22) yields

$$\varepsilon(t)=F_p(1-e^{-t})+\varepsilon_c e^{-t}+\frac{F_m(1-e^{\tau_s})}{\varepsilon_c-F_p+F_m}e^{-(t-\tau_s)}. \tag{8.25}$$

The oscillation period $T_{sc}$ is obtained by imposing that at the end of the period the cell recovers its initial deformation, that is, $\varepsilon(t=T_{sc})=\varepsilon_c$. Returning hereafter to dimensional variables, we obtain the following relationship between the contraction period $T_{sc}$ and the delay time $\tau_s$:

$$e^{T_{sc}/\tau_c}=1+\frac{e^{\tau_s/\tau_c}\left(e^{\tau_s/\tau_c}-1\right)}{\alpha(1-\alpha)}, \tag{8.26}$$

where $\alpha$ has been defined by expressing the aspiration force as $F_p=F_{\text{shiv}}+\alpha F_m$, where $F_{\text{shiv}}=2k_cR_c\varepsilon_{\text{shiv}}$ corresponds to the critical pressure required for shivering. We note that oscillations only arise if $F_{\text{shiv}}<F_p<F_{\text{shiv}}+F_m$, which corresponds to $0<\alpha<1$. Experimentally, we find that shivering happened when $500\text{ Pa}<\Delta P<1200\text{ Pa}$. Taking $\Delta P_c\approx 300$ Pa gives us the range of $\frac{F_p}{\pi R_c^2}\approx 200-900$ Pa. This range is compatible with our model if we set $\frac{F_{\text{shiv}}}{\pi R_c^2}=(200+900)/2$ Pa and $\frac{F_m}{\pi R_c^2}=(900-200)/2$ Pa, which yields $\varepsilon_{\text{shiv}}\approx 0.3$ and $F_m\approx 0.1\ \mu$N. The amplitude of the oscillations is the difference between the maximum and the minimum of $\varepsilon(t)$ given by Eqs. (8.24) and (8.25):

$$\Delta\varepsilon=\frac{F_m}{2k_cR_c}\left(1-e^{-\tau_s/\tau_c}\right). \tag{8.27}$$

Experimentally, $2R_c\Delta\varepsilon\approx 2\ \mu$m. According to Eq. (8.27) and taking $R_c\approx 15\ \mu$m and $\tau_c\approx 200$ seconds this corresponds to a delay time $\tau_s\approx 40$ seconds. Finally, Eq. (8.26) with $\alpha\approx 1/2$ yields $T_{sc}\approx 150$ seconds.

To afford a visual comparison between the predictions of this model and the experiments, we numerically simulate the behavior of a string of cells aspirated in a micropipette. Each cell in the string was modeled as a Kelvin–Voigt element. Additionally, each cell exhibits an active delayed contractility, as described by Eq. (8.20). The simplest numerical simulation considers a fixed number of cells inside the pipette (Fig. 8.3E). Shivering arises if the applied aspiration pressure is large enough to bring cell stretching beyond the contractility threshold, but not too large, so that with contractility forces on, cell stretching is reduced below the threshold, thus switching contractility off again. The period of such oscillations is proportional to the delay time $\tau_s$. With a fixed number of cells in the pipette, our model predicts regular oscillations, which we interpret to correspond to the frequent small contractions in the experiment, of period $T_{sc} = 4\tau_s \approx 200$ seconds.

The observed large contraction could be explained by considering synchronous contractions of several cells, which could be associated with cell reorganizations due to aggregate flow into the pipette. Thus we have considered a modified version of our numerical model where the number of cells inside the pipette increases over time due to aggregate flow, shown in Fig. 8.3F. Cell flow into the pipette induces heterogeneities in the oscillation pattern, yielding a simulated kymograph that more closely resembles the experimental one. In our experiments, we have noticed that shivering is more frequent in smaller size aggregates. This might be related to the ratio of the deformed volume of the aggregate body $\sim R_p^3$ to the total volume. A more systematic study should addressee this hypothesis. We also observe that interfering with myosin II activity, by 10 $\mu$M blebbistatin treatment, suppresses the contractions, and also makes the aggregate flow easier, as can be seen in the kymograph shown in Fig. 8.3G.

## 8.5 Permeability of cellular aggregates

Interstitial flow plays an important role in tissue morphogenesis, function, and pathology [61]. It is most prominently linked to the drainage of blood plasma leaking from capillaries toward lymphatic vessels. Interstitial flow occurs through the extracellular matrix. In tissues with little extracellular matrix or in model tissues lacking an extracellular matrix, interstitial flow is controlled by permeability at the intercellular junctions, which depends on the strength of intercellular adhesions.

Cellular aggregate permeability can be characterized by microfluidics. By placing a cellular aggregate in a specially designed chamber in a microfluidic channel, as sketched in Fig. 8.4A, an interstitial flow can be imposed through it. The aggregate permeability can be quantified by assuming the aggregate to behave as a porous medium following Darcy's law, that is,

$$Q = \frac{\Delta P}{R_h}, \tag{8.28}$$

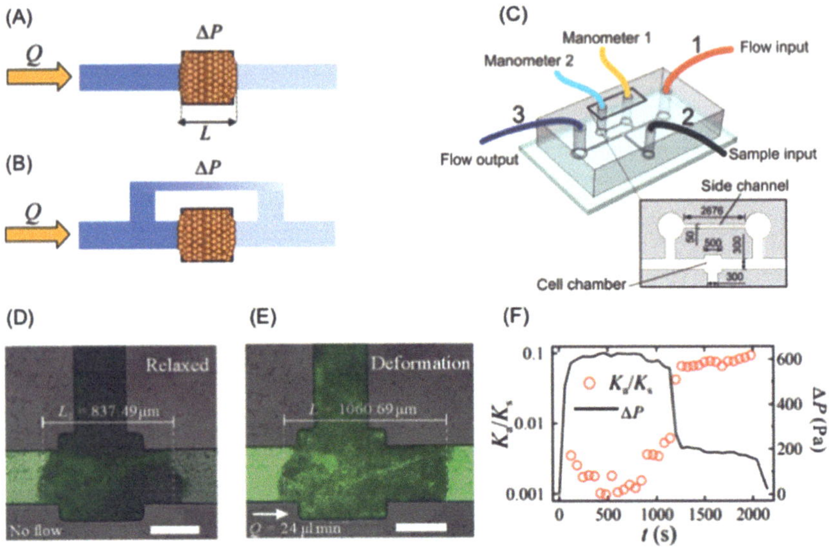


**Figure 8.4** Permeability of cellular aggregates. (A) Principle of characterization of aggregate permeability in a microfluidic channel and (B) usage of a side channel to prevent experimental artifacts. The practical realization of this idea is shown in (C), with dimensions in micrometer. (D, E) Experimental images of the measurement of permeability of a MCF-7 cellular aggregate in a microfluidic channel. A fluorescent tracer has been used to visualize the advancement of the flow through the aggregate, which goes from left to right. The scale bar corresponds to 200 $\mu$m. (F) Experimental measurement of the intrinsic permeability of a MCF-7 cellular aggregate, $K_a$, in the microfluidic system shown in (C). The intrinsic permeability of the side channel is here $K_s = 1.97 \times 10^{-10}$ m$^2$. In this experiment, aggregate fracture occurs at $t \approx 1200$ seconds, as indicated by the sudden drop of the pressure difference, causing $K_a$ to increase. *Images (C–F) are reproduced from Q.D. Tran, M. Marcos, D. Gonzalez-Rodriguez. Permeability and viscoelastic fracture of a model tumor under interstitial flow. Soft Matter 14 (2018) 6386–6392 with permission from the Royal Society of Chemistry.*

where Q is the flow rate through the aggregate for an imposed pressure difference $\Delta P$, and $R_h = \eta_l L/(K_a A)$ is the aggregate's total hydrodynamic resistance, which depends on the medium fluid viscosity $\eta_l$, the aggregate's length along the flow direction $L$ and cross-section $A$, and the aggregate's intrinsic permeability $K_a$. By imposing Q and measuring $\Delta P$ or vice versa, the intrinsic permeability can be determined.

In practice, experimental difficulties may prevent the realization of the setup in Fig. 8.4A. Tran et al. [62] attempted to realize this setup by imposing a flow rate through the aggregate using a syringe pump, and by measuring the resulting pressure difference across the aggregate by means of a manometer. Due to the low permeability of the MCF-7 breast cancer cell aggregates used in their study, which would require to impose a very small flow rate, the setup proved impractical. To overcome this difficulty, they proposed an alternative setup with a narrow side channel parallel to the aggregate chamber, as sketched in Fig. 8.4B and C. The side channel permeability is carefully chosen to allow sufficient continuous flow for proper operation of the system while at the same time not overshadowing the permeability of the sample. Using this setup, an aggregate permeability $K_a = (8 \pm 5) \times 10^{-13}\ \mathrm{m}^2$ was measured, as shown in Fig. 8.4F [62]. The large standard deviation reflects significant variability both between samples and over time, as well as a measurement uncertainty introduced by the permeability of the side channel. We also note that the permeability is associated with a characteristic length $\sqrt{K_a} \approx 1\ \mu\mathrm{m}$, comparable to the intercellular distance within the aggregate.

The picture of aggregate permeability as that of a porous medium governed by Darcy's law is a highly simplified one. In reality, aggregate deformation induced by flow will modify its permeability, as suggested by the experimental images in Fig. 8.4D and E. At moderate flow rates, permeability increases due to viscoelastic deformation, whereas larger flow rates induce the formation of cracks through the aggregate that highly permeabilize it. This latter phenomenon is known as "aggregate fracture" under flow. MCF-7 aggregates of a few hundred micrometers in size fracture under flow pressures of the order of 500 Pa [62,63]. We note that in this setup aggregate flow is prevented due to strong adhesion to the device walls, here treated with fibronectin. This differs from the micropipette experimental system, where the pipette walls are rendered nonadhesive. This is why pressure values of about 500 Pa do not induce fracture in the micropipette experiment. When discussing aggregate fracture, we must also keep in mind that the meaningful fracture threshold is prescribed by the intercellular adhesion energy, $G$. Indeed, it can be shown that the aforementioned pressure threshold is linked to an intercellular adhesion energy of the order of $G \approx 10^{-2}\ \mathrm{J\,m}^{-2}$, a value that has been reported for aggregates of different cell lines [36,62]. A simple relationship between pressure threshold $\Delta P_{\mathrm{frac}}$ and adhesion energy $G$ can be written for fast loading, under which the aggregate behaves elastically. In this case, $G \sim \Delta P_{\mathrm{frac}}(L_{\mathrm{frac}} - L_0) = \Delta P_{\mathrm{frac}}^2 L_0/E$. With $L_0 \approx 100\ \mu\mathrm{m}$ and $E \approx 1\ \mathrm{kPa}$, this simplified equation explains the relationship between the orders of magnitude of $\Delta P_{\mathrm{frac}}$ and $G$. Future work is still required to fully characterize aggregate permeability. Open questions are the dependence of aggregate permeability on deformation, the respective roles of intercellular adhesion energy and of the extracellular matrix, and a theoretical modeling of aggregate permeability experiments that includes poroelastic effects [64].

When studying a permeable sample, the equations describing micropipette aspiration discussed in Section 8.3.3 may need modification. Indeed, for a permeable aggregate, the flow velocity of the medium inside the pipette will no longer be equal to the velocity $v$ of advancement of the aggregate tongue, as was stated in Section 8.3.3 for an impermeable aggregate. Rather, the medium flow velocity will scale as $v_l \sim K_a \Delta P/(\eta_l L)$, where $L$ is the length of the aspirated tongue. The ratio between the viscous dissipation due to aggregate deformation, $f_v$, and viscous dissipation due to flow of medium, $f_l$, yields a new critical length, $L_c^{'} = 8K_a R_p^2 l_t / R_t^4$. Permeability effects can be neglected if $L \ll L_c^{'}$. For a typical cellular aggregate permeability $K_a < 10^{-12}$ m$^2$ in the experimental conditions described in Section 8.3.3, we obtain $L_c^{'} > 1$ m, which justifies neglecting permeability effects when studying cell aggregate viscoelasticity. We emphasize that $L \ll L_c^{'}$ means that the total aspiration pressure exerted on the aggregate is not affected by permeability, but it does not preclude the existence of interstitial flow, $v_l > v$. The condition for interstitial flow velocity to be significant compared to aggregate deformation speed is $L > L_{\text{perm}} = (K_a \eta)/(R_p \eta_t)$. When $L \gtrapprox L_{\text{perm}}$, part of the aspiration force is exerted on the aggregate as a shear stress of the interstitial fluid, which may yield to different physics of tissue deformation. In our experimental conditions, $L_{\text{perm}} \approx 1$ m and this effect is again negligible.

If the aggregate is sufficiently permeable, interstitial flow could be measured in the micropipette experiment by tracking the motion of debris downstream of the aspirated tongue. This would provide an estimate of the interstitial flow speed $v_l$, which is expected to depend on the aspirated length $L$. By measuring $\Delta v = v_l - v$ as a function of $L$, intrinsic permeability $K_a$ can be deduced by fitting the equation

$$\frac{1}{\Delta v} = \frac{\eta_l}{K_a \Delta P}(L_a + L), \tag{8.29}$$

where $L_a$ is an unknown constant accounting for the hydraulic resistance contributed by the aggregate body outside the pipette. Eq. (8.29) assumes the aggregate's permeability $K_a$ to be a constant, independent of the aggregate's deformation. An alternative hypothesis, which can be tested experimentally, is that $K_a$ increases with increasing $\Delta P$ due to increasing elastic deformation, that is, a poroelastic effect.

## 8.6 Conclusions and perspectives

In this chapter, we studied the flow of cellular aggregates into capillaries. Under constant stress conditions, the flow dynamics reveal the viscoelastic liquid behavior of these model tissues. At short time, we observe a fast deformation regime related to an elastic response. At long time, the tissue flows at a constant speed, thus behaving like a viscous liquid. We developed a methodology to extract elastic and viscous coefficients, as well as an effective surface tension, from the information contained in the aspiration and relaxation curves at constant pressure. In this description we consider the aggregates to be impermeable, a condition that is satisfied in our experiments but that may not always be valid. To address this question, we present a microfluidic system that allows measuring the permeability of cell aggregates and tissue explants. Microfluidic experiments confirm that permeability effects are negligible in our micropipette experiment conditions. However, porosity can be important in other tissue types and play a crucial role in their function, as has been observed recently in the case of cancer tumors [31]. We have thus discussed how to analyze micropipette aspiration experiments of tissues with significant permeability.

To address the increasing interest in understanding the role of rheology in morphogenesis and disease [19], the approaches developed in this chapter can readily be applied to embryonic tissues. In a recent study, micropipette aspiration has been applied to explants of *Xenopus* embryos at the gastrulation stage, in order to investigate mechanical differences between ectoderm and mesoderm tissues and to correlate tissue mechanics to acto-myosin activity [65]. A similar approach has been used to put in evidence that bacterial aggregates of *Neisseria meningitidis* actively modulate their rheological properties to increase their colonization efficiency into narrow capillaries [66].

Besides rheological measurements, the pipette system allows exploration of tissue mechanosensitivity under controlled external stress. In the example discussed in this chapter, the active nature of a cellular aggregate reveals itself as pulsatile contractions both at cellular and multicellular scales. Strikingly, pulsation only arises over a limited range of moderate external stress, whereas at larger stress the tissue contracts permanently. Then, activity manifests itself less conspicuously, as a change of the apparent rheological properties of the tissue. Combined with confocal microscopy, mechanisms of tissue intercalation can be studied when a controlled external stress is applied through micropipette aspiration, as has been used to study the role of growth factors such as FGF on vertex formation in mouse embryonic limb bud ectoderm [67].

The micropipette aspiration technique has been widely used in the past several decades for studies on single cells and biomimetic systems, such as giant unilamellar vesicles. However, its application to living tissues is at its dawn. The few examples discussed here show its potential for application to multicellular systems, ranging from in vitro cellular aggregates to organoids, embryonic tissue explants, and even in vivo tissues.

## Acknowledgments

We thank our colleagues who have worked with us on the research questions discussed in this chapter, most particularly Stéphane Douezan, Sylvie Dufour, Marcos, and Quang D. Tran.

## References

[1] N.M.E. Ayad, S. Kaushik, V.M. Weaver, Tissue mechanics, an important regulator of development and disease, Philos. Trans. R. Soc. B 374 (1779) (2019).
[2] S. Suresh, Biomechanics and biophysics of cancer cells, Acta Biomater. 3 (4) (2007) 413–438.
[3] H. Yu, J.K. Mouw, V.M. Weaver, Forcing form and function: biomechanical regulation of tumor evolution, Trends Cell Biol. (2010) 1–10.
[4] D. Mitrossilis, J. Fouchard, A. Guiroy, N. Desprat, N. Rodriguez, B. Fabry, et al., Single-cell response to stiffness exhibits muscle-like behavior, Proc. Natl. Acad. Sci. U. S. A. 106 (43) (2009) 18243–18248.
[5] L.A. Davidson, S.D. Joshi, H.Y. Kim, M. vonDassow, L. Zhang, J. Zhou, Emergent morphogenesis: elastic mechanics of a self-deforming tissue, J. Biomech. 43 (1) (2010) 63–70.
[6] M. von Dassow, J.A. Strother, L.A. Davidson, Surprisingly simple mechanical behavior of a complex embryonic tissue, PLoS One 5 (12) (2010) e15359.
[7] B. Guirao, Y. Bellaïche, Biomechanics of cell rearrangements in Drosophila, Curr. Opin. Cell Biol. 48 (2017) 113–124.
[8] J. Solon, A. Kaya-Copur, J. Colombelli, D. Brunner, Pulsed forces timed by a ratchet-like mechanism drive directed tissue movement during dorsal closure, Cell 137 (7) (2009) 1331–1342.
[9] M. Ben Amar, P. Qiuyang-Qu, T.T.K. Vuong-Brender, M. Labouesse, Assessing the contribution of active and passive stresses in *C. elegans* elongation, Phys. Rev. Lett. 121 (2018) 268102.
[10] A. Lardennois, G. sti, T. Ferraro, F. Llense, P. Mahou, J. Pontabry, et al., An actin-based viscoplastic lock ensures progressive body-axis elongation, Nature (2019) 1–21.
[11] B. Bénazéraf, PF, R.E. Baker, N. Denans, C.D. Little, O. Pourquié, A random cell motility gradient downstream of fgf controls elongation of an amniote embryo, Nature 466 (7303) (2010) 248–252.

[12] A.K. Lawton, A. Nandi, M.J. Stulberg, N. Dray, M.W. Sneddon, W. Pontius, et al., Regulated tissue fluidity steers zebrafish body elongation, Development 140 (3) (2013) 573–582.
[13] A. Mongera, A. Michaut, C. Guillot, F. Xiong, O. Pourquié, Mechanics of anteroposterior axis formation in vertebrates, Annu. Rev. Cell Dev. Biol. 35 (1) (2019) 259–283.
[14] D. Bi, X. Yang, M.C. Marchetti, M.L. Manning, Motility-driven glass and jamming transitions in biological tissues, Phys. Rev. X 6 (2) (2016) 021011–021013.
[15] I. Bonnet, P. Marcq, F. Bosveld, L. Fetler, Y. Bellaiche, F. Graner, Mechanical state, material properties and continuous description of an epithelial tissue, J. R. Soc. Interface 9 (75) (2012) 2614–2623.
[16] R.A. Foty, G. Forgacs, C.M. Pfleger, M.S. Steinberg, Liquid properties of embryonic tissues: measurement of interfacial tensions, Phys. Rev. Lett. 72 (14) (1994) 2298–2301.
[17] L. Oswald, S. Grosser, D.M. Smith, J.A. Käs, Jamming transitions in cancer, J. Phys. D: Appl. Phys. 50 (48) (2017) 483001.
[18] T. Savin, N.A. Kurpios, A.E. Shyer, P. Florescu, H. Liang, L. Mahadevan, et al., On the growth and form of the gut, Nature 476 (7358) (2011) 57–62.
[19] N.I. Petridou, C.P. Heisenberg, Tissue rheology in embryonic organization, EMBO J. 38 (20) (2019) 1–13.
[20] R.A. Foty, C.M. Pfleger, G. Forgacs, M.S. Steinberg, Surface tensions of embryonic tissues predict their mutual envelopment behavior, Development 122 (5) (1996) 1611–1620.
[21] M.L. Manning, R.A. Foty, M.S. Steinberg, E.M. Schoetz, Coaction of intercellular adhesion and cortical tension specifies tissue surface tension, Proc. Natl. Acad. Sci. U. S. A. 107 (28) (2010) 12517–12522.
[22] R. David, O. Luu, E.W. Damm, J.W.H. Wen, M. Nagel, R. Winklbauer, Tissue cohesion and the mechanics of cell rearrangement, Development 141 (19) (2014) 3672–3682.
[23] G. Forgacs, R.A. Foty, Y. Shafrir, M.S. Steinberg, Viscoelastic properties of living embryonic tissues: a quantitative study, Biophys. J. 74 (5) (1998) 2227–2234.
[24] E.H. Barriga, R. Mayor, Adjustable viscoelasticity allows for efficient collective cell migration, Semin. Cell Dev. Biol. 93 (2019) 55–68.
[25] N. Khalilgharibi, J. Fouchard, P. Recho, G. Charras, A. Kabla, The dynamic mechanical properties of cellularised aggregates, Curr. Opin. Cell Biol. 42 (2016) 113–120.
[26] A. Mongera, P. Rowghanian, H.J. Gustafson, E. Shelton, D.A. Kealhofer, E.K. Carn, et al., A fluid-to-solid jamming transition underlies vertebrate body axis elongation, Nature 561 (7723) (2018) 401–405.
[27] B. Wallmeyer, S. Trinschek, S. Yigit, U. Thiele, T. Betz, Collective cell migration in embryogenesis follows the laws of wetting, Biophys. J. 114 (2018) 213–222.
[28] S. Tripathy, E.J. Berger, Measuring viscoelasticity of soft samples using atomic force microscopy, J. Biomech. Eng. 131 (9) (2009).
[29] A. D'Angelo, K. Dierkes, C. Carolis, G. Salbreux, J. Solon, *In vivo* force application reveals a fast tissue softening and external friction increase during early embryogenesis, Curr. Biol. 29 (9) (2019) 1564–1571.e6.
[30] K. Doubrovinski, M. Swan, O. Polyakov, E.F. Wieschaus, Measurement of cortical elasticity in *Drosophila melanogaster* embryos using ferrofluids, Proc. Natl. Acad. Sci. U. S. A. 114 (5) (2017) 1051–1056.
[31] Y.L. Han, A.F. Pegoraro, H. Li, K. Li, Y. Yuan, G. Xu, et al., Cell swelling, softening and invasion in a three-dimensional breast cancer model, Nat. Phys. (2019) 1–10.

[32] A.D. Wessel, M. Gumalla, J. Grosshans, C.F. Schmidt, The mechanical properties of early drosophila embryos measured by high-speed video microrheology, Biophys. J. 108 (8) (2015).
[33] O. Campàs, T. Mammoto, R.A. Hasso, S. Sperling, D. O'Connell, A.G. Bischof, et al., Quantifying cell-generated mechanical forces within living embryonic tissues, Nat. Methods 11 (2) (2014) 183–189.
[34] W. Lee, N. Kalashnikov, S. Mok, R. Halaoui, E. Kuzmin, A.J. Putnam, et al., Dispersible hydrogel force sensors reveal patterns of solid mechanical stress in multicellular spheroid cultures, Nat. Commun. 10 (1) (2019) 144.
[35] E. Mohagheghian, J. Luo, J. Chen, G. Chaudhary, J. Chen, J. Sun, et al., Quantifying compressive forces between living cell layers and within tissues using elastic round microgels, Nat. Commun. (2018) 1–14.
[36] D. Gonzalez-Rodriguez, L. Bonnemay, J. Elgeti, S. Dufour, D. Cuvelier, F. Brochard-Wyart, Detachment and fracture of cellular aggregates, Soft Matter 9 (7) (2013) 2282–2290.
[37] A.R. Harris, J. Bellis, N. Khalilgharibi, T. Wyatt, B. Baum, A.J. Kabla, et al., Generating suspended cell monolayers for mechanobiological studies, Nat. Protoc. 8 (12) (2013) 2516–2530.
[38] A. Mgharbel, H. Delanoë-Ayari, J.-P. Rieu, Measuring accurately liquid and tissue surface tension with a compression plate tensiometer, HFSP J. 3 (2009) 213–221.
[39] T. Aoki, T. Ohashi, T. Matsumoto, M. Sato, The pipette aspiration applied to the local stiffness measurement of soft tissues, Ann. Biomed. Eng. 25 (1997) 581–587.
[40] K. Guevorkian, M.J. Colbert, M. Durth, S. Dufour, F. Brochard-Wyart, Aspiration of biological viscoelastic drops, Phys. Rev. Lett. 104 (2010) 218101.
[41] B. González-Bermúdez, G.V. Guinea, G.R. Plaza, Advances in micropipette aspiration: applications in cell biomechanics, models, and extended studies, Biophys. J. 116 (4) (2019) 587–594.
[42] K. Guevorkian, J.-L. Maître, Micropipette aspiration: a unique tool for exploring cell and tissue mechanics in vivoMethods in Cell Biology, vol. 139 Cell Polarity and Morphogenesis, Academic Press, 2017, pp. 187–201.
[43] R.M. Hochmuth, Micropipette aspiration of living cells, J. Biomech. 33 (1) (2000) 15–22.
[44] F. Guilak, J.R. Tedrow, R. Burgkart, Viscoelastic properties of the cell nucleus, Biochem. Biophys. Res. Commun. 269 (3) (2000) 781–786.
[45] E. Evans, A. Yeung, Apparent viscosity and cortical tension of blood granulocytes determined by micropipet aspiration, Biophys. J. 56 (1) (1989) 151–160.
[46] J.L. Maître, H. Turlier, R. Illukkumbura, B. Eismann, R. Niwayama, F. Nédélec, et al., Asymmetric division of contractile domains couples cell positioning and fate specification, Nature (2016) 1–14.
[47] A. Chaigne, C. Campillo, N.S. Gov, R. Voituriez, J. Azoury, C. Umaña-Diaz, et al., A soft cortex is essential for asymmetric spindle positioning in mouse oocytes, Nat. Cell Biol. 15 (8) (2013) 958–966.
[48] A. Nava, E. Mazza, M. Furrer, P. Villiger, W.H. Reinhart, In vivo mechanical characterization of human liver, Med. Image Anal. 12 (2) (2008) 203–216.
[49] Sutter Instruments. Pipette cookbook. https://www.sutter.com/PDFs/pipette_cookbook.pdf.
[50] Y.S. Chu, W.A. Thomas, O. Eder, F. Pincet, E. Perez, J.P. Thiery, et al., Force measurements in e-cadherin mediated cell doublets reveal rapid adhesion strengthened by actin cytoskeleton remodeling through rac and cdc42, J. Cell Biol. 167 (6) (2004) 1183.

[51] P.L. Ryan, R.A. Foty, J. Kohn, M.S. Steinberg, Tissue spreading on implantable substrates is a competitive outcome of cell-cell vs. cell-substratum adhesivity, Proc. Natl. Acad. Sci. U. S. A. 98 (8) (2001) 4323–4327.
[52] R. Foty, A simple hanging drop cell culture protocol for generation of 3D spheroids, J. Vis. Exp. 51 (2011) e2720.
[53] Y.C. Chen, X. Lou, Z. Zhang, P. Ingram, E. Yoon, High-throughput cancer cell sphere formation for characterizing the efficacy of photo dynamic therapy in 3D cell cultures, Sci. Rep. 5 (1) (2015) 12175.
[54] K. Guevorkian, D. Gonzalez-Rodriguez, C. Carlier, S. Dufour, F. Brochard-Wyart, Mechanosensitive shivering of model tissues under controlled aspiration, Proc. Natl. Acad. Sci. U. S. A. 108 (33) (2011) 13387–13392.
[55] D. Gonzalez-Rodriguez, K. Guevorkian, S. Douezan, F. Brochard-Wyart, Soft matter models of developing tissues and tumors, Science 338 (6109) (2012) 910–917.
[56] J.C.M. Mombach, D. Robert, F. Graner, G. Gillet, G.L. Thomas, M. Idiart, et al., Rounding of aggregates of biological cells: experiments and simulations, Phys. A: Stat. Mech. Appl. 352 (2) (2005) 525–534.
[57] K. Jakab, B. Damon, F. Marga, O. Doaga, V. Mironov, I. Kosztin, et al., Relating cell and tissue mechanics: implications and applications, Dev. Dyn. 237 (9) (2008) 2438–2449.
[58] Z. Dagan, S. Weinbaum, R. Pfeffer, An infinite-series solution for the creeping motion through an orifice of finite length, J. Fluid Mech. 115 (1982) 505–523.
[59] K. Piroird, C. Clanet, D. Quéré, Capillary extraction, Langmuir 27 (15) (2011) 9396–9402.
[60] P. Marmottant, Effective viscosity of cell aggregates: the role of fluctuations and stress, Proc. Natl. Acad. Sci. USA 12 (2009).
[61] M.A. Swartz, M.E. Fleury, Interstitial flow and its effects in soft tissues, Annu. Rev. Biomed. Eng. 9 (8) (2007) 229–256.
[62] Q.D. Tran, M. Marcos, D. Gonzalez-Rodriguez, Permeability and viscoelastic fracture of a model tumor under interstitial flow, Soft Matter 14 (2018) 6386–6392.
[63] Q.D. Tran, Marcos, D. Gonzalez-Rodriguez, Quantitative characterization of viscoelastic fracture induced by time-dependent intratumoral pressure in a 3D model tumor, Biomicrofluidics 13 (2019) 054107.
[64] D. Ambrosi, S. Pezzuto, D. Riccobelli, T. Stylianopoulos, P. Ciarletta, Solid tumors are poroelastic solids with a chemo-mechanical feedback on growth, J. Elast. 129 (1-2) (2017) 107–124.
[65] L. Kashkooli, D. Rozema, L. Espejo-Ramirez, P. Lasko, F. Fagotto. Ectoderm to mesoderm transition by downregulation of actomyosin contractility. *bioRxiv* (2019).
[66] D. Bonazzi, V. Lo Schiavo, S. Machata, I. Djafer-Cherif, P. Nivoit, V. Manriquez, et al., Intermittent pili-mediated forces fluidize neisseria meningitidis aggregates promoting vascular colonization, Cell 174 (1) (2018) 143–155.
[67] J. Wen, H. Tao, K. Lau, H. Liu, C.A. Simmons, Y. Sun, et al., Cell and tissue scale forces coregulate fgfr2-dependent tetrads and rosettes in the mouse embryo, Biophys. J. 112 (10) (2017) 2209–2218.

CHAPTER NINE

# Viscoelasticity of multicellular systems caused by collective cell migration: multiscale modeling considerations

**Ivana Pajic-Lijakovic and Milan Milivojevic**
Faculty of Technology and Metallurgy, Belgrade University, Belgrade, Serbia

## 9.1 Introduction

A deep insight into the relation between viscoelasticity and collective cell migration is essential for understanding wide range of biological processes such as embryo morphogenesis, wound healing and cancer invasion [1–5]. Long-time viscoelasticity depends on (1) the volume fraction and configuration of migrating cells and the rate of its change, (2) the viscoelasticity of migrating cell clusters, and (3) the viscoelasticity of surrounding resting cells [1,2,6–8]. A pragmatic approach to such a multidisciplinary field sometimes overshadows some serious problems which arise from the complexity of multiscale interactions, so adequate interpretation should include theoretical considerations at both mesoscopic and macroscopic levels. The framework for further interpretation of the problem can be the phenomenological theory of the dynamical behavior of continua, which considers the dissipative effects that accompany deformation. To interpret such complex dissipative effects on various scales in biological systems and connect them with the viscoelasticity of multicellular systems, the irreversible thermodynamics approach is needed. We study this problem at mesoscopic and macroscopic levels by applying both the basic principles of irreversible thermodynamics [9,10] and rheology based on mechanical coupling modes [11,12].

The key parameter that influences the viscoelasticity is the size, fluidity, and thickness of the biointerface between migrating and resting cell subpopulations. The size of the biointerface is related to the configuration of migrating cells. Multicellular systems can form various cellular configurations depending

*Viscoelasticity and Collective Cell Migration.*
DOI: https://doi.org/10.1016/B978-0-12-820310-1.00002-1

on cell type and loading conditions. Mikami et al. [13] discussed collective cell migration of stratified epithelial cells toward the wound in the form of monolayer sheets. All epithelial cells within the sheet migrate, maintaining cell–cell adhesions [14]. Migrating cell sheets slide over the surrounding cell layers in the resting state [15]. A number of sheets and their sizes depend on the size, shape, and depth of injury. Cell organization in the form of migrating cell sheets and their sliding over the surrounding unperturbed cell layers of epithelium pointed to an ordered lamellar structure. Friedl and Alexander [16] considered collective cell migration during cancer invasion and metastases. They concluded that some tumor types can migrate within partially connected strands through surrounding tissue, while others can migrate in the forms of monolayer sheets or small cell clusters. Some other cell types can also migrate within small clusters through surrounding tissue [1]. Consequently, three types of migrating cell configurations can be distinguished: (1) ordered lamellar type and less ordered types such as (2) partially connected cell strands, and (3) dispersed small cell clusters. Shellard and Mayor [17] reported that larger migrated cluster behaves as a supracellular unit by establishing a supracellular cytoskeleton organization.

The sliding phenomenon is related to slip effects that influence the biointerface fluidity [18]. Slip effects are connected with establishing the velocity gradient surrounding the migrating clusters. Pronounced slip effects point out that surrounding cells stay in resting state and their velocity is equal to zero. The fluidity depends on the state of cell–cell adhesion contacts under stress near the migrating cluster. Rozbicki et al. [19] and Shellard and Mayor [17] considered a movement of larger cellular clusters and experimentally determined two regions: (1) ordered region of the cluster and (2) surrounding perturbed boundary layers. Circular movement of cells is observed within the perturbed layers. This circular movement of cells within the perturbed layers ensures the decrease of the viscosity. The multiscale nature of the biointerface dynamics represents the product of residual stress accumulation (1) within the core regions of migrating clusters during their movement through dense surrounding (internal effects), (2) the collisions of the velocity fronts (external effects). The accumulated stress can induce disordering of migrating cell clusters and consequently migrating-to-resting cell state transition which leads to change the biointerface size. The collision of the velocity fronts also induces a local increase of compressive stress and can lead to stagnant zone formation and decrease of the biointerface size. A collision of velocity fronts is caused by uncorrelated motility. Uncorrelated motility has been related to perturbations of cell signaling and delay of gene expression [3].

Gene expression induces time delay in cell response to various mechanical and biochemical stimuli [3]. This time delay might be relevant for cell coupling because what cells acquire at the present time is the information of surrounding cells some time ago. These perturbations can induce that (1) cells in the same population respond to different signals and/or (2) cells behave differently in response to the same signals [5] that could lead to uncorrelated motility. The uncorrelated motility caused by a collision of velocity fronts is significant even in 2D [20]. Cumulative effects of these mechanical and biochemical processes influence the configuration of migrating cells and the rate of its changes and on that base the viscoelasticity of multicellular systems on various space and timescales [2,7,8].

Motivated by these recent findings, we pointed to the relation between the biointerface size, fluidity, and thickness on long-time viscoelasticity of multicellular surfaces through mechanical coupling modes. Formulation of constitutive models at mesoscopic and macroscopic levels from the standpoints of thermodynamics and rheology was discussed as a function of volume fraction and configuration of migrating cells depending on mechanical and biochemical perturbations.

## 9.2 Phenomenological description of long-time rearrangement of multicellular surfaces under external stress

Two-model systems were considered in the context of viscoelasticity caused by collective cell migration. One of them is cell aggregate micropipette aspiration, while the other is cell aggregate uniaxial compression between parallel plates and aggregate rounding after compression (Fig. 9.1).

Micropipette aspiration induces excitation of an aggregate surface part $\Delta V_{sr}$, while the aggregate compression between parallel plates induces excitation of the whole aggregate surface region $V_{sr} = A_{sr}(t)L_{sr}(t)$ (where $A_{sr}(t)$ is the aggregate surface and $L_{sr}(t)$ is the thickness of the aggregate surface region). Both systems were considered at a timescale of hours which corresponds to collective cell migration.

Aggregate surface rearrangement (1) under uniaxial compression and (2) during aggregate rounding after compression satisfied the condition that the volume of the perturbed surface region is constant. Cell aggregate surface $A_{sr}(t)$ increases and the thickness of the surface region $L_{sr}(t)$

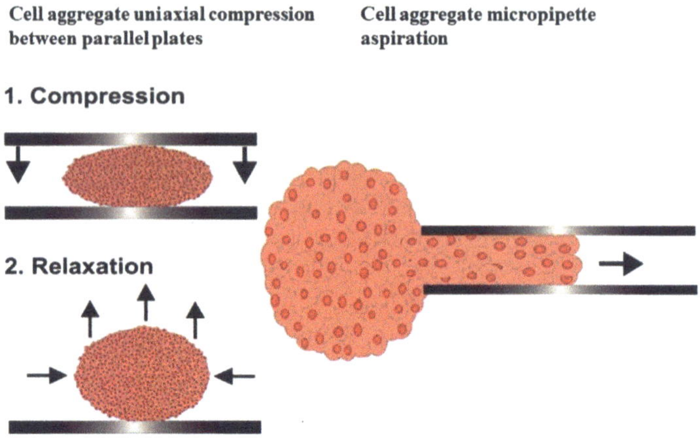


**Figure 9.1** Considered model systems are (1) cell aggregate uniaxial compression and free relaxation after compression and (2) cell aggregate micropipette aspiration.

decreases under compression, while the volume $V_{sr}$ stays approximately constant. On the other hand, aggregate surface $A_{sr}(t)$ decreases and the thickness of the surface region $L_{sr}(t)$ increases during the aggregate rounding after compression, while the volume $V_{sr}$ stays constant. The number of cells in the surface region is approximately constant, that is, $N \approx const.$

Micropipette aspiration induces excitation of $N$ cells locally within a surface part. One part of exciting cells is instantaneously introduced into pipette, while the other part undergoes collective cell migration toward the pipette depending on the applied pressure. Guevorkian et al. [21] applied pressure locally in the range of 0.5–1.2 kPa. The lower pressure of 0.5 kPa induces activation of perturbed cells and their entrance into the pipette via collective cell migration during 120 minutes. However, under higher pressure, one part of perturbed cells is instantaneously inserted into the pipette. These cells undergo collective cell migration within the pipette, while the other part of already perturbed cells enters into the pipette.

Consequently, the population of perturbed cells under externally applied stress in both cases can be treated thermodynamically as the closed system and consider as the canonical ensemble at a macroscopic level. However, at the mesoscopic level, surface domains can be treated as thermodynamic open subsystems due to the movement of migrating cell clusters from one domain to the other. Externally applied stress $\tilde{\sigma}_{ext}$ induced internally generated stress $\Delta\tilde{\Pi}_{eff}(q_1, q_2, \ldots q_n)$ unhomogeneously distributed within the multicellular surface, while $q_1, q_2, \ldots q_n$ represent exciting degrees of freedom. The internal stress can induce resting-to-migrating cell state transition depending on the intensity and duration of applied stress. Corresponding cause–consequence relation has been described based on Edward's statistic [7]. Edward's statistic is developed for the rearrangement of grain systems under externally applied stress [22,23]. Cellular systems are much complex and need much more degrees of freedom. However, the general thermodynamical principles can be applied in the form:

$$\exp(-\tilde{\sigma}_{ext}\Delta f^{-1}(\lambda_E \tilde{Z})) = \int \exp(-\Delta\tilde{\Pi}_{eff}(q_1, q_2, \ldots q_n)\Delta f^{-1}(\lambda_E \tilde{Z}))dq_1 dq_2 \ldots dq_n \quad (9.1)$$

where $f^{-1}(\lambda_E \tilde{Z})$ is the inverse map and $\lambda_E \tilde{Z}$ represents the stress "quanta" by the analogy with thermal quanta equal to $k_B T$, $k_B$ is Boltzmann constant, and $T$ is temperature. Externally applied stress $\tilde{\sigma}_{ext}$ corresponds to free energy in classical thermodynamics, while the internally generated stress $\Delta\tilde{\Pi}_{eff}(q_1, q_2, \ldots q_n)$ corresponds to Hamiltonian in classical thermodynamics.

Cell surface rearrangement caused by collective cell migration has been treated as a thermodynamic system close to equilibrium at the macroscopic level [8,24]. However, cell surface rearrangement considered at a mesoscopic level has been treated as a thermodynamical system far from equilibrium [2,6]. It is in accordance with the fact that internal fluctuations, which are significant during system rearrangement at the mesoscopic level, can be neglected at the macroscopic level [25,26].

Linear nature of the cell rearrangement obtained experimentally at a macroscopic level [27,28] has been modeled by applying Eyring transition state theory [8,24,28]. However, thermodynamic fluctuations significant at the mesoscopic level can induce nonlinear behavior of the systems at the mesoscopic level [2,6].

A significant difference in cell stiffness between migrating and resting cell clusters indicates that volume fraction of migrating cells and their distribution can influence long-time viscoelasticity of multicellular surfaces [2,6]. This aspect of cell surface rearrangement has been treated by the analogy with physics in the form of a multiphase system, that is, pseudo-blend [2,6]. Migrating cells can form three types of configuration: (1) monolayer sheets (lamellar structure), (2) small clusters, and (3) and various combinations of these two types. The lamellar structure of pseudo-blend corresponds to parallel mode coupling, while the highly dispersed system in the form of small clusters points to series mode coupling. Mixed-mode

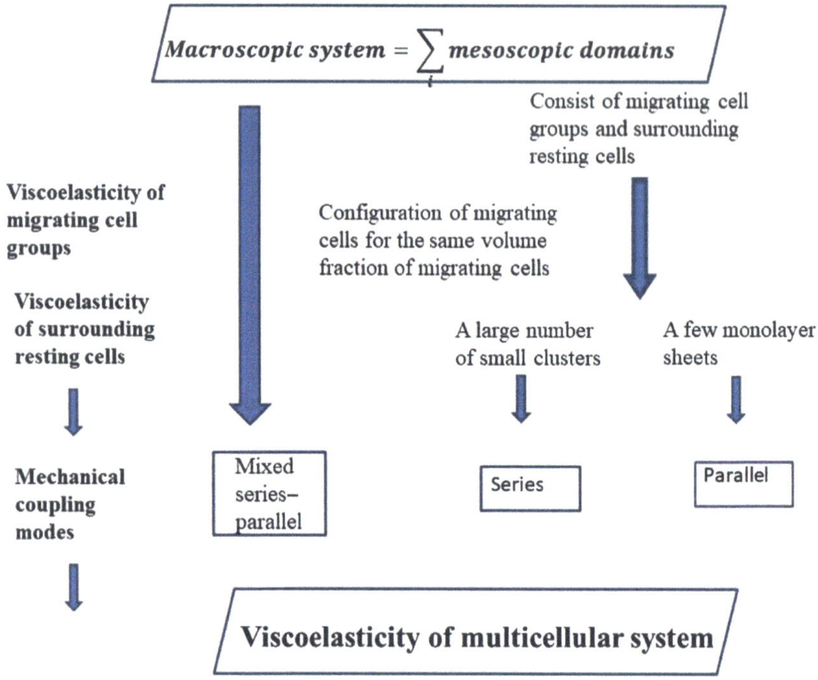


Figure 9.2 Schematic presentation of the modeling strategy at mesoscopic and macroscopic levels.

coupling, which accounts for both fractions of cells (migrating and resting) coupled in series and in parallel, is suitable for considering configuration changes of migrating cells and its influence on viscoelasticity at the macroscopic level. However, at the mesoscopic level, we will consider surface domains that correspond to cell migration in (1) the form of small clusters, modeled by series mode coupling, and (2) the form of monolayer sheets, modeled by parallel mode coupling.

The schematic presentation of the modeling strategy at mesoscopic and macroscopic levels is shown in Fig. 9.2.

## 9.3 Long-time viscoelasticity at a mesoscopic level—constitutive modeling

Viscoelasticity caused by collective cell migration at a mesoscopic level should account for the contributions: (1) migrating cell pseudo-phase, (2) resting cell pseudo-phase, and (3) interface between migrating and resting cell "pseudo-phases." Consequently, the formulation of the constitutive model should include the following steps [6]:

1. Formulation of the constitutive model for migrating cell subpopulation as:

$$\tilde{\sigma}_m(r,t) = \tilde{\sigma}_m(\tilde{\varepsilon}_m(r,t)). \tag{9.2}$$

where $\tilde{\sigma}_m(r,t)$ and $\tilde{\varepsilon}_m(r,t)$ are the local stress and the strain within migrating cell clusters. The complex modulus of migrating pseudo-phase can be expressed as $G_m^*(r,\omega) = \frac{F(\tilde{\sigma}_m(r,t))}{F(\tilde{\varepsilon}_m(r,t))}$ (where $F[^\circ]$ is the Fourier operator and $\omega$ is angular velocity equal to $\omega = \frac{2\pi}{t}$).

2. Formulation of the constitutive model for resting cell subpopulation as:

$$\tilde{\sigma}_r(r,t) = \tilde{\sigma}_r(\tilde{\varepsilon}_r(r,t)) \tag{9.3}$$

where $\tilde{\sigma}_r\ (r,t)$ and $\tilde{\varepsilon}_r\ (r,t)$ are the local stress and the strain within resting cell clusters, respectively. The complex modulus of resting pseudo-phase can be expressed as $G_r^*(r,\omega) = \frac{F(\tilde{\sigma}_r(r,t))}{F(\tilde{\varepsilon}_r(r,t))}$.

Joanny and Prost [29] and Notbohm et al. [20] defined the active, migrating state of cells by (1) passive, resting stress $\tilde{\sigma}_r(r,t)$ and (2) the additional contribution to stress due to cell activation $\tilde{\sigma}_A(r,t)$. Consequently, the stress of migrating cell cluster has been expressed as $\tilde{\sigma}_m(r,t) = \tilde{\sigma}_r(r,t) + \tilde{\sigma}_A(r,t)$. They described cell passive contribution in the context of rheological elements—spring and dashpot. Notbohm et al. [20] described cell passive, resting state as purely reversible by spring element as $\tilde{\sigma}_r(r,t) = E_P\tilde{\varepsilon}_r(r,t)$ [where $E_P$ is the elastic modulus of cells in the passive (resting) state]. However, Joanny and Prost [29] considered cell passive state as more dissipative. They proposed a modified rheological element consists of dashpot and additional contribution caused by cell orientation. It is expressed as $\tilde{\sigma}_r(r,t) = \eta_P \frac{d\tilde{\varepsilon}_r(r,t)}{dt} + \nu\tilde{h}(r,t)$ [where $\eta_P$ is the viscosity of cells in the passive (resting) state, $\tilde{h}(r,t)$ is the cell orientation field which depends on cell polarization, and $\nu$ is the constitutive parameter]. The first term on the right-hand side is a dashpot element, while the second one accounts for cell orientation. Cell activation has been related to the difference between the chemical potential of adenosine triphosphate (ATP) and its hydrolysis products $\Delta\mu(r,t)$. Joanny and Prost [29] described stress contribution caused by cell activation as $\tilde{\sigma}_A(r,t) = \xi\Delta\mu(r,t)\tilde{I}$ (where $\xi$ is the activity coefficient, and $\tilde{I}$ is the unit tensor). Notbohm et al. [20] related $\Delta\mu(r,t)$ to the concentration of active myosin. The chemical potential difference increases with ATP concentration. Cell activation is related to contractions of actin–myosin stress fibers caused by ATP hydrolysis. Detail rheological analysis of these modeling approaches is provided by Pajic-Lijakovic and Milivojevic [6].

3. Formulation of internal dynamics at the biointerface. Interface dynamics is related to the mechanical coupling between pseudo-phases. The mechanical coupling is a function of the interface size and thickness. The local interface size is expressed as:

$$\Gamma_i(r,t) = \frac{V_m(r,t)}{A_m(r,t)} \tag{9.4}$$

where $\Gamma_i(r,t)$ is the local biointerface size, $V_m(r,t)$ is the average volume of migrating cell cluster, and $A_m(r,t)$ is the average surface of the migrating cell cluster. Lower interface size corresponds to the lamellar structure of the pseudo-blend in which cells migrate within few monolayer sheets [30,31]. Higher interface size (for the same volume fraction of migrating cells) corresponds to cell migration within a large number of small migrating clusters dispersed within the resting cell pseudo-phase [16]. The lamellar structure of the pseudo-blend leads to parallel mode coupling between pseudo-phases, while the highly dispersed structure of pseudo-blend leads to series mode coupling between pseudo-phases. Biointerface can be sharp or finite. Finite biointerface is the more realistic case, but we considered and compared both cases based on theoretical analysis.

## 9.3.1 Series and parallel mode coupling at a mesoscopic level: the sharp biointerface

The sharp biointerface represents a product of pronounced slip effects. For this condition a multicellular system can be modeled as a two-phase system. The biointerface size has been related to viscoelasticity based on mechanical coupling modes [2,6]. Series mode coupling corresponds to cell migration within small cell clusters, while parallel mode coupling corresponds to cell migration in the form of monolayer sheets. Volume fractions of migrating and resting cell pseudo-phases for the sharp biointerface satisfy the condition:

$$\phi_m(r,t) + \phi_r(r,t) = 1 \tag{9.5}$$

where $r$ is the space coordinate which corresponds to a part of the multicellular system, $\phi_m(r,t)$ is the local volume fraction of migrating cells and

$\phi_r(r,t)$ is the local volume fraction of resting cells. For parallel coupling mode the stress and strain can be expressed by [12] model proposed for two-phase polymer blends as:

$$\begin{aligned}\tilde{\sigma}_T(r,t) &= \phi_m(r,t)\tilde{\sigma}_m(r,t) + (1-\phi_m(r,t))\tilde{\sigma}_r(r,t)\\ \tilde{\varepsilon}_T(r,t) &= \tilde{\varepsilon}_m(r,t) = \tilde{\varepsilon}_r(r,t)\end{aligned} \tag{9.6}$$

where $\tilde{\sigma}_T(r,t)$ is the local surface stress within multicellular system and $\tilde{\varepsilon}_T(r,t)$ is the local surface strain within multicellular system.

For series mode coupling the stress and strain were expressed as [12]:

$$\begin{aligned}\tilde{\sigma}_T(r,t) &= \tilde{\sigma}_m(r,t) = \tilde{\sigma}_r(r,t)\\ \tilde{\varepsilon}_T(r,t) &= \phi_m(r,t)\tilde{\varepsilon}_m(r,t) + (1-\phi_m(r,t))\tilde{\varepsilon}_r(r,t)\end{aligned} \tag{9.7}$$

Series mode coupling leads to the softer tissue domain than that obtained for the parallel mode coupling for the same volume fraction of migrating cells due to intensive energy dissipation within the biointerface [2,6].

#### *9.3.1.1 Configuration changes of migrating cells for the sharp biointerface*

Cell surface rearrangement considered at a mesoscopic level has been treated as a thermodynamical open system far from equilibrium [2,6]. The change of the volume fraction of migrating cells for sharp biointerface at a mesoscopic level was expressed by the modified model for viscoelastic phase separation proposed for viscoelastic solid by Tanaka [32]. The theoretical basis is now applied to the case of cellular mixtures [2,6]. The model represents the modification of phase model B proposed for thermodynamical systems far from equilibrium [9]. It is expressed as:

$$\frac{\partial \phi_m(r,t)}{\partial t} = \vec{\nabla}\left[D_{eff}\left(\phi_m(r,t)\,\vec{\nabla}\frac{\delta F(r,t)}{\delta\phi_m(r,t)}\right) + \left(\vec{\nabla}\cdot\tilde{\sigma}_m(r,t) - \vec{\nabla}\cdot\tilde{\sigma}_r(r,t)\right)\right] \tag{9.8}$$

where $D_{eff}$ is the effective diffusion coefficient, $F(r,t)$ is the free energy functional which accounts for the action of dynamic surface tension (where $F(r,t) = \int \gamma(r,t)d^2r$ and $\gamma(r,t)$ is the surface tension). The first term on the right-hand side represents the thermodynamical affinity that accounts for the cumulative effects of biochemical processes such as cell adhesion, signaling, and gene expressions, while the stress difference (the second term) represents the mechanical driving force for cell migration. The volume fraction of migrating cells also influences the local stress depending on the mechanical coupling mode (Eqs. 9.6 and 9.7). Long-time cell rearrangement at a mesoscopic level can satisfy one of the following conditions: (1) unsteady-state dynamics and (2) steady-state dynamics. The schematic presentation of the model for the sharp biointerface at a mesoscopic level is shown in Fig. 9.3.

Unsteady-state dynamics can enhance cell migration for $\frac{\partial \phi_m(r,t)}{\partial t} > 0$ or reduce cell migration for $\frac{\partial \phi_m(r,t)}{\partial t} < 0$ [2,6]. Uncorrelated motility induces collisions of velocity fronts that can result in the formation of stagnant zones. Stagnant zones are unstable. They exist for some time and disappear and cells establish an ordered trend of rearrangement again.

Steady-state dynamics corresponds to the condition $\frac{\partial \phi_m(r,t)}{\partial t} = 0$ and $\phi_m(r)$ is time-independent. For this condition, corresponding complex modulus of multicellular surface domains $G_T^*(r,\omega) = \frac{F(\tilde{\sigma}_T(r,t))}{F(\tilde{\varepsilon}_T(r,t))}$ (where $F[°]$ is the Fourier operator) can be expressed for series mode coupling and for parallel mode coupling. Complex modulus for (1) parallel mode coupling is expressed as $\frac{G_T^*(r,\omega)}{G_r^*(r,\omega)} = \phi_m(r)\frac{G_m^*(r,\omega)}{G_r^*(r,\omega)} + (1-\phi_m(r))$ and (2) series mode coupling is expressed as $\frac{G_T^*(r,\omega)}{G_r^*(r,\omega)} = \frac{1}{(1-\phi_m(r)) + \frac{\phi_m(r)}{\frac{G_m^*(r,\omega)}{G_r^*(r,\omega)}}}$. The modulus $G_T^*(r,\omega)$ should satisfy the conditions: (1) for $\phi_m(r) = 0$, the modulus $G_T^*(r,\omega) = G_r^*(r,\omega)$ and (2) for $\phi_m(r) = 1$, the modulus ratio is equal to $G_T^*(r,\omega) = G_m^*(r,\omega)$. Pajic-Lijakovic and Milivojevic [6] reported that the modulus ratio increases slowly for series mode coupling such that

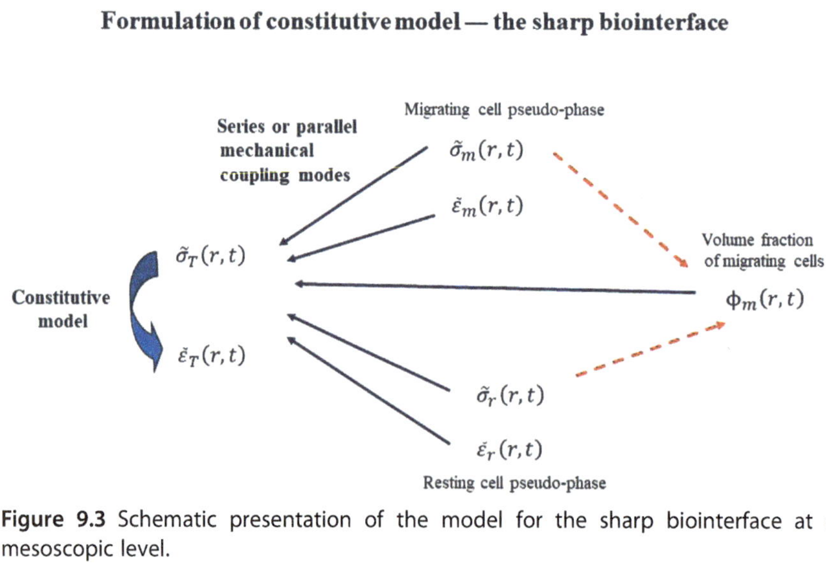


**Figure 9.3** Schematic presentation of the model for the sharp biointerface at a mesoscopic level.

$\frac{G_T^*(r,\omega)}{G_r^*(r,\omega)} < 5$ for $\phi_m(r) \leq 0.8$ and then increases rapidly with the volume fraction of migrating cells. The modulus ratio increases more intensive and linearly with the volume fraction of migrating cells for parallel mode coupling compared with the series mode coupling.

## 9.3.2 Series and parallel mode coupling at a mesoscopic level: the finite biointerface

The finite biointerface is realized by reducing slip effects which leads to an increase in the velocity gradient through the interface. For this condition, three pseudo-phases should be accounted for modeling consideration [18]. The finite biointerface represents boundary layers around migrating clusters which corresponds to a lamellar structure between the migrating and the biointerface pseudo-phases. Constitutive model of the finite biointerface can be expressed as $\tilde{\sigma}_{int}(r,t) = \tilde{\sigma}_{int}(\tilde{\varepsilon}_{int}(r,t))$ (where $\tilde{\sigma}_{int}(r,t)$ is the local stress and $\tilde{\varepsilon}_{int}(r,t)$ is the local strain within the finite biointerface). The corresponding complex modulus can be expressed as $G_{int}^*(r,\omega) = \frac{F(\tilde{\sigma}_{int}(r,t))}{F(\tilde{\varepsilon}_{int}(r,t))}$ (where $F[^\circ]$ is the Fourier operator and $\omega$ is angular velocity equal to $\omega = \frac{2\pi}{t}$). The complex modulus $G_{int}^*(r,\omega)$ satisfies the condition $G_r^*(r,\omega) \leq G_{int}^*(r,\omega) \leq G_m^*(r,\omega)$ depending on the slip effects.

Volume fractions of all pseudo-phases should satisfy the condition:

$$\phi_m(r,t) + \phi_r(r,t) + \phi_{int}(r,t) = 1 \tag{9.9}$$

where $\phi_{int}(r,t)$ is the local volume fraction of cells within the biointerface. For parallel coupling mode between migrating cell pseudo-phase and the biointerface, the stress and strain can be expressed as:

$$\begin{aligned} \tilde{\sigma}_{m-int}(r,t) &= \phi_m^*(r,t)\tilde{\sigma}_m(r,t) + \phi_{int}^*(r,t)\tilde{\sigma}_{int}(r,t) \\ \tilde{\varepsilon}_{m-int}(r,t) &= \tilde{\varepsilon}_m(r,t) = \tilde{\varepsilon}_{int}(r,t) \end{aligned} \tag{9.10}$$

where $\tilde{\sigma}_{m-int}(r,t)$ is the local stress arising by coupling the biointerface and migrating cell pseudo-phase; $\tilde{\sigma}_{int}(r,t)$ is the local stress at the biointerface; $\tilde{\varepsilon}_{m-int}(r,t)$ is the local strain arising from coupling the biointerface and migrating cell pseudo-phase; $\tilde{\varepsilon}_{int}(r,t)$ is the local strain at the biointerface; $\phi_m^*(r,t)$ is the volume fraction of migrating cells, which is equal to $\phi_m^*(r,t) = \frac{\phi_m(r,t)}{1-\phi_r(r,t)}$; and $\phi_{int}^*(r,t)$ is the volume fraction of cells, which is equal to $\phi_{int}^*(r,t) = \frac{\phi_{int}(r,t)}{1-\phi_r(r,t)}$. Thus defined $m-int$ pseudo-phase can be further coupled with the resting cell pseudo-phase. This coupling can be modeled as a series or parallel depending on the configuration of migrating cells. For cell migration within a small migrating cell clusters enveloped by biointerface layers, series mode coupling is suitable and expressed as:

$$\begin{aligned} \tilde{\sigma}_T(r,t) &= \tilde{\sigma}_{m-int}(r,t) = \tilde{\sigma}_r(r,t) \\ \tilde{\varepsilon}_T(r,t) &= (\phi_m(r,t) + \phi_{int}(r,t))\tilde{\varepsilon}_{m-int}(r,t) + \phi_r(r,t)\tilde{\varepsilon}_r(r,t) \end{aligned} \tag{9.11}$$

where $\phi_r(r,t)$ is the volume fraction of resting cells, which is equal to $\phi_r(r,t) = 1 - \phi_m(r,t) - \phi_{int}(r,t)$. For cell migration within a few monolayer sheets enveloped by biointerface layers, parallel mode coupling is suitable and expressed as:

$$\begin{aligned} \tilde{\sigma}_T(r,t) &= (\phi_m(r,t) + \phi_{int}(r,t))\tilde{\sigma}_{m-int}(r,t) + \phi_r(r,t)\tilde{\sigma}_r(r,t) \\ \tilde{\varepsilon}_T(r,t) &= \tilde{\varepsilon}_{m-int}(r,t) = \tilde{\varepsilon}_r(r,t) \end{aligned} \tag{9.12}$$

Complex cause–consequence relations between the biointerface size, fluidity, and thickness influence long-time viscoelasticity of multicellular systems.

#### 9.3.2.1 Configuration changes of migrating cells for the finite biointerface

The change of the volume fraction of migrating cells for the finite biointerface at a mesoscopic level can be expressed by modified model (Eq. 9.8). In this case, two equations are needed, one for the volume fraction of migrating cells $\phi_m(r,t)$ and the other for the volume fraction of cells within the biointerface. The model can be formulated by Pajic-Lijakovic and Milivojevic [18] as:

$$\begin{aligned} \frac{\partial \phi_m(r,t)}{\partial t} &= \vec{\nabla}\left[D_{eff}\left(\phi_m(r,t)\,\vec{\nabla}\frac{\delta F(r,t)}{\delta \phi_m(r,t)}\right) + \left(\vec{\nabla}\cdot\tilde{\sigma}_m(r,t) - \vec{\nabla}\cdot\tilde{\sigma}_{int}(r,t)\right)\right] \\ \frac{\partial \phi_{int}(r,t)}{\partial t} &= \vec{\nabla}\left[D_{eff}\left(\phi_{int}(r,t)\,\vec{\nabla}\frac{\delta F(r,t)}{\delta \phi_{int}(r,t)}\right) + \left(\vec{\nabla}\cdot\tilde{\sigma}_{int}(r,t) - \vec{\nabla}\cdot\tilde{\sigma}_r(r,t)\right)\right] \end{aligned} \tag{9.13}$$

The volume fractions $\phi_m(r,t)$ and $\phi_{int}(r,t)$ also influence the local stress depending on the mechanical coupling mode (Eqs. 9.10–9.12). Long-time cell rearrangement at a mesoscopic level can satisfy one of the following conditions: (1) unsteady-state dynamics or (2) steady-state dynamics. The schematic presentation of the model for the finite biointerface is shown in Fig. 9.4.

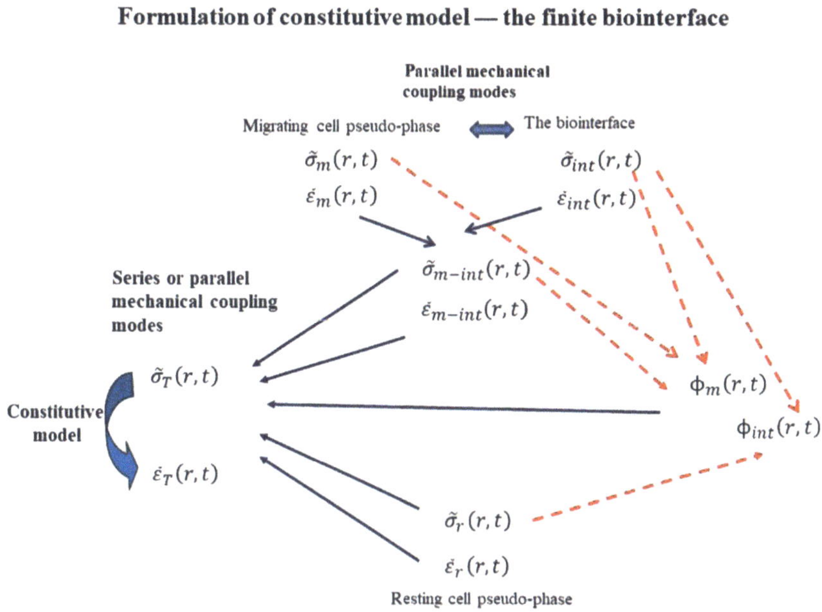


**Figure 9.4** Schematic presentation of the model for the finite biointerface at a mesoscopic level.

Unsteady-state dynamics can (1) enhance cell migration for $\frac{\partial \phi_m(r,t)}{\partial t} > 0$ and intensive perturbations within a biointerface for $\frac{\partial \phi_{int}(r,t)}{\partial t} > 0$ and (2) reduce cell migration for $\frac{\partial \phi_m(r,t)}{\partial t} < 0$ and ordered cell rearrangement within a biointerface for $\frac{\partial \phi_{int}(r,t)}{\partial t} < 0$.

Steady-state dynamics corresponds to the condition: $\frac{\partial \phi_m(r,t)}{\partial t} = 0$, $\phi_m(r)$ is time-independent, as well as $\frac{\partial \phi_{int}(r,t)}{\partial t} = 0$ and $\phi_{int}(r)$ is time-independent. For these conditions the complex modulus for parallel coupling between migrating cell clusters and the finite biointerface can be expressed as $\frac{G^*_{m-int}(r,\omega)}{G^*_r(r,\omega)} = \phi^*_m(r)\frac{G^*_m(r,\omega)}{G^*_r(r,\omega)} + \phi^*_{int}(r)\frac{G^*_{int}(r,\omega)}{G^*_r(r,\omega)}$ (where $\phi^*_m(r) = \frac{\phi_m(r)}{1-\phi_r(r)}$ and $\phi^*_{int}(r) = \frac{\phi_{int}(r)}{1-\phi_r(r)}$ while the modulus ratio $\frac{G^*_{m-int}(r,\omega)}{G^*_r(r,\omega)}$ is in the range $\frac{G^*_{m-int}(r,\omega)}{G^*_r(r,\omega)} \in [1, \frac{G^*_m(r,\omega)}{G^*_r(r,\omega)}]$). When $\phi^*_{int}(r) \to 0$ then $\phi^*_m(r) \to 1$ and $\frac{G^*_{m-int}(r,\omega)}{G^*_r(r,\omega)} \to \frac{G^*_m(r,\omega)}{G^*_r(r,\omega)}$ (the sharp biointerface).

The complex modulus of the surface domain for (1) parallel mode coupling is expressed as $\frac{G^*_T(r,\omega)}{G^*_r(r,\omega)} = (\phi_m(r) + \phi_{int}(r))\frac{G^*_{m-int}(r,\omega)}{G^*_r(r,\omega)} + (1 - \phi_m(r) - \phi_{int}(r))$ and (2) series mode coupling is expressed as $\frac{G^*_T(r,\omega)}{G^*_r(r,\omega)} = \frac{1}{(1-\phi_m(r)-\phi_{int}(r)) + \frac{\phi_m(r)+\phi_{int}(r)}{\frac{G^*_{m-int}(r,\omega)}{G^*_r(r,\omega)}}}$ [18]. The modulus $G^*_T(r,\omega)$ should satisfy the conditions: (1) for $\phi_m(r) = 0$ and $\phi_{int}(r) = 0$ the modulus $G_T^*(r,\omega) = G_r^*(r,\omega)$ and (2) for $\phi_m(r) + \phi_{int}(r) = 1$ the modulus ratio is equal to $G_T^*(r,\omega) = G_{m-int}^*(r,\omega)$.

### 9.3.3 Interface size influences viscoelasticity by mechanical coupling modes

The biointerface size is a key factor that influences the rheological behavior of multicellular surfaces through the mechanical coupling modes [2]. It depends on cell type and microenvironmental conditions and can be perturbed by uncorrelated motility. Larger interface size that corresponds to a large number of small migrating cell clusters dispersed within resting pseudo-phase is modeled by series mechanical coupling between migrating and resting cell pseudo-phases. Smaller interface size that corresponds to cell migration within a few monolayer sheets is modeled by parallel mode coupling. The increase of the interface size leads to the weakening of the multicellular surfaces due to additional energy dissipation that significantly influences their viscoelasticity [2,6]. This weakening can be quantified by the modulus ratio $\frac{G_T^*(r,\omega)}{G_r^*(r,\omega)}$. The lower value of the modulus ratio corresponds to the softer multicellular surface. Migrating cell clusters are stiffer than resting ones due to an accumulation of contractile energy [33]. Consequently, the complex modulus $G_m^*(r,\omega)$ is higher than the complex modulus $G_r^*(r,\omega)$. The modulus ratio $\frac{G_m^*(r,\omega)}{G_r^*(r,\omega)}$ can be in the range $\frac{G_m^*(r,\omega)}{G_r^*(r,\omega)} \in [10, 100]$ depending on cell types and microenvironmental conditions. Pajic-Lijakovic and Milivojevic [2,6] supposed that the modulus ratio between migrating and surrounding resting cells approximately is equal to $\frac{G_m^*(r,\omega)}{G_r^*(r,\omega)} \approx 100$. Then, the modulus ratio $\frac{G_T^*(r,\omega)}{G_r^*(r,\omega)}$ for the sharp interface can be estimated for the corresponding volume fraction of migrating cells $\phi_m$. For the volume fraction of migrating cells $\phi_m = 0.2$, the modulus ratio for parallel mode coupling and the sharp interface is equal to $\frac{G_T^*(r,\omega)}{G_r^*(r,\omega)} = 20$, while the modulus ratio for series mode coupling is much lower and equal to $\frac{G_T^*(r,\omega)}{G_r^*(r,\omega)} = 1$. For the volume fraction of migrating cells $\phi_m = 0.5$, the modulus ratio for parallel mode coupling and the sharp interface is equal to $\frac{G_T^*(r,\omega)}{G_r^*(r,\omega)} = 45$, while the modulus ratio for series mode coupling is $\frac{G_T^*(r,\omega)}{G_r^*(r,\omega)} = 3$. For the volume fraction of migrating cells $\phi_m = 0.8$, the modulus ratio for parallel mode coupling and the sharp interface is equal to $\frac{G_T^*(r,\omega)}{G_r^*(r,\omega)} = 80$, while the modulus ratio for series mode coupling is $\frac{G_T^*(r,\omega)}{G_r^*(r,\omega)} = 6$. The finite interface case for supposed no-slip effects is related to an increase of the biointerface fluidity. Factors such as (1) reduction of slip effects (quantified by increasing the velocity gradient through the interface) under constant thickness of the biointerface and (2) decrease of the biointerface thickness (quantified by decreasing the ratio $\frac{\phi_{int}(r)}{\phi_m(r)}$) under the same slip effects lead to increase of the shear rate and intensify energy dissipation (quantified by increasing $\frac{G^*_{m-int}(r,\omega)}{G^*_r(r,\omega)}$) result in the system weakening [18].

## 9.4 Long-time viscoelasticity at a macroscopic level—constitutive modeling

Long-time cell rearrangement at a macroscopic level depends on (1) resting-to-migrating cell state transition and vice versa and (2) configuration changes of migrating cells from small clusters to monolayer sheets. Obtained configuration changes of migrating cells should be related to the viscoelasticity of multicellular surfaces based on mechanical coupling modes. Pajic-Lijakovic and Milivojevic [2,6] considered cell surface rearrangement at a mesoscopic level and proposed: (1) series mode coupling for the surface domains in which cell migrate in the form of small clusters and (2) parallel mode coupling for the surface domains in which cell migrate in the form of monolayer sheets. Pajic-Lijakovic and Milivojevic [7] expanded this consideration obtained at a mesoscopic level to the macroscopic level by formulating mixed, series-parallel, mode coupling. Mixed-mode coupling accounts for both fractions of cells (migrating and resting) coupled in series and in parallel. The volume fraction of migrating cells for the sharp biointerface is equal to $\phi_m(t)$, while the finite biointerface is equal to $\phi_{meff}(t) = \phi_m(t) + \phi_{int}(t)$. To simplify modeling consideration at a macroscopic level, we used the notation $\phi_m(t)$ for both types of biointerfaces.

### 9.4.1 Mixed series-parallel mode coupling at a macroscopic level

For mixed-coupling mode the multicellular surface stress and strain are formulated by modifying the model proposed by Kolarik et al. [11] for polymer blends. The model for the mixed-coupling mode is expressed as:

$$\begin{gathered}\tilde{\sigma}_{surface}(t) = \phi_P(t)\tilde{\sigma}_P(t) + (1 - \phi_P(t))\tilde{\sigma}_S(t) \\ \tilde{\varepsilon}_{surface}(t) = \tilde{\varepsilon}_P(t) = \tilde{\varepsilon}_S(t)\end{gathered} \tag{9.14}$$

where $\tilde{\sigma}_{surface}(t)$ is the surface stress tensor, $\tilde{\sigma}_P(t)$ is the stress part for the corresponding parallel coupling, $\tilde{\sigma}_S(t)$ is the stress part for the corresponding series coupling, $\tilde{\varepsilon}_{surface}(t)$ is the surface strain tensor, $\tilde{\varepsilon}_P(t)$ is the strain part for the corresponding parallel coupling, $\tilde{\varepsilon}_S(t)$ is the strain part for corresponding series coupling, $\phi_P(t)$ is the volume fractions of cells coupled in parallel equal to $\phi_P(t) = \phi_{P_m}(t) + \phi_{P_r}(t)$ (where $\phi_{P_m}$ is the volume fraction of migrating cells in the form of monolayer sheets coupled in parallel and $\phi_{P_r}(t)$ is the surrounding resting cells coupled in parallel). The volume fraction of cells coupled in parallel satisfies the condition $\phi_P(t) = 1 - \phi_S(t)$ (where $\phi_S(t)$ is the volume fraction of cells coupled in series). The stress tensor $\tilde{\sigma}_P(t)$ is equal to:

$$\phi_P(t)\tilde{\sigma}_P(t) = \phi_{P_m}(t)\tilde{\sigma}_{Pm}(t) + \phi_{P_r}(t)\tilde{\sigma}_{Pr}(t) \tag{9.15}$$

where $\tilde{\sigma}_{Pm}(t)$ and $\tilde{\sigma}_{Pr}(t)$ are the stresses for migrating and resting parts coupled in parallel. The strain tensor $\tilde{\varepsilon}_P(t)$ is equal to:

$$\tilde{\varepsilon}_P(t) = \tilde{\varepsilon}_{Pm}(t) = \tilde{\varepsilon}_{Pr}(t) \tag{9.16}$$

where $\tilde{\varepsilon}_{Pm}(t)$ and $\tilde{\varepsilon}_{Pr}(t)$ are the local strains for migrating and resting parts coupled in parallel. The stress tensor $\tilde{\sigma}_S(t)$ is equal to:

$$\tilde{\sigma}_S(t) = \tilde{\sigma}_{Sm}(t) = \tilde{\sigma}_{Sr}(t). \tag{9.17}$$

where $\tilde{\sigma}_{Sm}(t)$ and $\tilde{\sigma}_{Sr}(t)$ are the stresses for migrating and resting parts coupled in series. The strain tensor $\tilde{\varepsilon}_S(t)$ is equal to:

$$\phi_S(t)\tilde{\varepsilon}_S(t) = \phi_{S_m}(t)\tilde{\varepsilon}_{Sm}(t) + \phi_{S_r}(t)\tilde{\varepsilon}_{Sr}(t). \tag{9.18}$$

where $\tilde{\varepsilon}_{Sm}(t)$ and $\tilde{\varepsilon}_{Sr}(t)$ are the local strains for migrating and resting parts coupled in series. Volume fraction of cells coupled in series $\phi_S(t)$ is equal to $\phi_S(t) = \phi_{S_m}(t) + \phi_{S_r}(t)$ (where $\phi_{S_m}(t)$ is the volume fraction of migrating cells in the form of small clusters and $\phi_{S_r}(t)$ is the volume fraction of surrounding resting cells coupled in series). For mixed-coupling mode the volume fraction of migrating cells satisfies the condition $\phi_m(t) = \phi_{S_m}(t) + \phi_{Pm}(t)$, while the volume fraction of resting cells satisfies the condition $\phi_r(t) = \phi_{S_r}(t) + \phi_{Pr}(t)$. Cell rearrangement at a macroscopic level has been treated as a canonical ensemble of $N$ cells in the surface region equal to $N = N_r(t) + N_{Sm}(t) + N_{Pm}(t)$ (where $N_r(t)$ is the number of resting cells, $N_{Sm}(t)$ is the number of migrating cells in the form of small clusters, and $N_{Pm}(t)$ is the number of migrating cells in the form of monolayer sheets). Consequently, the volume fraction of resting cells is equal to $\phi_r(t) = \frac{N_r(t)}{N}$, the volume fraction of migrating cells in the form of small clusters is equal to $\phi_{S_m}(t) = \frac{N_{Sm}(t)}{N}$, and the volume fraction of migrating cells in the form of monolayer sheets is equal to $\phi_{Pm}(t) = \frac{N_{Pm}(t)}{N}$. The volume fraction of resting cells in the surrounding of $\phi_{Sm}(t)$ is equal to $\phi_{Sr}(t) = \frac{N_{Sr}(t)}{N}$, while the volume fraction of resting cells in the surrounding of $\phi_{Pm}(t)$ is equal to $\phi_{Pr}(t) = \frac{N_{Pr}(t)}{N}$. Resting pseudo-phase is in the surrounding of migrating pseudo-phase for both configurations. Higher volume fraction $\phi_{Pm}(t)$ corresponds to (1) higher volume fraction $\phi_{Pr}(t)$ and (2) lower volume fractions $\phi_{Sm}(t)$ and $\phi_{Sr}(t)$. When volume fraction is (1) $\phi_{Pm}(t) \rightarrow 0$, it means that $\phi_{Pr}(t) \rightarrow 0$ and/or (2) $\phi_{Sm}(t) \rightarrow 0$ it means that $\phi_{Sr}(t) \rightarrow 0$. Consequently, Pajic-Lijakovic and Milivojevic [7] correlated the volume fraction of resting pseudo-phase with the volume fraction of migrating pseudo-phase for both configurations in the form of additional condition as:

$$\frac{\phi_{Pr}(t)}{\phi_{Sr}(t)} = \frac{\phi_{Pm}(t)}{\phi_{Sm}(t)} \tag{9.19}$$

which offers the possibility for formulating the volume fractions of resting cells for both configurations as $\phi_{Sr}(t) = \phi_r(t)\frac{1}{1 + X(t)}$ and $\phi_{Pr}(t) = \phi_r(t)\frac{X(t)}{1 + X(t)}$, while the model parameter $X(t)$ is equal to $X(t) = \frac{\phi_{Pm}(t)}{\phi_{Sm}(t)}$.

Configuration changes of a migrating cell pseudo-phase take place to minimize the interface size between resting and migrating cell pseudo-phases. Higher interface size leads to intensive mechanical and biochemical perturbations, which is caused by uncorrelated motility [7,24]. The initial phase of cell migration which corresponds to $\phi_m(t) \ll 1$ can be described thermodynamically as nucleation of new "migrating" phase. This condition corresponds to cell migration in the form of small clusters, that is, $\phi_m(t) = \phi_m(t)^{low} \ll 1$ and $\phi_m(t)^{low} \approx \phi_{Sm}(t)$ while $\phi_{Pm}(t) \rightarrow 0$. An increase in the volume fraction of migrating cells leads to an increase in the interface size. To reduce the interface size as well as the surface free energy, migrating cells can form monolayer sheets instead of small clusters. Consequently, for the high volume fraction of migrating cell $\phi_m(t) = \phi_m(t)^{high} \rightarrow 1$, cell motion in the form of monolayer sheets becomes dominant, that is, $\phi_m(t)^{high} \approx \phi_{Pm}(t)$ while $\phi_{Sm}(t) \rightarrow 0$.

## 9.4.2 Configuration changes of migrating cells at a macroscopic level

Cell surface rearrangement has been treated as a thermodynamical closed system near equilibrium at the macroscopic level [7,24]. Linear nature of long-time rearrangement, obtained experimentally at a macroscopic level, has been modeled by applying Eyring transition state theory by

Marmottant et al. [28] and Pajic-Lijakovic and Milivojevic [7,24]. Pajic-Lijakovic and Milivojevic [2] treated the configuration changes of migrating cells as two-step processes. Both steps are reversible. The first step represents the resting-to-migrating cell state transition and vice versa, while the second one represents the configuration changes of migrating cells from dispersed small clusters to monolayer sheets and vice versa. The two-step model has been expressed as:

$$R \underset{k_1}{\overset{k_2}{\rightleftarrows}} Ms \underset{k_3}{\overset{k_4}{\rightleftarrows}} Mp \tag{9.20}$$

where $R$ represents the resting pseudo-phase, $Ms$ is the migrating cells in the form of small clusters coupled in series, and $Mp$ is the migrating cells coupled in parallel, $k_1$ is the rate of $R \rightarrow Ms$ cell state transition, $k_2$ is the rate of $Ms \rightarrow R$ cell state transition, $k_3$ is the rate of $Ms \rightarrow Mp$ cell state transition, and $k_4$ is the rate of $Mp \rightarrow Ms$ cell state transition. Kinetic constants should be related to the driving forces of phase transitions.

The two-step model is formulated by Pajic-Lijakovic and Milivojevic [7] as:

$$\begin{aligned}
\frac{d\phi_r(t)}{dt} &= -k_1\phi_r(t) + k_2\phi_{Sm}(t) \\
\frac{d\phi_{Sm}(t)}{dt} &= k_1\phi_r(t) - (k_2 + k_3)\phi_{Sm}(t) + k_4\phi_{Pm}(t) \\
\frac{d\phi_{Pm}(t)}{dt} &= k_3\phi_{Sm}(t) - k_4\phi_{Pm}(t).
\end{aligned} \tag{9.21}$$

with an initial condition: at $t = 0$, all cells within the surface are in resting state, that is, the volume fractions are equal to $\phi_r(0) = 1$ and $\phi_{Sm}(0) = \phi_{Sm}(0) = 0$. The solution of model equations is expressed [34] as:

$$\begin{aligned}
\phi_r(t) &= \phi_r(0)\left[1 - k_1\left(\frac{k_1 + k_3}{\lambda_1\lambda_2} - \frac{k_3 - \lambda_1 + k_4}{\lambda_1(\lambda_2 - \lambda_1)}\exp(-\lambda_1 t) + \frac{k_4 - \lambda_2 + k_3}{\lambda_2(\lambda_2 - \lambda_1)}\exp(-\lambda_2 t)\right)\right] \\
\phi_{Sm}(t) &= \phi_r(0)k_1\left[\frac{k_4}{\lambda_1\lambda_2} - \frac{k_4 - \lambda_1}{\lambda_1(\lambda_2 - \lambda_1)}\exp(-\lambda_1 t) + \frac{k_4 - \lambda_2}{\lambda_2(\lambda_2 - \lambda_1)}\exp(-\lambda_2 t)\right] \\
\phi_{Pm}(t) &= \phi_r(0)k_1k_3\left[\frac{1}{\lambda_1\lambda_2} - \frac{1}{\lambda_1(\lambda_2 - \lambda_1)}\exp(-\lambda_1 t) + \frac{1}{\lambda_2(\lambda_2 - \lambda_1)}\exp(-\lambda_2 t)\right]
\end{aligned} \tag{9.22}$$

where the model parameters $\lambda_1$ and $\lambda_2$ can be determined from the relations $\lambda_1\lambda_2 = k_2k_4 + k_1k_3 + k_1k_4$ and $k_1 + k_2 + k_3 + k_4$. Model parameters $\lambda_1$ and $\lambda_2$ have real solutions for the condition $(k_2k_4 + k_1k_3 + k_1k_4)^2 - 4(k_1 + k_2 + k_3 + k_4) \geq 0$. The volume fraction of resting cells consists of two subpopulations: (1) resting cells in surrounding of small migrating cell clusters and (2) resting cells in surrounding of migrating monolayer sheets. State of the multicellular surface depends on the values of kinetic constants $k_1$, $k_2$, $k_3$, and $k_4$.

The choice of kinetic constants defines the various scenarios of long-time cell surface rearrangement and the corresponding equilibrium configurations of migrating cells described by volume fractions $\phi_{Sm}^{eq}$, $\phi_{Pm}^{eq}$, $\phi_{Sr}^{eq}$, and $\phi_{Pr}^{eq}$.

Equilibrium volume fractions were determined for the condition $t \to \infty$ and expressed as:

$$\begin{aligned}
\phi_{Sm}^{eq} &= k_1 \frac{k_4}{\lambda_1\lambda_2} \\
\phi_{Sr}^{eq} &= k_1 k_3 \frac{1}{\lambda_1\lambda_2} \\
\phi_{Sr}^{eq} &= \left(1 - k_1 \frac{k_1 + k_3}{\lambda_1\lambda_2}\right) \frac{1}{1 + X_{eq}} \\
\phi_{Sr}^{eq} &= \left(1 - k_1 \frac{k_1 + k_3}{\lambda_1\lambda_2}\right) \frac{X_{eq}}{1 + X_{eq}}
\end{aligned} \tag{9.23}$$

where $X_{eq} = k_3$. This model is formulated for 3D systems. However, it can be also applied to 2D systems. In this case surface fractions of cells should be used instead of volume fractions in the modeling equations.

### 9.4.3 Kinetic constants formulation

Resting-to-migrating cell state transition represents the first step of cell long-time rearrangement. It is induced by local stress accumulation and described by kinetic constants $k_1$ and $k_2$. Change in migrating cells configuration (from dispersed small clusters to monolayer sheets) represents the second step of cell long-time rearrangement. It is induced by the cohesiveness in-homogeneity and described by kinetic constants $k_3$ and $k_4$. External stress $\tilde{\sigma}_{ext}$ provokes internal stress accumulation $\Delta\tilde{\Pi}_{eff}$ (effective "mechanical driving force") which leads to resting-to-migrating cell state transition (the first step) (Eq. 9.20). In the second step, in-homogeneity of cell cohesiveness $\Delta E_{int}$ provokes cells biomechanical response related to cell signaling $\Delta E_{eff}$ (effective "biochemical driving force") which leads to configuration changes of migrating cells from small clusters to monolayer sheets. Kinetic constants should be formulated in the context of the transition state theory by formulating corresponding energy barriers. A mechanical barrier is formulated in Appendix 1, while the biochemical barrier is formulated in Appendix 2.

The rate constants $k_1$ and $k_2$ are expressed as:

$$k_1 = k^I \exp(-(\tilde{\sigma}_{ext} - \Delta\tilde{\Pi}_{eff})\cdot f^{-1}(\lambda_E \tilde{Z}))$$
$$k_2 = k^I \exp(-(\tilde{\sigma}_{ext} + \Delta\tilde{\Pi}_{eff})\cdot f^{-1}(\lambda_E \tilde{Z})) \tag{9.24}$$

where $k^I$ is the frequency and the effective mechanical driving force is equal to $\Delta\Pi_{effij} = \Delta\Pi_{r-mij} - \Delta\Pi_{Pij}$, $= \Delta\tilde{\Pi}_{r-mij}$ is the stress difference between migrating and resting cell pseudo-phases at the biointerface, $\Delta\tilde{\Pi}_{Pij}$ is the perturbation stress component which is approximately equal to the unit stress, that is, $\Delta\Pi_{Pij} \approx \lambda_E Z_{ij}$, $\lambda_E Z_{ij}$ is the stress quanta. The number of stress quanta for resting-to-migrating cell state transition is $n^+_{stress} = (\tilde{\sigma}_{ext} - \Delta\tilde{\Pi}_{eff})\cdot f^{-1}(\lambda_E \tilde{Z})$, while the number of quanta for migrating-to-resting cell state transition is equal to $n^-_{stress} = (\tilde{\sigma}_{ext} + \Delta\tilde{\Pi}_{eff})\cdot f^{-1}(\lambda_E \tilde{Z})$. The rate constant $k_1$ for resting-to-migrating cell state transition is supposed to be equal to $k_1 \sim t_p^{-1}$ (where $t_p$ is the cell persistence time). McCann et al. [35] reported that cell persistence time corresponds to several tens of minutes.

The ratio $\frac{k_1}{k_2}$ is equal to $\frac{k_1}{k_2} = \exp(\Delta n_{stress})$ (where $\Delta n_{stress} = n^-_{stress} - n^+_{stress} \geq 0$). Three cases of the ratio $\frac{k_1}{k_2}$ could be considered depending on $\Delta n_{stress}$:

1. $\Delta n_{stress} > 0$ (and $= \Delta\Pi_{r-mij} > \Delta\Pi_{Pij}$), which corresponds to $k_1 > k_2$;
2. $\Delta n_{stress} \approx 0$ (and $= \Delta\Pi_{r-mij} \approx \Delta\Pi_{Pij}$), which corresponds to $k_1 \approx k_2$; and
3. $\Delta n_{stress} < 0$ (and $= \Delta\Pi_{r-mij} < \Delta\Pi_{Pij}$), which corresponds to $k_1 < k_2$.

The rates $k_3$ and $k_4$ are expressed as:

$$k_3 = k^{II} \exp\left(-\frac{(\Delta E_{int} - \Delta E_{eff})}{k_B T_{eff}}\right)$$
$$k_4 = k^{II} \exp\left(-\frac{(\Delta E_{int} + \Delta E_{eff})}{k_B T_{eff}}\right) \tag{9.25}$$

where $k^{II}$ is the frequency; $\Delta E_{int}$ is the biochemical barrier equal to the spreading coefficient [36], that is, $\Delta E_{int} = \gamma_r \Delta A_m - (\gamma_m + \gamma_{int})\Delta A_m$; $\gamma_r$ is the surface tension of resting cell pseudo-phase; $\gamma_m$ is the surface tension of migrating cell pseudo-phase; $\gamma_{int}$ is the interfacial tension; $\Delta A_m$ is the surface change of migrating cell clusters; $\Delta E_{eff}$ is the effective "biochemical driving force, that is, the cohesiveness difference between the pseudo-phases" which leads to configuration change of migrating cells from small clusters to monolayer sheets; and $T_{eff}$ is the effective temperature. Kinetic constants $k_3$ and $k_4$ account for the cohesiveness difference. The effective "biochemical driving force" is equal to $\Delta E_{eff} = \Delta E_S - \Delta E_P$ (where $\Delta E_S$ accounts for cell signaling which manages cohesiveness of migrating and resting cell pseudo-phases near the interface and $\Delta E_P$ is the energy perturbation). A detailed description of the effective biochemical driving force formulation is given in Appendix 2.

As was shown, the stress barrier for the first-step process was defined relative to a unit of stress. Similarly, the energy barrier for the second-step process should be defined relative to a unit of energy. This unit of energy should be related to cellular mobility as the main cellular characteristic responsible for long-time rearrangement. By the analogy with a thermal unit of energy expressed as $\Delta E_T = k_B T$, we define the cellular long-time rearrangement energy unit as $\Delta E_C = k_B T_{eff}$. Concept of effective temperature has been applied for considering rearrangement of various thermodynamical systems (near to equilibrium and far from equilibrium) from glasses and sheared fluids to granular systems [37]. Pajic-Lijakovic and Milivojevic [7] applied this concept to the long-time rearrangement of dense cellular systems. The number of interfacial energy quanta for configuration changes of migrating cells from small clusters to monolayer sheets is equal to $n_{int}^{+} = \frac{\Delta E_{int} - \Delta E_{eff}}{k_B T_{eff}}$, while the number of interfacial energy quanta for configuration changes of migrating cells from monolayer sheets to small clusters is equal to $n_{int}^{-} = \frac{\Delta E_{int} + \Delta E_{eff}}{k_B T_{eff}}$.

The ratio $\frac{k_3}{k_4}$ is equal to $\frac{k_3}{4} = \exp(\Delta n_{int})$ (where $\Delta n_{int} = n_{int}^{-} - n_{int}^{+} \geq 0$). Three causes of the ratio $\frac{k_3}{k_4}$ can be considered depending on $\Delta n_{int}$:

1. $\Delta n_{int} > 0$ (and $\Delta E_S > \Delta E_P$), which corresponds to $k_3 > k_4$;
2. $\Delta n_{int} \approx 0$ (and $\Delta E_S \approx \Delta E_P$), which corresponds to $k_3 \approx k_4$; and
3. $\Delta n_{int} < 0$ (and $\Delta E_S < \Delta E_P$), which corresponds to $k_3 < k_4$.

The ratio $\frac{k_1}{k_3}$ is equal to $\frac{k_1}{k_3} = \exp(n_{int}^{+} - n_{stress}^{+})$. Three causes of the ratio $\frac{k_1}{k_3}$ can be considered:

1. if $n_{int}^{+} > n_{stress}^{+}$ indicates that $k_1 > k_3$,
2. if $n_{int}^{+} \approx n_{stress}^{+}$ indicates that $k_1 \approx k_3$, and
3. if $n_{int}^{+} < n_{stress}^{+}$ indicates that $k_1 < k_3$.

The ratio $\frac{k_2}{k_4}$ is equal to $\frac{k_2}{k_4} = \exp(n_{int}^{-} - n_{stress}^{-})$. Three causes of the ratio $\frac{k_2}{k_4}$ could be considered:

1. if $n_{int}^{-} > n_{stress}^{-}$ indicates that $k_2 > k_4$,
2. if $n_{int}^{-} \approx n_{stress}^{-}$ indicates that $k_2 \approx k_4$, and
3. if $n_{int}^{-} < n_{stress}^{-}$ indicates that $k_2 < k_4$.

Consequently, two conditions are necessary for defining the single state of the multicellular surface. Two constants are related per single condition. We formulated 12 conditions placed within three groups. Combinations within groups cannot define a new system state. The number of states can be expressed as $n_{state} = C_2^{12} - C_2^{3} = 54$ (where $n_{state}$ is the number of states, $C_2^{12}$ is the combination of 12 conditions of second class, and $C_2^{3}$ is the combination of 3 conditions of the second class).

These three groups of cases determine the influence of uncorrelated motility on resting-to-migrating cell state transition and vice versa as well as configuration changes of migrating cells through mechanical perturbations $\Delta\tilde{\Pi}_{Pij}$ and biochemical perturbations $\Delta E_P$. Two cases (first group—case 1 and second group—case 1) correspond to low mechanical and biochemical perturbations, respectively. Two cases (first group—case 2 and second group—case 2) correspond to medium mechanical and biochemical perturbations. Two cases (first group—case 3 and second group—case 3) correspond to large mechanical and biochemical perturbations. Uncorrelated motility accounts for internal and external effects. Internal effects are caused by stress accumulation within migrating cell clusters during their intercalation through the dense cellular environment [2]. External effects represent the consequence of the collision of velocity fronts which is significant even in 2D [20]. Uncorrelated motility represents the main cause of change the relaxation rate of aggregate shape after uniaxial compression from cycle to cycle [24]. This modeling consideration accounts for uncorrelated motility and its impact on dynamics of configuration changes.

### 9.4.4 Complex modulus of multicellular surfaces expressed for equilibrium configuration of migrating cells

Detailed consideration of time-dependent rheological response of multicellular systems requires additional knowledge of constitutive models for migrating cell clusters and surrounding resting cells. We will present the simplified consideration of current equilibrium states based on the equilibrium configurations of migrating cells, expressed by volume fractions $\phi_{Sm}^{eq}$, $\phi_{Pm}^{eq}$, $\phi_{Sr}^{eq}$, and $\phi_{Pr}^{eq}$. Consequently, the modeling results can be characterized from the standpoint of the surface stiffness in the context of mixed-coupling mode at a macroscopic level. The surface stiffness can be quantified by (1) the complex modulus of the cell surface $G^*_{surface}(\omega) = \frac{F(\tilde{\sigma}_{surface}(t))}{F(\tilde{\varepsilon}_{surface}(t))}$, (2) the complex modulus of the migrating pseudo-phase $G^*_m(\omega) = \frac{F(\tilde{\sigma}_m(t))}{F(\tilde{\varepsilon}_m(t))}$, and (3) the complex modulus of the resting pseudo-phase $G^*_r(\omega) = \frac{F(\tilde{\sigma}_r(t))}{F(\tilde{\varepsilon}_r(t))}$. The cell surface complex modulus for the mixed-coupling mode under the equilibrium condition described by $\phi^{eq}_{Sm}$, $\phi^{eq}_{Pm}$, $\phi^{eq}_{Sr}$, and $\phi^{eq}_{Pr}$ can be expressed from Eqs. (9.21)–(9.23) as:

$$\frac{G^*_{surface}(\omega)}{G^*_r(\omega)} = \phi^{eq}_{Pm}\frac{G^*_m(\omega)}{G^*_r(\omega)} + \phi^{eq}_{Pr} + \left(\phi^{eq}_{Sm} + \phi^{eq}_{Sr}\right)^2 \frac{1}{\frac{\phi^{eq}_{Sm}}{\frac{G^*_m(\omega)}{G^*_r(\omega)}} + \phi^{eq}_{Sr}} \tag{9.26}$$

Model equation (Eq. 9.26) should satisfy the following conditions: (1) if $\phi^{eq}_{Pm} \to 0$ and $\phi^{eq}_{Sm} \to 0$ while $\phi^{eq}_{Pr} + \phi^{eq}_{Sr} = 1$, then $\frac{G^*_{surface}(\omega)}{G^*_r(\omega)} \to 1$ and (2) if $\phi^{eq}_{Pr} \to 0$ and $\phi^{eq}_{Sr} \to 0$ while $\phi^{eq}_{Pm} + \phi^{eq}_{Sm} = 1$, then $\frac{G^*_{surface}(\omega)}{G^*_r(\omega)} = \frac{G^*_m(\omega)}{G^*_r(\omega)}$.

Characteristic equilibrium states for cellular configurations obtained during cell surface rearrangement are estimated by Eq. (9.23), while the corresponding surface stiffness can be predicted by Eq. (9.26). Some characteristic equilibrium states for corresponding low, medium, and large mechanical and biochemical perturbations are presented in Table 9.1.

Two experimental systems (1) cell aggregate rounding after uniaxial compression and (2) cell aggregate micropipette aspiration were discussed in the context of postulated characteristic cases presented as the result of this modeling consideration at a macroscopic level.

### 9.4.5 Discussion of multicellular surfaces responses under stress conditions

Cell aggregate shape changes during its rounding, after uniaxial compression takes place during successive relaxation cycles. The relaxation rates per cycles are not random but they have the tendency to gather around two or three values indicating organized cell migrating pattern. Every relaxation rate can be related to a corresponding cell migrating state. Three scenarios were considered base on experimental data by Mombach et al. [27]: (1) most of the cells migrate all the time, which correspond to the highest value of the relaxation rate $k_m$; (2) some cell clusters migrate while the others (at the same time) stay in resting state, which correspond

**Table 9.1** Characteristic problems of the viscoelasticity for the equilibrium configurations of migrating cells at a macroscopic level.

| Problem no. | Kinetic constant proportions | Volume fractions of cells at equilibrium | Complex modulus ratio | Intensity and type of perturbations |
|---|---|---|---|---|
| 1 | $k_1 \prec\prec k_2$<br>$k_3 \prec\prec k_4$<br>$k_4 \approx k_2$ | $\varphi_r^{eq} \to 1$ | $\frac{G*(\omega)}{G_r*(\omega)} \to 1$ | Large mechanical and large biochemical |
| 2 | $k_1 \approx k_2 \approx k_3 \approx k_4$ | $\varphi_r^{eq} = \varphi_{Sm}^{eq} \approx \varphi_{Pm}^{eq}$ | $\frac{G*(\omega)}{G_r*(\omega)} = 0.33 \frac{G_m*(\omega)}{G_r*(\omega)}$ | Medium mechanical and medium biochemical |
| 3 | $k_3 \prec\prec k_4$<br>$k_4 \geq k_1$<br>$k_1 \geq k_2$ | $\varphi_{Sm}^{eq} \succ\succ \varphi_{Pm}^{eq} \varphi_{Sr}^{eq} \approx \varphi_r^{eq}$ | $\frac{G*(\omega)}{G_r*(\omega)} \approx \frac{\left(\varphi_{Sm}{}^{eq} + \varphi_{Sr}{}^{eq}\right)^2}{\varphi_{Sr}^{eq}}$ | Low or medium mechanical and large biochemical |
| 4 | $k_1 \succ\succ k_2$<br>$k_3 \succ\succ k_4$<br>$k_3 \approx k_1$ | $\varphi_r^{eq} \to 0$ | $\frac{G*(\omega)}{G_r*(\omega)} \to \frac{G_m*(\omega)}{G_r*(\omega)}$ | Low mechanical |
| 5 | $k_3 \succ\succ k_4$<br>$k_3 \geq k_1$<br>$k_1 \geq k_2$ | $\varphi_{Sm}^{eq} \prec\prec \varphi_{Pm}^{eq} \varphi_{\mathrm{Pr}}^{eq} \approx \varphi_r^{eq}$ | $\frac{G*(\omega)}{G_r*(\omega)} = \varphi_{Pm}{}^{eq} \frac{G_m*(\omega)}{G_r*(\omega)} \left(\varphi_{Pm}^{eq} \succ 0.33\right)$ | Low or medium mechanical and low biochemical |

to the median value of the relaxation rate $k_t$; and (3) most of the cells stay in the resting state, which correspond to the relaxation rate $k_r \approx 0$. These three scenarios have been discussed based on the proposed model [7,24].

The first scenario corresponds to the large volume fraction of migrating cells that can be realized under low or medium mechanical perturbations and low biochemical perturbations. The characteristic case 5 presented in Table 9.1 was the most suitable. A second scenario in which some cell clusters migrate while the others (at the same time) stay in resting state corresponds to medium mechanical and biochemical perturbations. This configuration of migrating cells is more disordered than the one proposed for the first scenario. For this condition, case 2 should be the most suitable (Table 9.1). The third scenario corresponds to case 1 (Table 9.1) realized for large mechanical and biochemical perturbations.

Instead of cell aggregate uniaxial compression between parallel plates, aggregate micropipette aspiration is another widely used experimental system for considering a collective cell migration and its influence on long-time viscoelasticity. Experimental data by Guevorkian et al. [21] was considered based on the proposed model. Activation of perturbed cells and their entrance into the pipette via collective cell migration is induced by the pressure of 0.5 kPa for 120 minutes. They detected short-time contractions with the period of 3–4 minutes and long-time contractions with the period of 10–17 minutes. Short-time contractions correspond to single-cell contractions, while the long-time contractions account for collective phenomena of cell rearrangement. Cell entrance into the pipette could be estimated within two-time regimes: (1) initial regime for $t \in [0, 50 \text{ minutes}]$ and (2) final regime for $t \in [50, 120 \text{ minutes}]$ [7]. The long-time contractions are more intensive in the final regime compared with the initial regime pointed to the disordered configuration of migrating cells. Both regimes should be characterized by the large volume fraction of migrating cells within the pipette. Consequently, the initial regime can be identified as the characteristic case 5 (Table 9.1) suitable for larger volume fractions of migrating cells and more ordered configurations. The final regime corresponds to the more disordered configuration of migrating cells and has been identified as case 2 (Table 9.1).

Bearing in mind that mathematical models represent only the simplified description of this complex stochastic phenomenon, we pointed out the importance of understanding of the cause–consequence relation between: (1) mechanical and biochemical perturbations generated during collective cell migration, (2) configuration changes of migrating cells, and (3) long-time viscoelasticity of multicellular surfaces.

## 9.5 Conclusion

A significant difference in cell stiffness between (active) migrating and (passive) resting cell pseudo-phases indicates that volume fraction of migrating cells and their distribution influence the long-time viscoelasticity of multicellular surfaces. This aspect of cell surface rearrangement has been treated by the analogy with physics in the form of a multiphase system. The number of pseudo-phases depends on the characteristics of the biointerface. For the sharp biointerface the multicellular system can be treated as the two-phase system, while for the finite biointerface, it can be treated as the three-phase system. The factors such as size, thickness, and slip effects of the biointerface significantly influence long-time viscoelasticity. Size of the biointerface depends on the volume fraction and configuration of migrating cells and influences the viscoelasticity through mechanical coupling modes. Slip effects of the biointerface influence the fluidity of perturbed boundary layers around the migrating clusters. Migrating cells can form three types of configuration: (1) monolayer sheets (lamellar structure), (2) small clusters, and (3) and various combinations of these two types. Various types of configurations represent the product of stochastic effects such as biochemical and mechanical perturbations. The lamellar structure of pseudo-blends corresponds to parallel mode coupling, while the highly dispersed system in forms of small clusters points to series mode coupling considered at a mesoscopic level. Series-parallel mode coupling which accounts for various combinations of these two basic configuration types is suitable for modeling the viscoelasticity at a macroscopic level.

Configuration changes of migrating cells were treated as: (1) thermodynamics systems far from equilibrium at the mesoscopic level and modeled by modified model for viscoelastic phase separation and (2) thermodynamic system close to equilibrium at the macroscopic level and modeled by two-step Eyring model. Two-model systems were considered in the context of (1) configuration changes, (2) viscoelasticity, and (3) their interrelations. One is cell aggregate uniaxial compression between parallel plates and aggregate rounding after compression, while the other is cell aggregate micropipette aspiration.

This modeling consideration pointed out the necessity of additional experiments to (1) relate cell configurations with the rheological response of multicellular surfaces under various stress conditions, (2) characterized the biointerface in the context of its fluidity and thickness, and (3) clarify the impact of uncorrelated motility on the cell rearrangement.

## Acknowledgment

This work was supported by the Ministry of Education, Science and Technological Development of the Republic of Serbia (Contract No. 451-03-68/2020-14/200135).

## Declaration of interest

The author reports no conflict of interest. Both the authors contributed equally to the paper.

## Appendix 1

Part of surface cells loses the equilibrium number of adhesion complexes during the aggregate compression andbecomes activated (the first step of Eq. 9.21). We supposed that cell aggregate uniaxial compression cannot induce changes of single-cell volumes. External stress relates to the internal stress (force moments) of the cells. Consequently, internal stress function $\Delta\tilde{\Pi}_{eff}$ is equilibrated with the external stress $\tilde{\sigma}_{exp}$ such that $\sigma_{expij} = \Delta\Pi_{effij}$ during aggregate compression. Internal stress can be expressed as: $\Delta\Pi_{eff}$

$$\Delta\tilde{\Pi}_{eff} = \frac{1}{V^{\alpha}} \sum_{\alpha,\beta} \vec{f}^{\alpha\beta} \vec{r}^{\alpha\beta} \tag{9.A1.1}$$

where $V^{\alpha}$ is the volume of the single cell, $\vec{f}^{\alpha\beta}$ is the force acting on the cell $\beta$ from cell $\alpha$, and $\vec{r}^{\alpha\beta}$ is the vector from the center of $\alpha$ to the center of $\beta$. The entropy of cell rearrangement can be expressed as:

$$S = S(V_{surface}, \tilde{\sigma}_{ext}, N) \tag{9.A1.2}$$

where $S$ is the surface entropy and $V_{surface}$ is the surface volume. Edwards and Grinev [22] and Edwards [23] considered rearrangement of granular systems under external stress and introduced thermodynamic intensive variable $\tilde{Z}$ by the analogy with temperature defined as:

$$\frac{1}{\tilde{Z}} = \frac{\partial S}{\partial \tilde{\sigma}_{in}} \tag{9.A1.3}$$

where $\tilde{\sigma}_{in}$ is the internal stress caused under external stress $\tilde{\sigma}_{ext}$. At first, it is a quite different system from the one considered here. But, more careful examination reveals that many of their properties can be defined in the same terms for dense cell populations.

## Appendix 2

The biochemical barrier for configuration changes of migrating cells (the second step of Eq. 9.21) is expressed as the spreading coefficient. Spreading coefficient accounts for the cohesiveness of both pseudo-phases and interfacial tension [36]. Consequently, the corresponding energy barrier could be expressed as:

$$\Delta E_{int} = \gamma_r \Delta A_m - (\gamma_m + \gamma_{int}) \Delta A_m \tag{9.A2.1}$$

where $\gamma_r$ is the equilibrium surface tension of resting cell pseudo-phase, $\gamma_m$ is the equilibrium surface tension of migrating cell pseudo-phase, $\gamma_{int}$ is the interfacial tension, and $\Delta A_m$ is the surface change of migrating cell clusters. Surface tension represents the specific measure of the surface free energy.

The effective driving force for the second step $\Delta E_{eff}$ (i.e., driving energy) comes from cell signaling. This biochemical driving force can be expressed as:

$$\Delta E_{eff} = \Delta E_S - \Delta E_P \tag{9.A1.2}$$

where $\Delta E_P$ is the energy perturbation and $\Delta E_S$ accounts for cell signaling, which manage cohesiveness of migrating and resting cell pseudo-phases near the interface expressed as [7]:

$$\Delta E_S = k_B T_{eff} \Delta H \tag{9.A1.3}$$

where $k_B$ is the Boltzmann constant, $T_{eff}$ is the effective temperature expressed based on generalization of Einstein's relation [37]. Consequently, the effective temperature can be expressed as:

$$k_B T_{eff} = \frac{D_{eff}}{\mu'} \tag{9.4}$$

where $D_{eff}$ is the diffusivity of migrating cells and $\mu'$ is the mobility of velocity fronts. The variable $\Delta H$ represents Shannon information entropy that has been applied to cell signaling and expressed by Rhee et al. [38]:

$$\Delta H = H(I) - H(I/R) \tag{9.A1.5}$$

where $H(I)$ can be interpreted to be the overall uncertainty about the input $I$, and $H(I/R)$ is the residual uncertainty about $I$ after the value of the response $R$ is known. Energy perturbation $\Delta E_P$ could be equal to:

$$\Delta E_P = k_B T_{eff} \tag{9.A1.6}$$

The perturbations of cell signaling can biochemically reduce the action of the driving energy and inhibit conformation changes of migrating cells.

## References

[1] E.H. Barriga, K. Franze, G. Charras, R. Mayor, Tissue stiffening coordinates morphogenesis by triggering collective cell migration in vivo, Nature 554 (2018) 523–527.
[2] I. Pajic-Lijakovic, M. Milivojevic, Long-time viscoelasticity of multicellular surfaces caused by collective cell migration – multi-scale modelling considerations, Sem. Cell Dev. Biol. 93 (2019) 87–96.
[3] G. Petrungaro, L. Morelli, K. Uriu, Information flow in the presence of cell mixing and signalling delays during embryonic development, Sem. Cell Dev. Biol. 93 (2019) 26–35.
[4] E.H. Barriga, R. Mayor, Adjustable viscoelasticity allows for efficient collective cell migration, Sem. Cell Dev. Biol. 93 (2019) 55–68.
[5] G.B. Blanchard, A.G. Fletcher, L.J. Schumacher, The devil is in the mesoscale: mechanical and behavioural heterogeneity in collective cell movement, Sem. Cell Dev. Biol. 93 (2019) 46–54.
[6] I. Pajic-Lijakovic, M. Milivojevic, Viscoelasticity of multicellular surfaces, J. Biomech. 60 (2017) 1–8.
[7] I. Pajic-Lijakovic, M. Milivojevic, Functional epithelium remodeling in response to applied stress under in vitro conditions, Appl. Bionics Biomech. 2019 (2019). Available from: https://doi.org/10.1155/2019/4892709.
[8] I. Pajic-Lijakovic, M. Milivojevic, Jamming state transition and collective cell migration, J. Biol. Eng. 13 (73) (2019) 1–12.
[9] T. Ala-Nissila, K. Majaniemi, K. Elder, Phase-field modeling of dynamical interface phenomena in fluids, Novel Methods in Soft Matter Simulations, 640 (2004) 357–388.
[10] H. Tanaka, A viscoelastic model of phase separation, Phys. Rev. 56 (4) (1997) 4451–4462.
[11] J. Kolarik, L. Fambri, A. Pegoretti, A. Penati, Ternary polymer blends: prediction of mechanical properties for various phase structures, Polym. Adv. Technol. 11 (2000) 75–81.

[12] M. Takayanagi, S. Uemura, M. Shunsuke, Application of equivalent model method to dynamic rheo-optical properties of crystalline polymer, J. Polym. Sci. C. 5 (1964) 113–122.
[13] T. Mikami, K. Yoshida, H. Sawada, M. Esaki, K. Yasumura, M. Ono, Inhibition of Rho-associated kinases disturbs the collective cell migration of stratified TE-10 cells, Biol. Res. 48 (2015) 1–15.
[14] E. Scarpa, R. Mayor, Collective cell migration in development, J. Cell Biol. 212 (2) (2016) 143–155.
[15] T. Kuwabara, D.G. Parkins, D.G. Cogan, Sliding of the epithelium in experimental corneal wounds, Invest. Ophthalmol. 15 (1976) 4–14.
[16] P. Friedl, S. Alexander, Cancer invasion and the microenvironment: plasticity and reciprocity, Cell 147 (2011) 992–1009.
[17] A. Shellard, R. Mayor, Supracellular migration – beyond collective cell migration, J. Cell Sci. 132 (2019) jcs226142.
[18] I. Pajic-Lijakovic, M. Milivojevic, Viscoelasticity of multicellular systems caused by collective cell migration: dynamics at the biointerface, Eur. Biophys. J. 49 (2020). Available from: https://doi.org/10.1007/s00249-020-01431-2.
[19] E. Rozbicki, M. Chuai, A.I. Karjalainen, F. Song, H.M. Sang, R. Martin, et al., Myosin II-mediated cell shape changes and cell intercalation contribute to primitive streak formation, Nat. Cell Biol. 17 (4) (2015) 397–408.
[20] J. Notbohm, S. Banerjee, K.J.C. Utuje, B. Gweon, H. Jang, Y. Park, et al., Cellular contraction and polarization drive collective cellular motion, Biophys. J. 110 (2016) 2729–2738.
[21] K. Guevorkian, D. Gonzalez-Rodriguez, C. Carlier, S. Dufour, F. Brochard-Wyart, Mechanosensitive shivering of model tissues under controlled aspiration, PNAS 108 (33) (2011) 13387–13392.
[22] S. Edwards, D.V. Grinev, Statistical mechanics of vibration-induced compaction of powders, Phys. Rev. E 58 (1988) 4758–4762.
[23] S.F. Edwards, The full canonical ensemble of a granular system, Physica A 353 (2005) 114–118.
[24] I. Pajic-Lijakovic, M. Milivojevic, Successive relaxation cycles during long-time cell aggregate rounding after uni-axial compression, J. Biol. Phys. 43 (2017) 197–209.
[25] D. Requera, J.M. Rubi, J.M.G. Vilar, The mesoscopic dynamics of thermodynamics systems, J. Phys. Chem. B 109 (2005) 21502–21551.
[26] J.M. Rubi, Mesoscopic non-equilibrium thermodynamics, Contrib. Sci. 11 (2015) 147–151.
[27] J.C.M. Mombach, D. Robert, F. Graner, G. Gillet, G.L. Thomas, M. Idiart, et al., Rounding of aggregates of biological cells: experiments and simulations, Physica A 352 (2005) 525–534.
[28] P. Marmottant, A. Mgharbel, J. Kafer, B. Audren, J.P. Rieu, J.C. Vial, et al., The role of fluctuations and stress on the effective viscosity of cell aggregates, PNAS 106 (2009) 17271–17275.
[29] J.F. Joanny, J. Prost, Active gels as a description of the actin-myosin cytoskeleton, HFSP J. 3 (2) (2009) 94–104.
[30] A.G. Clark, D.M. Vignjajevic, Models of cancer cell invasion and the rule of micro-environment, Curr. Opin. Cell Biol. 36 (2015) 13–22.
[31] J. Ranft, M. Basan, J. Elgeti, J.F. Joanny, J. Prost, F. Julicher, Fluidization of tissues by cell division and apoptosis. PNAS 107 (49) (2010) 20863–20868.
[32] H. Tanaka, A viscoelastic model of phase separation, Phys. Rev. 56 (4) (1997) 4451–4462.
[33] J.R. Lange, B. Fabry, Cell and tissue mechanics in cell migration, Exp. Cell. Res. 319 (2013) 2418–2423.

[34] V.I. Korobov, V.F. Ochkov, Chemical Kinetics With Mathcad and Maple, Springer-Verlag, Wien, 2011.
[35] C. McCann, P.W. Kriebel, C.A. Parent, W. Losert, Cell speed, persistence and information transmission during signal relay and collective migration, J. Cell Sci. 123 (10) (2010) 1724–1731.
[36] W.D. Harkins, Feldman, Films. The spreading of liquids and the spreading coefficient, J. Am. Chem. Soc. 44 (1922) 2665–2685.
[37] J. Casas-Vazquez, D. Jou, Temperature in non-equilibrium states: a review of open problems and current proposals, Rep. Prog. Phys. 66 (2003) 1937–2023.
[38] A. Rhee, R. Cheong, A. Levchenko, The application of information theory to biochemical signalling systems, Phys. Biol. 9 (2012) 045001.

CHAPTER TEN

# Recent advances in imaging of cell elasticity

**Teckla Akinyi[1], Pol Grasland-Mongrain[2], Manish Bhatt[1], Stefan Catheline[3] and Guy Cloutier[1,4]**

[1]Laboratory of Biorheology and Medical Ultrasonics, University of Montreal Hospital Research Center, Montréal, QC, Canada
[2]ENS de Lyon, Université Claude Bernard, CNRS, Laboratoire de Physique, Lyon, France
[3]LabTAU, INSERM unit 1032, Université de Lyon, Lyon, France
[4]Department of Radiology, Radio-oncology and Nuclear Medicine, and Institute of Biomedical Engineering, University of Montreal, Montréal, QC, Canada

## 10.1 Introduction

Biological cells have a complex chemical and mechanical environment upon which function and structural integrity rely. In performing their normal functions, cells adhere and spread on the extracellular matrix (ECM) to form tissues and collectively or individually migrate to the injured site during immune response. Depending on the functional state or in response to external stimuli, the mechanical properties of cells can change. These changes are mitigated by the rearrangement of subcellular structures and their ability to convert mechanical stresses to biochemical, bioelectrical, or morphological changes. The cell is thus a mechanical structure with the ability to probe external forces, detect internal mechanics of substructures, and generate responsive and active forces [1]. Section 10.1 cursorily introduces the basic structure of the cell, highlighting key cellular components that are responsible for its mechanical properties and responsiveness.

Maintenance of cellular structures and appropriate cellular responses to external stresses largely depend on the deformability of the cells; the ability to survive the mechanical stress of motility or invasion without rupturing. This is investigated through the elasticity of the whole cell, which in turn informs on its developmental state and pathophysiology. Changes in cell elasticity infer on diseases, which can clinically manifest as a loss in function or change in the physical state of the organ or tissue. The clinical

*Viscoelasticity and Collective Cell Migration.*
DOI: https://doi.org/10.1016/B978-0-12-820310-1.00010-0

relevance of cellular elasticity is thus discussed by giving examples of studies reporting an observed change in elasticity such as the increase in Young's modulus of the zona pellucida of the ovum after fertilization [2], in malaria, asthma, and sickle cell anemia where cellular stiffening was observed [3], or the decrease in elasticity as shown in bladder and cancer cells [4]. Beyond characterizing cell development and function, the significance of tracking cellular elasticity in the pharmaceutical industry is also presented.

Elasticity describes the relationship between stress (force) and strain (deformation). This is the tendency of a material to return to its original shape after being deformed by mechanical stress, graphically presented as a stress–strain curve. The curve/ratio provides information on how much a material can resist forces of stress (stretch). A high elastic modulus implies it takes a lot of force to stretch the material a small amount (the material is then described as being stiff), whereas a low modulus value denotes deformability. Thus it is evident that stiffness is the resistance of a body to deformation upon exertion of an external force and elasticity is a measure of stiffness, reported in the unit of Pascals (Pa). Often, elasticity is reported interchangeably with stiffness. In Section 10.2, methods used to assess cell elasticity, the accompanying physical concepts and current efforts for quantifying a single cell are discussed. Since the average cell dimension is in the micrometer range (50–150 μm typically) [5], imaging techniques need to acquire high spatial resolution planes or volumes to phenotype the whole cell or subcellular structures. The newly introduced technique of optical microelastography, which offers high spatial and temporal resolutions, is presented.

The interpretation of experimental methods used to quantify mechanical properties of cells relies on making assumptions about the cell surface, shape, rheology model, and boundary conditions. The simplest model assumes isotropy and homogeneity of a spherical cell; however, experimental results both from biomechanical and biochemical analyses point toward a more heterogeneous and anisotropic cell, for example, the clustering of integrins versus a more even distribution along the cell membrane [6]. Choosing the right mechanical model is increasingly becoming a key milestone in developing robust methods and more accurate analysis tools for single-cell elastography. However not all tools rely on model analysis, for example, wave-based elastography, which directly characterizes the propagation of a wave through the cell to assess its mechanical property. Consequently, Section 10.3 explores the necessity of and considerations in rheological modeling of a single cell. These computational models can provide more accurate interpretation of experimental results, offering simulation analysis for comparison.

Imaging cellular elasticity has made great strides over the past few decades; the chapter concludes with exploring some miscellaneous techniques that aim to provide wholesome and definitive tools to capture cellular biomechanics. Among the recent trends are assessing both viscosity and elasticity, and providing space and time resolutions to capture dynamic processes of subcellular structures. Assessment of both viscosity and elasticity is necessary because cells are viscoelastic; that is, they exhibit both viscous and elastic properties as a natural consequence of their composition, made of semi-solid organelles, the cytoskeleton, and the nucleus suspended in a fluid cytoplasm. Viscosity is described by the time lag between stress and strain, or the opposition to a movement to maintain a static position when a force is applied. Cells will flow like a fluid and have the ability to recover their shape when the deforming stress is withdrawn. It is thus necessary to develop wholesome and definitive tools that would also quantify the viscosity. Accuracy in such measurements is limited by the reliance on verifying experimental results through computational modeling. Modeling techniques involve gross approximations based on assumptions. To optimize results in cellular elastography, model-independent methods have been suggested to the community. Moreover, the applicability of such techniques to clinical applications depends on their efficiency in processing cells. High-throughput techniques are of interest, and imaging methods used in single-cell assessment are being adapted to increase processing rates. The ability to process millions of cells in a single biological sample should propel elastography to more accurate results as it reduces the incubation time in buffer solutions or media. The ability to provide micrometer-based spatial resolution and millisecond-based temporal resolution could also revolutionize the field by giving access to dynamic changes in biomechanics associated with biological processes.

The walkthrough in this chapter purposes to provide a basis for the unfamiliar reader to understand the basic cell structures responsible for its mechanical properties, available techniques for quantifying single-cell elasticity as well as considerations in choosing a measurement method. The current and future interests of biorheologists, as well as clinical relevance are discussed.

## 10.2 Cell structure: key architectural players in elasticity

Body parts are a group of organs (bones, muscles, and vessels) that are constructed by different tissues. The tissues are comprised of grouped cells held together by the ECM. The cell is the basic unit of life, which consists of the nucleus, cytoplasm, and membrane. This hierarchical nature of the living body suggests that deformation or mechanical stress to body parts is translated to structural rearrangements of the cell thus potentially affecting its function. Dynamic changes in the architecture and molecular composition of a cell during function, dysfunction or external attack will consequent in changes in mechanical properties of the cell. These mechanical changes on the cytoplasm, membrane, or nucleus can be predictive of developmental failure or markers of healthy versus diseased cells. This section cursorily highlights the key architectural players that modulate the response to mechanical forces and pathway by which mechanical stresses are transmitted.

### 10.2.1 Cell membrane

The cell membrane is a double-layer of a complex of phospholipids, cholesterol, and proteins embedded with channels that make it semipermeable. It plays two major roles: (1) provides structural integrity by creating a boundary between the materials inside and outside the cell, and (2) contains passage tunnels for channeling biomolecules into and out of the cell crucial for its homeostasis [7]. The permeability of the membrane as well as its rigidity determine the cell shape and appropriate cellular filtering of biomolecules.

Physical properties of the membrane have been of interest for decades, propelled by elasticity theories [8–10], that highlight the role of membrane elasticity in cell functions like adhesion, motility, cell division, wound healing, and intracellular trafficking. Here, we briefly discuss key structural organization of membrane proteins that contribute to its mechanical properties during normal functioning. The cell membrane is the surface upon which cells adhere to the ECM. Acting as a first point of contact, the flexibility of the ECM regulates receptor adaptability; if the ECM is flexible, rapid deformations are sensed while sustained stresses are dissipated before they reach the cell [11]. Mechanical stresses transmitted to individual cells are distributed via their adhesion to the ECM.

Physiologically, adhesion is crucial in understanding how bacteria, parasites, and viruses interact with a cell. Furthermore, adhesive properties of individual cells determine the cell shape, the structural integrity of tissues, the communication through adhesive junctions, and cell migration [11–14]. The sites of adhesion are provided by cell surface receptors called integrins, which are embedded in the cell membrane where they are activated and participate in binding to the ECM and in cell–cell adhesion [15–19]. The cluster of integrins on the membrane receives the mechanical signal from the ECM and transmits it across the membrane and to the cytoskeleton via a link to actin filaments [20,21].

Both actin filaments and integrins, which are crucial for cell mechanics, are anchored in the membrane and provide a pathway into the cytoskeletal (CSK), defining the cell membrane as a heterogeneous structure [20,22–24]. Actin structures are associated with the membrane in the form of branched and unbranched filaments that regulate protrusion when cells adhere and crawl on a substrate during cell spreading, as well as the ability of the cell to pull itself forward [25,26]. For example, during cell migration, protrusive and contractile forces required for the cell to push and squeeze depend on the rearrangement of actin filaments into subcellular structures on the membrane [27]. Additionally, actin filaments contribute to this motion by forming a crosslink permeated by myosin motors that are essential in cortex regeneration [26]. Moreover, the connection of the membrane to the CSK provides the resistance of the membrane and balances its tension through dynamic mechanical changes [28,29].

### 10.2.2 Cytoplasm

Within the cytoplasm is the CSK; this is a network of protein fibers that include actin filaments and microtubules, which support the cell shape as well as anchor organelles in a viscous fluid suspension called the cytosol. Mechanical strength of the cell majorly depends on actin filaments that have been shown to dominate in modulating cellular elasticity and by extensional cellular morphology [30–33]. Tensional forces are generated in the cytoplasm as actomyosin slides in their CSK. As for the microtubules, their biomechanical effects on the cell are secondary in that their interaction with actin filaments resists and balances inward directed tensional forces [29,30]. Moreover, such tensional forces are balanced by CSK fibers from individual cells, which are interconnected and through adhesion to the ECM reinforce the mechanical strength of the tissue. CSK forces play an integral role in cell motility by housing and positioning motor proteins crucial in transporting information between the extracellular environment and the nucleus [14]. The rearrangements of actin filaments also mitigate the organization of organelles and corresponding molecules depending on the needs of the cell.

### 10.2.3 Nucleus

For the eukaryotic cell, the defining feature and largest organelle is the nucleus. It acts as the functional epicenter of the cell containing the genetic material organized as DNA molecules and nucleoproteins to form chromosomes. The nucleus is isolated from the rest of the organelles by a double membrane that is selectively permeable and structurally supported by the nucleoskeleton. Through the LINC (linker of the nucleoskeleton and cytoskeleton) complexes, forces are transmitted between the cytoskeleton and the nucleus. The physical connection between the CSK and nucleoskeletal has been hypothesized as crucial in intracellular nuclear movement and positioning, in CSK organization and cell migration, as well as in triggering force induced changes in nuclear structures. Albeit in its infancy, research in nucleo-CSK coupling and the role of the proteins in the LINC complex is deemed essential in elucidating how the nucleus processes mechanical forces [34] and in understanding the effects on gene expression.

From the above review, it can be concluded that the effect of mechanical stresses on the cellular functionality and shape depends on the mechanical properties of the ECM, the organization of membrane proteins and CSK filaments, cytoplasmic tensional forces, and the cyto-nucleoskeletal coupling. Proper distribution and transmission of mechanical forces and the ability of the cell to perform its function while maintaining structural integrity; that is, without shearing or breaking, is the basis of cellular elasticity.

### 10.2.4 Clinical relevance of cellular elasticity

Correct physiology and appropriate mechanotransduction—how cells sense and transform mechanical signals into biochemical processes, depend on the elasticity of the cell components. Dysfunction at any point in cellular mechanics may consequent in cell, tissue, and organ pathophysiology or developmental abnormalities resulting in diseases spanning a wide range of fields in health [35–39]. A few examples of elasticity changes manifested clinically in diseases include decreased lung elasticity in emphysema [40], breast and bladder cancer cells [41,42], narrowing of airways in asthma [43], enhanced rigidity and adhesion of red cells to the endothelium in malaria and sickle cell disease [44,45], increased stiffness of brain cells and tissues in Alexander disease [46], increased stiffness in ECM of the lung in idiopathic pulmonary fibrosis [47], reduced stiffness in emphysematous lung tissues and chronic obstructive pulmonary disease [48,49], arterial stiffening in hypertension [50], reduced elasticity in gastrointestinal cancer, malaria, and human breast epithelial cancerous cells [51,52], increased stiffness in myofibrillar myopathies [53], and body weight loads and urine pressure causing cell necrosis in diabetic foot ulcers and kidney disorders [28].

The correlation of cellular elasticity with morphology has gained momentum in the pharmaceutical industry with calls for the creation of a unique area termed as "mechanopharmacology" [54]. This would focus on the consideration of cellular mechanics in identifying drug targets, particularly drug action on the cell [55,56]. Knowledge of cellular biomechanics has two major contributions to pharmaceutical developments: (1) development of pharmacological agents that are biomechanically similar to the target organ and tissue, mainly the creation of substrates that span the known pathophysiological range of cells in which to develop drugs [57], and (2) identification of cellular stiffness as a biomarker of therapeutic activity. Studies that have exploited stiffness as a biomarker include but are not limited to: antimitotic microtubule targeting agents that have been correlated to corrected whole-cell stiffness on cancer cells [58]; reduction of inflammation of endothelial cells by the simvastatin drug, as illustrated by elasticity changes [56]; and observed increase in cell membrane permeability (reduced elasticity modulus) by glycyrrhizin, in the development of drug delivery systems [59].

Moreover, investigation into the elasticity of pathogens is instrumental in elucidating pathogen–host interaction, particularly pathogen adherence in drug screening. The elastic properties of the bonds between the pathogen and host have been shown to be responsible for robust adhesion and cooperation [60–65]. Such mechanical properties have been used to investigate adhesion of *Borrelia burgdorferi* in Lyme disease [66], binding of the Ebola virus to the host cell membrane [67], the membrane properties of *Escherichia coli* [68], and the effect of membrane stiffening of red blood cells on deformations resulting from adhesion of malaria parasites [69].

Cellular elasticity can also be used to phenotype mutations of said pathogens and potential immune evasion. In *Candida albicans*, a common human fungal pathogen, mutations of the wild type have been reported to have decreased cell wall elasticity [70], whereas wild type 042pet *E. coli* is reported to be more rigid that the mutant 042aap [71].

Pointedly, most of the clinical manifestations of diseases are related to changes in tissues and by extension of cellular mechanics [35,72]. Scientists have responded to this by focusing studies beyond genetic testing and to physical phenotyping of tissues and cells, with hopes to elucidate the effects of mechanical stresses on biological functions. To this end, molecular biologists now focus on the pathways in which mechanical forces are converted to biochemical signals in the field of mechanotransduction, which is the process by which cells convert mechanical stimuli to biochemical signals [21,34,35,73,74]. On the other hand, physicists and engineers apply rheological methods (the science of deformations and flow) to understand cellular mechanics.

Rheology primary deals with the relationships between stresses and deformations and has been applied over different industries from chemical processing (plastics, paints, and lubricants) to geological seismology and most notably biotechnology (cellular and biomaterials) [75]. A wide array of techniques have been developed with length scales to probe cellular elasticity, viscosity, and viscoelasticity. By inducing external mechanical stresses, the response from the cell can be tracked and used to phenotype its properties, development, and pathophysiology. Albeit not all novel technologies as atomic force microscopy (AFM) or optical tweezers have been around for decades, recent developments to push the spatiotemporal scales would potentially enable the capture of dynamic processes in addition to biomechanical phenotypes. Advancements in such techniques are presented more thoroughly in the following section.

## 10.3 Estimation of cell elasticity

Viscoelasticity is a fundamental property to better understand biological cells. This parameter is indeed related to the cell anatomy, functionality, and pathological state. Due to its importance, various techniques have been developed to estimate it.

### 10.3.1 Cellular deformation

Most widespread techniques to estimate cell viscoelasticity are based on cell deformation. Cells can be deformed in many ways, including micropipettes or microplates [76], microfluidic deformation [77,78], magnetic bead twisting [79], and tweezers or stretchers [80]. Micropipettes, microplates, and microfluidic tubes can deform the cell by various ways, notably aspiration, squeezing, tubular constrains, etc. More information on optical, acoustic, and magnetic tweezers is provided later in this chapter. Active or passive cell deformation can be observed from optical images under a microscope. Global elasticity (or whole-cell elasticity) can then be estimated from the applied force intensity and the observed deformation. Viscosity can also be deduced by observing deformation along time.

However, mapping both viscosity and elasticity is more difficult because the local deformation depends strongly on the internal distribution of stress, which is unknown. Scientists can use different models to estimate this distribution, but the assumptions behind the model have a strong influence on the final estimation. As an illustration, mouse oocyte zona pellucida elasticity was measured in two independent experiments [81,82]. The first team found a value of 3−14 ($\pm$ 2) kPa depending on the model, and the second group reported a value of 42.2 $\pm$ 2 kPa. Hence, the results strongly depend on the model and images are hardly quantitative [83] but offer approximations. Despite this drawback, cell deformation is widely used nowadays, thanks to its simplicity and adaptability to different setups. Also, spatial and temporal resolution can be excellent, as they mainly depend on the camera pixel resolution and frame rate.

### 10.3.2 Cellular deformation using tweezers

An important cellular deformation technique lies in the use of optical tweezers or stretchers [84−86]. It is based on the difference in refractive index between different parts of the sample (or with the surroundings). Due to this difference in optical property, a focused laser beam can trap a part of the cell and move it. One can then deform the whole cell to estimate its global elasticity [85]. However, it is rarely used to map elasticity from local deformations [86], as it needs an appropriate modeling of the cell mechanical behavior (similarly to the cell deformation with micropipettes); and it is complicated to trap only a point of the cell, requiring to add external optical beads (making the technique very similar to active microrheology, described later). The technique presents nevertheless the advantage to be contactless and quite easy to implement. It is often used to measure red blood cells global elasticity [87–89]. Cell tweezers were also developed using acoustic [87,88] and magnetic fields [89,90].

### 10.3.3 Atomic force microscopy

AFM has widely been used to map cell elasticity [91–93] and has become with time one of the standard to assess cell viscoelasticity. With this technique, the cell is poked with a small cantilever. Cantilever displacement is measured through the deviation of a laser beam. Measurements are done point by point, allowing to reach a submicrometric resolution. Viscoelasticity can then be deduced from the temporal response of the cantilever penetration at each point of the medium. Similarly to cell deformation techniques, AFM needs an appropriate physical modeling, as even if the applied force can be precisely measured, the stress distribution in the sample is unknown. As measurements highly depend on the chosen model, AFM requires a precise calibration of the cantilever shape: change in shape due to probe aging, in particular, can have an important impact. Finally, due to the mechanical measurement point by point, acquisitions are rather slow, taking typically at least a few minutes to achieve a reasonable number of data points. One must be careful that no biological process impacting cell viscoelasticity occurs during the acquisition time—this is not always negligible: for example, a study using a micropipette indentation showed that mouse oocyte elasticity was decreasing by a factor of two in less than 10 seconds during probing [94]. The long period of acquisition with AFM also requires the cell to be fixed to the support, a process which can alter cell characteristics. Also, although some models allow to measure viscoelasticity in the bulk, this technique essentially makes surface measurements [95] (Fig. 10.1).

### 10.3.4 Microrheology

Microrheology is a technique initially developed to assess viscoelastic properties of soft matter [96,97]. This technique uses micrometer sized spheres as particle probes. Each particle is then individually tracked over time—typically for a few minutes. We usually distinguish "passive microrheology," where particles are moving due to Brownian noise, and "active microrheology," where particles are moved by an external force (magnetic field, optical tweezers, etc.). Note that "active microrheology" is

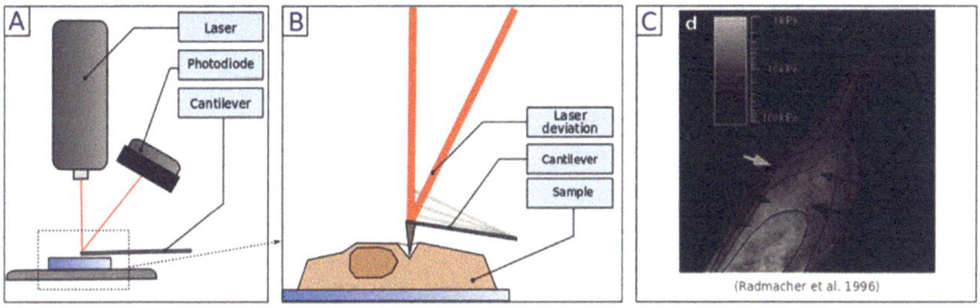


**Figure 10.1** (A) and (B) Schematic of an AFM experiment. A cantilever is poking a cell attached to a substrate. This deviates a laser detected by a photodiode. (C) The deviation behavior gives information on the cell elasticity at each point. *AFM*, atomic force microscopy. *From M. Radmacher, M. Fritz, C.M. Kacher, J.P. Cleveland, P.K. Hansma, Measuring the viscoelastic properties of human platelets with the atomic force microscope, Biophys. J. 70 (1996) 556–567.*

sometimes named from the external force type: for example, "magnetic twisting cytometry." In magnetic twisting cytometry (MTC), microbeads are allowed to bind to cell surfaces for 10–15 minutes. Introducing a strong magnetic field (1000G for 10 microseconds) allows magnetization and alignment of magnetic moments of the cell bound beads. Mechanical stress is then applied using a weaker twisting magnetic field that is perpendicular to the original magnetic field. Finally, measurement of the average bead rotation induced by the twisting magnetic field allows the estimation of mechanical properties of the cell [79,98,99].

The particle density has to remain low to facilitate tracking, and more importantly, to limit particle interactions due to the Van der Waals force or other forces that would influence particle displacement. Knowing the physical force which makes them to move and the mean square displacement of each particle allows the estimation of both elasticity and viscosity of the sample at the particle locations. Hence, microrheology cannot provide detailed maps of a sample viscoelasticity, but rather measurements at discrete particle locations.

Nonetheless, microrheology has been applied to assess mechanical properties of biological cells [100]. Scientists have been able to estimate both elasticity and viscosity in different zones of the cells, such as the interphase, nucleus, and cytoplasm [101]. However, applying this technique to biological samples makes the assumption that the particle displacement is only due to physical processes. This assumption can be invalid for active cells and/or long measurement times, making this technique less reliable. Another drawback lies in the invasive nature of the method: particles have to be injected inside the cell for tracking purpose without impacting its functionality (Fig. 10.2).

## 10.3.5 Brillouin scattering microscopy

Brillouin scattering microscopy has been introduced and further developed recently to estimate cell elasticity [102,103]. This technique entails transmitting a laser beam in the studied cell. Internal mechanical waves inside the cell shift the laser frequency. Measurement of this shift allows to estimate bulk elasticity modulus at the laser beam location. By moving the laser beam point by point provides 2D and even 3D measurements. In 2015 Scarcelli et al. [102] showed that the bulk modulus varies proportionally with the Young modulus in NIH 3T3 cells with various concentration of sucrose. This might not be true for all samples, as even in a homogeneous isotropic linear elastic material, the physical relationship between these two magnitudes involves a third parameter (Poisson's ratio). This method, as with most microrheology technologies, is rather sample dependent: for example, in hydrated materials, water content dominates Brillouin spectroscopy measurements instead of stiffness [104]. However, although it can reach a good spatial resolution ($0.5 \times 0.5 \times 2\ \mu m^3$ in Scarcelli et al.), the point-by-point measurement makes the acquisition of a large number of points rather slow, typically a few minutes (Fig. 10.3).

## 10.3.6 Microelastography

Shear wave elastography is a technique developed a few decades ago to map organ viscoelasticity, through ultrasound imaging [105,106] or magnetic resonance imaging [107]. It has been successfully applied in many

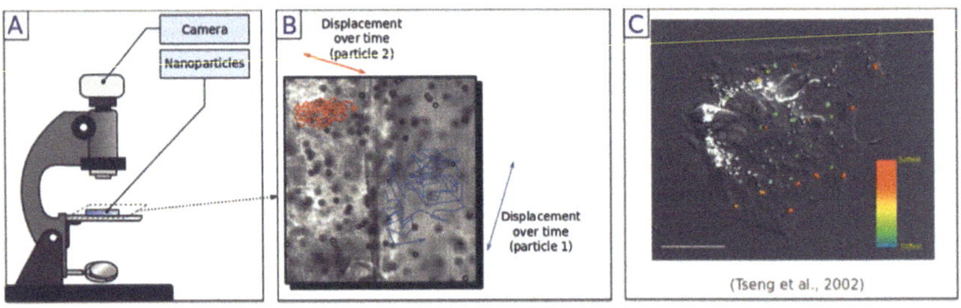


**Figure 10.2** (A) and (B) Schematic of a microrheology experiment. Particles are injected in the sample. A camera tracks the particle along time. (C) The mean square displacement of particles allow to evaluate local viscoelasticity. *From Y. Tseng, T.P. Kole, D. Wirtz, Micromechanical mapping of live cells by multiple-particle-tracking microrheology, Biophys. J. 83 (2002) 3162–3176.*

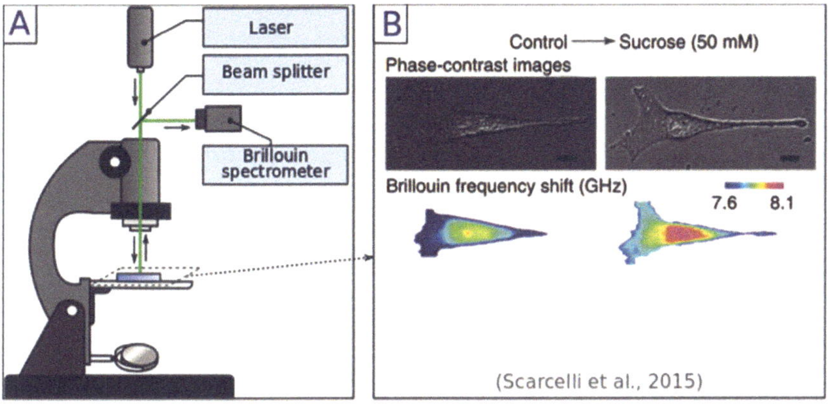


**Figure 10.3** (A) Schematic of a Brillouin spectroscopy experiment. A laser beam is emitted on the sample. Longitudinal waves inside the cell shift the laser frequency. A Brillouin scattering then measures the frequency shift, related to the local longitudinal elasticity modulus. (B) A map of the Brillouin frequency shift can be produced with a high spatial resolution. *From G. Scarcelli, W.J. Polacheck, H.T. Nia, K. Patel, A.J. Grodzinsky, R.D. Kamm, et al., Noncontact three-dimensional mapping of intracellular hydromechanical properties by Brillouin microscopy, Nat. Methods, 12 (2015) 1132.*

organs, including the liver [108], breast [109], and the thyroid [110]. This technique is based on tracking of a type of mechanical wave, the shear wave, in the studied sample. As the shear wave speed is related to the sample elasticity under certain assumptions, one can map elasticity by measuring the speed of the shear wave everywhere in the sample. Besides, shear wave amplitude also decreases with viscosity, so this parameter can be also estimated from the shear wave attenuation over distance [111]. Concretely, in shear wave elastography, a time-lapse of the sample is imaged, usually with ultrasound or magnetic resonance imaging. A shear wave is induced in the sample, through a mechanical or acoustical vibration, and images are captured as the wave propagates through the sample. The displacement between images is calculated with a tracking algorithm. Elasticity and viscosity are then mapped in the sample from the displacement propagation as shear waves.

This technique has been recently applied at the cellular scale [112], by achieving three main milestones:

1. the generation of shear waves at high frequency (15 kHz), at least 10–300 times higher than the usual shear wave frequency used in elastography;

2. the high-speed detection of the waves; indeed, in soft matters, shear waves usually propagate at a speed of a few meters per second, so tracking the shear waves in a sample of a few tens of micrometer requires an imaging frequency higher than 100,000 kHz. In the reported experiments of Grasland-Mongrain et al., a 200,000 frames per second (fps) optical camera was used; and
3. the use of correlation-based techniques to calculate displacement propagation; as the shear wave front was not well defined in reported experiments, a robust 2D optical-flow-based speckle tracking method had to be used.

With these three challenges resolved, the technique could be used to map the elasticity of $\approx 80\,\mu m$ diameter mouse oocytes. The authors reported a spatial resolution of about 10 μm, for a measurement acquired in less than 1 microseconds. Despite this relatively low spatial resolution, the measurement speed is an important advantage to follow dynamic cellular processes. Also, the time resolution avoids artefacts due to changes in cell biomechanics during measurements. Scientists are currently trying to improve the spatial resolution by using different types of shear waves and by improving elasticity estimation algorithms. Theoretically, the spatial resolution limit should be the one of optics; so it is anticipated that a submicrometer resolution could be achieved. One important point of the technique is the absence of viscoelastic model required to reconstruct cell biomechanics, contrary to cell deformation techniques. But this goes with a difficulty: cell elastography primarily provides wave speed information, which can be related to cell elasticity assuming some a priori on shear wave propagation and cell property. Additionally, the method has not yet been developed to measure cell viscosity (Fig. 10.4).

### 10.3.7 Summary

In summary, depending on the application, estimating the viscoelasticity of a whole cell in a few locations can be sufficient, without the need to locally map this parameter with a good spatial resolution. Also, some techniques are able to estimate both elasticity and viscosity of the sample, whereas some only provide one of these parameters, or a combination. A single method is currently able to map cell elasticity at a temporal resolution on the order of a millisecond. The choice of the measurement method can vary depending on the results of interest. The table below offers a summary of discussed methods (Table 10.1).

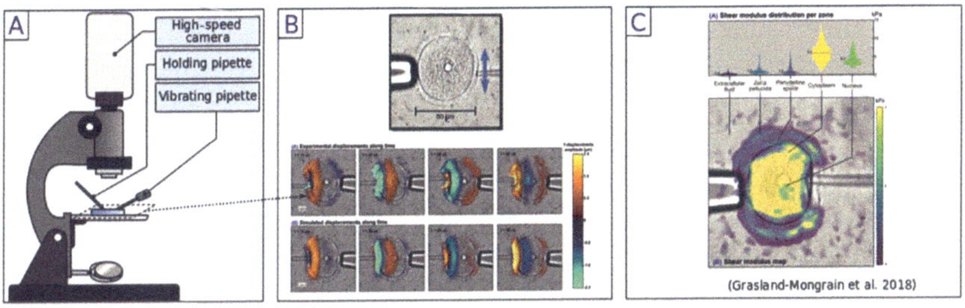


**Figure 10.4** (A) Schematic of a microelastography experiment. A pipette is vibrating the sample. The vibration propagation is observed by a high-speed camera. (B) A computer algorithm is then estimating the local shear wave speed, related to the cell elasticity. (C) Map of the cell elasticity. *From P. Grasland-Mongrain, et al., Ultrafast imaging of cell elasticity with optical microelastography, Proc. Natl Acad. Sci. U. S. A. 115 (2018) 861–866.*

**Table 10.1** State-of-the-art rheology technologies to assess cell viscoelasticity and description of performance.

| | **Setup simplicity and price** | **Spatial resolution** | **Temporal resolution** | **Measured physical parameters** |
|---|---|---|---|---|
| Cell deformation [76,79,82] | ++ | <μm | ≈second | Deformation |
| Optical tweezer [85,86,113] | + | <μm | ≈second | Deformation |
| Atomic force microscopy [91–93] | − | ≈μm | ≈minute | Cantilever deviation |
| Nanoparticles tracking [114] | + | >10 μm (no | mapping possible) | >minute |
| Particle mean square displacement | | | | |
| Brillouin scattering [102,103] | − | ≈μm | ≈second | Brillouin frequency shift (related to the longitudinal modulus) |
| Microelastography [112] | − | >10 μm | ≈millisecond | Shear wave speed |

## 10.4 Rheological modeling of a single cell

This section is highlighting some of the models applied for the interpretation of experimental data used to quantify cellular elasticity. Modeling of living cells attempts to quantify mechanical properties and responses when perturbed by an external load or stress. There are principally two schools of thought on the physical distribution of forces by cells: the micro/nano-structural approach and continuum models. Structural approaches consider discreet load-bearing elements, with the cytoskeleton as the main component. Continuum models, on the other hand, consider load-bearing elements that are infinitesimally small relative to the size of the cell. Biomedical models allow for the quantification of experimental data by making assumptions on the cell surface and shape, with the simplest model assuming homogeneity and isotropic distribution of mechanical stresses.

Noteworthy, is that although some general models have been developed, there is not a "one ring to rule them all" universal model to quantify mechanical properties of living cells due to their dynamic nature and ability to adjust to different microenvironments. Moreover, rheological techniques use different types of forces, magnitudes, and loading rates that are eliciting different responses from the cells [115,116]. Rheologists should thus choose a model that appropriately fits testing conditions, is physiologically relevant, and is descriptive of specific cells. After briefly introducing basic principles of each model, recent insights such models have provided are highlighted in the context of cellular elastography.

### 10.4.1 Structure-based models

#### *10.4.1.1 Tensegrity model*

Micro/nano-structural approaches deem the cytoskeleton as the main structural component with the widely accepted composition of microfilaments, microtubules, and intermediate filaments. The organization and association of this meshwork influences cellular tension and alternations responsible for cell shape and movement. The complexity of this structure in its most basic form is architecturally modeled as six wooden bars that are not in contact but are connected to a series of 24 elastic strings which hold up the rigid bars in an approximate sphere. The integrity of this structure is observed when a force is exerted from above flattening it out or anchored at the bottom spreading it out, without breaking. When compressional forces are released, the structure spontaneously reverts to its original shape, whereas the rigidity of wooden bars remains unchanged. The reliance of shape integrity on internal tension gave this model its name—tensegrity for tensional integrity [117–119].

This model, adapted from construction architecture, is applied to biological cells by making a number of assumptions [118,120–123]. These include: (1) mechanical load of cells are transferred across points where they anchor to their support system, that is the ECM or to other cells; (2) the cytoskeleton is responsible for the internal tension of a cell; (3) at molecular scales, local force interactions are dominant to gravitational effects, which means that the gravity is negligible; and (4) a compressible nucleus allows the extension of tensegrity to the nucleoskeletal, that is the whole cell can adhere to the ECM or to other cells and stretch in cell spreading without breaking or shearing.

Using this 30 structure model, an expression for the elasticity is derived based on a coarse-grain procedure. The working principle of the derivation is that external forces $E$ on the model are equal to the strain energy and follow the relationship presented in the equation below [124]:

$$E \approx \frac{\sigma_c^0 \varphi_c}{3} \approx \frac{\sigma_s^0 \varphi_s}{3}. \tag{10.1}$$

The above equation can be applied to experimentally generate upper and lower bounds of cellular elasticity, where $\sigma_c^0$ is the initial tensile stress of the strings, $\varphi_c$ is the relative density of the strings, $\sigma_s^0$ is the initial compression stress in the wooden struts, and $\varphi_s$ is the relative density of the wooden struts. This is achieved by identifying the tensile strings with the CSK actin filaments and the wooden struts with the microtubules [125,126].

Years of experimental proof have seen tensegrity been adapted in different studies, especially for investigating CSK mechanics in adherent cells and in determining the structural origin of cellular viscoelasticity. The tensegrity model has supported experimental efforts in characterizing the response of CSK elements on global CSK behavior [127], dynamics of cell reorientation in collective cell alignment and movement [128], and quantification of cellular elasticity providing strong evidence that the CSK is the primary determinant of the cell elastic response [125,126,129,130] (Fig. 10.5).

Despite the ability of the tensegrity model to accurately predict elastic characteristics of cells, it has a few limitations that include: failure to

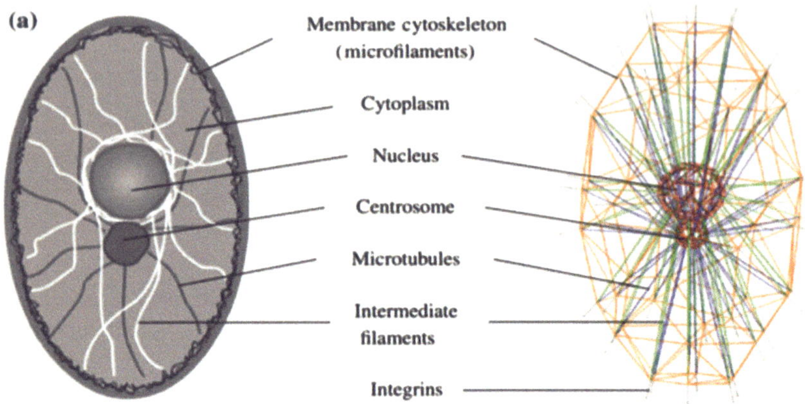


**Figure 10.5** (A) Tensegrity model of the cell, with the comparator of the cell structure to the model, highlighting key structural elements. *From L. Zhang, X. Feng, S. Li, Review and perspective on soft matter modeling in cellular mechanobiology: cell contact, adhesion, mechanosensing, and motility, Acta Mech. 228 (2017) 4095–4122.*

consider effects from thermal and nonthermal fluctuations, effects from the nuclear membrane and the fluid-like cytosol and compressibility of the microtubules, which are modeled as rigid struts. In the simplicity of the tensegrity model lies its elegance and flaws in characterizing complex hierarchal biological structures. Nonetheless, efforts are in place to develop more robust and multiscale tensegrity models incorporating considerations such as relaxing the rigidity of the struts, and introducing compressibility and hyperelasticity of the bars thereby accounting for nonlinear responses [131,132]. Needs are also to consider models that incorporate the nucleoskeletal capturing the hierarchical cell [122].

#### *10.4.1.2 Percolation model*

A model adapted from the mathematics and material science fields, namely the percolation theory, describes the random organization of elastic clusters. The emphasis of this model is in the connectivity of basic units, which are essential in information transfer [133]. Above a critical concentration of subunits or a percolation threshold, the structure develops elastic properties capable of response to mechanical stimuli. This threshold highly depends on the random geometry of the arrangement giving the structure great flexibility in assemblies, consequently the elasticity of the structure is dynamic [116,134]. The constituent subunits can assume any shape and randomly connect, thus changing global elasticity while adjusting to the environment. Biologically, interconnected networks are plenty within the cell inclusive of the nucleoskeletal filaments, CSK, or ECM. This model views the CSK as an entangled random mesh of network forming macromolecules. Applying the percolation model to the assembly of such networks explains morphogenic motions that are driven by the matrix. That is, locally assembled interconnected filaments and microtubules are dynamic and constantly rearranging while maintaining a global elasticity.

The percolation model does not compete with the tensegrity model, on the contrary the two can be thought of as complementary. As discussed in Ref. [133], tensegrity structures could develop from assembly of percolation structures. Moreover, there is the shared characteristic of interconnected ness and dynamic subunits contributing to a global sensitivity that allows uninterrupted propagation of stimuli. Unfortunately, this model has remained highly conceptual with limited quantitative results or correlation to elasticity [135]. Efforts to quantify elasticity based on a percolation, percolation-like model, or composite model include: calculation of the shear modulus of membrane skeleton based on a random incomplete lattice of spectrin tetramers [134], use of a random graph model of the spectrin CSK to investigate changes in red blood cells [136], and modeling skin lattice based on percolation to investigate tissue aging, which is correlated to elasticity loss [137].

## 10.4.2 Continuum-based models

Instead of focusing on particular constitutive structures as the main determinant of elasticity, continuum models focus on the continuous distribution of forces and strains within the whole cell, thus providing less details [115]. These models are, however, easier and direct in extracting mechanical properties and correlating them with experimental observations. The simplest model of the cell assumes isotropy and homogeneity; however, experimental results pointing to a more viscous nucleus compared to the cytoplasm have redirected model analysis to more robust options that capture heterogeneity, anisotropy, and viscoelasticity. Below such models are discussed.

### *10.4.2.1 Liquid drop models/cortical shell liquid models*

These models are based on the assumption that cells behave like a liquid drop. That is, they adapt a spherical shape when suspended, deform continuously as response to suction pressure into a micropipette acquiring a smaller diameter, and steadily recover the initial spherical shape upon release [138]. These models were first developed for the rheology of neutrophils in micropipette aspiration. Among these models are the simple and compound Newtonian liquid drop, shear thinning liquid drop model, and the Maxwell liquid drop model. The first three are satisfactorily suited for modeling large deformations. For the simple Newtonian liquid drop model, the cell is assumed to be a homogenous viscous liquid with a cell cortex that is anisotropic and with statistic tension. For example, leukocytes adopt a spherical liquid drop shape when suspended [138]. The constitutive equation of this model approximates the cell membrane and its cytoskeleton as a liquid cortical layer with negligible viscosity and a constant surface tension until an expansion limit is achieved [76,139,140]. Such studies have revealed the overall elasticity and viscosity to vary continuously with the degree of deformation [141].

The compound Newtonian model takes into account the nonhomogeneity of the interior of the cell; with a nucleus that is stiffer and more viscous than the cytoplasm [141–143]. Eukaryotic cells that are composed of a cell membrane, a cytoplasm, and a nucleus can be modeled with this approach. The model divides the cell into a three layered structure; the ectoplasm which is under persistent tension, the endoplasm as the softest region which is fluid-like, and the core layer composed of a condensed nucleus. For the structurally heterogeneous cell to behave as the simple Newtonian model, a few considerations should be accounted for: the time-scales of the cytoplasm and nucleus as well as the initial deformation ratio of the nucleus to the cytoplasmic shell should be comparable [144–146]. Time-dependent behavior of passive neutrophils due to large deformations can be explained by the compound liquid drop model in this manner [145]. There could be an infinite number of possibilities to fit a particular rheological model to study cell recovery depending upon the cell type and the experimental setup, meaning that only one kind of experiments cannot sufficiently identify all viscoelastic properties that are concerned. Among the success of this method is that in its limitations it is understood that dealing with the cell as a whole is inadequate when studying heterogeneous structures and a more accurate model would deal with the structures (cortex, cytoplasm, and nucleus) individually [115,147]. Like Newtonian models, shear thinning is based on micropipette aspiration. The underlying principle relies on the apparent viscosity of the cytoplasm that decreases with increasing mean shear rate, that is, the aspiration pressure. In the Newtonian liquid model, the aspiration speed is constant under constant aspiration pressure, however, the observed acceleration right before the whole cell is sucked in takes the form of a power law [148]. Combined with a finite-element-model, the shear thinning explanation is promising as it performs well in different experimental conditions capturing the complex cellular geometry during aspiration and other nonlinear relationships [115,149,150]. Shear thinning is successful in providing a comprehensive picture of cells during micropipette aspiration, however, other measurements with small scale dynamics do not exhibit shear thinning, limiting the above models to modeling larger deformations in cells.

To account for short-time smaller deformations, the Maxwell model which assumes a prestressed cortical shell containing a Maxwell fluid is better suited to explain the initial rapid entry during the micropipette aspiration of the cell [151]. Here, the cortical shell represents the near-surface actin layer of the cell, whereas the Maxwell fluid constitutes average mechanical properties of the cytoplasm with the main difference being that a Maxwell model contains an elastic component. This model was first utilized to study the recovery behavior of passive leukocytes in micropipette aspiration under small deformations [151]. Later, this model was also extended for large leukocyte deformations using a finite element method analysis to show that the Maxwell liquid model could not accurately describe the experimental data without allowing elasticity and viscosity values of the cytoplasm to increase computationally as the cell gets sucked into the pipette [141,152]. The constitutive equation is as below [151], where a dashpot and spring in series capture mechanical properties in $k$ as an elastic constant and $\mu$ as a viscous constant. The parameter $\tau_{ij}$ is the deviatoric stress constant, $\dot{\tau}_{ij}$ is its spatial derivative, and $\dot{\gamma}_{ij}$ is the engineering strain constant [153]. Noteworthy is that the Maxwell model degenerates into a Newtonian model as the elastic constant $k$ approaches infinity, implying that as the cell is being aspirated it transitions from a Maxwell to a Newtonian fluid with increasing elasticity and viscosity moduli [115]. This model is given by:

$$\tau_{ij} + \frac{\mu}{k}\dot{\tau}_{ij} = \mu\dot{\gamma}_{ij}. \tag{10.2}$$

Small deformation behavior in cells is fundamentally different than the large deformation of flow. In the above equation, if $k$ approaches to infinity, the Maxwell fluid degenerates into a Newtonian fluid, meaning that the cytoplasm undergoes a transformation from the Maxwell to the Newtonian behavior during micropipette aspiration, with increased elasticity and viscosity moduli. If the cell holding time is very rapid (5−7 seconds), the sausage shape deformed cells can exhibit a rapid initial elastic recovery [141]. However, if cells are to be held for a longer duration, the recovery becomes slower and the Newtonian liquid behavior dominates [145,146]. Perhaps the most significant contribution of the Maxwell model is in identifying cells as viscoelastic.

#### *10.4.2.2 Solid models*

Solid models assume that cells are made of one or more homogeneous layers and the incorporated material models may range from incompressible elastic to a viscoelastic solid depending upon experimental conditions. Cells are considered to be in a solid phase and the homogeneity assumption simplifies experimental data analysis as several mechanical parameters vanish from the mathematical equations. In micropipette experiments, endothelial cells and chondrocytes could not flow into the pipette even when the suction pressure was beyond the critical suction pressure, giving it a solid-like behavior [154,155]. Furthermore, nonspherical shapes in such experiments can be attributed to the reassembly of the cytoskeleton elements in response to the shear stress making it behaves like a solid rather than a fluid. Solid models are discussed separately for linear elastic and linear viscoelastic materials in the following subsections.

##### Linear elastic solid model

In a linear elastic model, cells are assumed to be in a solid phase with homogeneous and elastic properties. This model is a simplification of the viscoelastic solid model in which the time factor is ignored. The elasticity of the cell can then be represented in terms of its shear modulus $G$. Under the assumption of linear elasticity, the shear modulus is related to the Young's modulus $E$ by the equation $E = 2(1 + \nu)G$, where $\nu$ is the Poisson's ratio approximately equal to 0.5 for an incompressible material. Even though such a linear elastic model may not be adequate for describing cell mechanics, it serves as a basis for the viscoelastic solution.

In micropipette aspiration experiments, the radius of the cell is much larger than the radius of the pipette, and so, the cell is approximated as an incompressible elastic half-space. If the applied suction pressure is $\Delta P$, the aspiration length of the cell is $L$, the radius of the pipette is $R_p$, and the ratio of the pipette wall thickness to the pipette radius is $\phi_P$ ($\phi_P$ = 2.0−2.1 when the ratio is equal to 0.2−1.0), then the shear modulus $G$ can be computed as [154]:

$$G = \frac{\phi_P \Delta P}{2\pi\left(L/R_p\right)}. \tag{10.3}$$

For AFM indentation of adherent cells, if $F$ is the indentation force, $\theta$ the indentation angle, and $\delta$ is the indentation depth of the AFM tip, then the shear modulus can be computed as [156]:

$$G = \frac{F(1-\nu)\tan\theta}{1.4906\delta^2}. \tag{10.4}$$

The indentation of adherent cells can also be performed using a cytoindenter or a cell poker. For a cell poker with a cylindrical tip whose surface curvature and cell thickness are larger compared with the cylindrical diameter, the linear elastic solution to compute the shear modulus can then be expressed as below, where $R_I$ is the radius of the indenter:

$$G = F\frac{1-\nu}{4R_I}. \tag{10.5}$$

Finally, for MTC experiments [79,98,99], a finite element method can be utilized to achieve an analytical solution [157]. The spherical magnetic field is assumed to be fully adhered to the cell surface, and the bead and the substrate are assumed to be rigid. The expression to compute the shear modulus can be given as:

$$G = \frac{T}{k_s\alpha}\left(\frac{1}{\phi}\right) \text{or } G = \frac{TR}{k_s\beta}\left(\frac{R}{d}\right), \tag{10.6}$$

where $T$ is the applied torque, $k_s$ is the shape factor for the bead, $\phi$ is the angular rotation of the bead, $R$ is the radius of the bead, $d$ is the displacement of the bead, and $\alpha$ and $\beta$ are constants that depend on the cell height and the amount of embedded beads.

Linear elastic models are useful for simplified computations and thus for determining the estimate of the shear modulus. However, cells are often surrounded by fluids such as lipid membrane, which make them exhibit both fluid-like and solid-like properties. Thus linear viscoelastic models are utilized to integrate both of these material phase properties.

### Linear viscoelastic solid model

The linear viscoelastic solid model can be considered equivalent to having a Maxwell model in parallel with a spring element. Such a model can be used to study the small-strain deformation of cells, such as human leukocytes in micropipette aspiration experiments. The creep response $L(t)$ for a micropipette experiment can be given by [158]:

$$\frac{L(t)}{R_p} = \frac{\phi_P \Delta P}{2\pi k_1}\left[1 + \frac{k_1}{k_1 + k_2} e^{-\frac{t}{\tau}}\right] H(t), \tag{10.7}$$

where $k_1$ and $k_2$ are elastic constants, $H(t)$ is the Heaviside function, and $\tau$ is the characteristic creep time given as:

$$\tau = \mu \frac{k_1 + k_2}{k_1 k_2}, \tag{10.8}$$

with $\mu$ being the viscous constant. For a cell poker experiment with a cylindrical tip, the response of an adherent cell to a creep indentation $\delta(t)$ is given as [115,159]:

$$\delta(t) = \frac{F}{8R_I k_1}\left[1 + \left(\frac{k_1}{k_1 + k_2} - 1\right) e^{-\frac{t}{\tau}}\right] H(t). \tag{10.9}$$

These linear solid models were first applied for modeling leukocytes. However, leukocytes were found to be more accurately described using liquid drop models. Several anchorage-dependent cells such as endothelial cells, osteoblasts, chondrocytes, and cell nuclei follow standard linear solid models [155]. Both linear elastic and linear viscoelastic models are utilized for cell modeling or for investigating changes to cellular mechanics under different requirements and different conditions. Applications include models that can be utilized to determine properties of different types of cells. These models can also be used to study cellular deformation [160,161].

#### 10.4.2.3 Power law structural damping model

The dynamic behavior of cells is often studied in conditions where they undergo dynamic forces. Generally, MTC and AFM techniques are utilized for conducting dynamic tests on adherent cells, where a low amplitude sinusoidal force signal results in a sinusoidal displacement at the same frequency but exhibiting a phase lag in its steady state. Alcaraz et al. [162] developed a mathematical model for computing the complex shear modulus in an oscillatory AFM experiment by implementing a correction term to account for the contribution of the hydrodynamic drag force that may be significant due to viscous friction. In MTC experiments, the complex shear modulus may be computed from the classical elasticity equations with the condition that a linear viscoelasticity model is valid.

However, in both AFM and MTC experiments, the real part of the complex modulus of cells may depend on the frequency (ranging typically from $10^{-2}$ to 100 Hz) that can be defined by a weak power law with an exponent between 0.1 and 0.4.

The complex part of the shear modulus also exhibits a power law frequency dependency at low frequencies (<10 Hz); but at higher frequencies, the Newtonian viscous component overcomes it. This behavior was observed in several human cells including bronchial and alveolar epithelial cell lines [162], airway smooth muscle cells, lung epithelial cells, neutrophils, and mouse cells such as embryonic carcinoma cells and pulmonary macrophages [98]. Such nonlinear behavior cannot be accurately determined by spring-dashpot models (Maxwell, Newtonian, linear models), since the spring-dashpot system may always overestimate the exponent of the power law. Thus a power law structural damping model based on experimental findings was proposed to observe rheological behavior of adherent cells [163]. According to this empirical model, the complex shear modulus of adherent cells can be expressed as:

$$G^*(\omega) = G' + iG'' = G_0\left(\frac{\omega}{\omega_0}\right)^{\alpha}(1 + i\eta)\Gamma(1-\alpha)\cos\frac{\pi\alpha}{2} + i\omega\mu. \quad (10.10)$$

Here, $\omega$ is the angular frequency, $\alpha$ is the exponent of the power law ($0 < \alpha < 1$), $\eta$ is the structural damping coefficient defined as $\eta = \tan\frac{\pi\alpha}{2}$, $\Gamma(.)$ denotes the gamma function, and $\mu$ is the Newtonian viscous term. $G_0$ and $\omega_0$ are two scaling factors for the storage modulus and the angular frequency, respectively. Later, it was shown that such a power law based rheological behavior can also be described using fractional derivatives [164]. For example, a fractional Kelvin-Voigt rheological model can be described as:

$$\tilde{\sigma}_{KVM}(t) = G_s\tilde{\varepsilon}_{KVM}(t) + \eta_\alpha D^{\alpha}(\tilde{\varepsilon}_{KVM}(t)), \quad (10.11)$$

where $\tilde{\sigma}_{KVM}(t)$ denotes stress, $G_s$ is the elastic shear modulus, $\tilde{\varepsilon}_{KVM}(t)$ corresponds to strain, and $\eta_\alpha$ describes the effective modulus. The fractional derivate is given by $D^{\alpha}\tilde{\varepsilon}(t) = \frac{d^{\alpha}\tilde{\varepsilon}(t)}{dt^{\alpha}}$, where its order $\alpha$ is the damping coefficient of the system structural changes. For simulating oscillatory behavior of airway smooth muscle cells, following expressions can be derived from the above equation [164]:

$$G'(\omega) = G_s + \eta_\alpha\omega^{\alpha}\cos\left(\frac{\pi\alpha}{2}\right), \text{ and} \quad (10.12)$$

$$G''(\omega) = \eta_\alpha \omega^\alpha \sin\left(\frac{\pi\alpha}{2}\right). \tag{10.13}$$

Furthermore, adherent cells exhibit a proportionality behavior with the contractile stress, that is, their dynamic moduli increase with an increase in the contractile stress. The power law structural model is also utilized to quantify cell to cell variation in power law rheology [165] or to study temporal evolution of a single-cell rheology [166].

#### *10.4.2.4 Biphasic model*

Solid models discussed in previous sections consider cells to be in a single-phase material. However, many cells are often in two phases even in a steady state. For example, the cytoplasm consists both in solid polymeric contents and an interstitial fluid. Thus the two phases must be treated separately in such conditions. Thus a biphasic model [167] was introduced to consider the viscoelasticity in cells that are made of constituents in two phases, and for which, the liquid phase may diffuse through the solid phase. This theory was applied to study musculoskeletal cell mechanics and to model a single-chondrocyte subject to a flat punch indentation. In such a model, the solid phase is considered to be linearly elastic and the fluid phase is assumed to be an in viscid fluid.

The biphasic theory implies that cells and tissues exhibit viscoelastic behavior due to momentum exchange between the liquid and solid phases. Due to this complex behavior and the irregular geometry, developing a perfect analytical solution for a biphasic theory is very demanding. Nonetheless, the stress in the solid and liquid phases for the biphasic model can be computed as:

$$\sigma^s = -\phi^s p\mathbf{I} + \lambda_s tr(\varepsilon)\mathbf{I} + 2\mu_s\varepsilon, \tag{10.14}$$

$$\sigma^f = -\phi^f p\mathbf{I}. \tag{10.15}$$

Herein, $\sigma^s$ and $\sigma^f$ are the Causchy stress tensor for the solid and liquid phases, respectively, $\phi^s$ and $\phi^f$ are the solid and fluid volumetric functions ($\phi^s + \phi^f = 1$), $p$ is the fluid pressure, $\mathbf{I}$ is the identity tensor, $\lambda_s$ and $\mu_s$ are Lamé constants for the solid phase, and $tr(\varepsilon)$ is the trace of the Cauchy's infinitesimal strain tensor. A study based on a microindentation approach showed that a nonlinear biphasic model best represents the time-dependent mechanical behavior of articular cartilages [168]. The biphasic properties of single-bone cells and cartilages were utilized to characterize the dynamic environment of the surrounding ECM around chondrocytes in studying mechanotransduction of chondrocytes and the mechanobiology of the cartilage [169,170]. However, the fluid barrier behavior of the plasma membrane is not explored in detail, which is an important step in further developing more realistic cell models.

## 10.5 Trends in viscoelastography

Thus far, the viscoelastic nature of cells has been alluded to. Despite early working confirming this nature, microrheology measurements of viscoelasticity are scarce with most methods only capturing the elasticity. This is because at long time-scales (≥1 ms), the elastic response dominates [163,171,172]. It is thus necessary to develop techniques with higher temporal resolution (higher probing frequencies) to achieve a more comprehensive mechanical phenotype of the intact cell. Additionally, this would be advantageous for elucidating dynamic processes that occur in a small time scale. Demonstratively, Rigato et al. [171] successfully conducted high frequency (1−100 kHz) microrheology experiments using AFM. High-speed AFM is adapted with a miniature piezoelectric element with a resonant frequency of 200 kHz, and cells are visualized by transmission microscopy. Active rheology performed across the frequency range revealed that storage and loss moduli of fibroblasts increased in a similar way up to about 300 Hz after which, the loss modulus had a faster rate and finally became greater than the storage modulus above 30 kHz. Following, low and high viscoelastic regimes were identified, characterizing a purely viscous behavior or a fluid response with a transition frequency of 84 kHz for fibroblasts [171]. These results give access to the characterization of mechanical phenotypes of cells, both static and dynamic states, as well as the morphology. The authors demonstrated the utility of high-frequency imaging by charactering cells with a manipulated CSK architecture, confirming the role of the CSK in cellular biomechanics showing that cells with different CSK arrangements have unique mechanical properties. Moreover, they demonstrated the utility of high-frequency imaging in distinguishing the phenotype of benign and malignant breast cancer cells, posing viscoelasticity as a powerful biomarker.

The promise of frequency-based rheology is further demonstrated by Grasland-Mongrain et al. [112], who developed an ultrafast imaging system with optical microscopy. By inducing a high-frequency (15 kHz) shear wave within mammalian oocytes, the wave propagation could be followed using an ultrafast camera (200,000 frames per second). From the image cineloop, the elastic modulus of the whole cell was extracted by tracking the displacement of the wave, frame to frame, allowing the generation of elastic maps. The high spatial resolution gave access to zonal classification allowing mapping elasticity of the extracellular fluid, zona pellicuda, cytoplasm, and nucleus, providing more support for the mechanical heterogeneity of cells. Imaging whole cells without any chemical manipulation greatly opens the potential for longitudinal imaging, demonstrated by the same authors who imaged the oocyte at different maturation stages. Although results are in two dimensions, the method is extendable to three dimensions by focusing the camera at different depths. Additionally, the method may extract viscous properties by incorporating shear wave attenuation tracking capabilities [173]. Verified by simulation results, this method is reasonably accurate with high spatial and temporal resolutions.

To optimize such high-frequency rheological techniques, reducing the number of approximations would be beneficial. Model-independent rheology has been suggested to quantify the viscoelasticity of biological tissues by Kazemirad et al. [111] based on the propagation of shear waves. Following Navier's equation that governs the wave motion, a solution to the inverse wave propagation problem was analytically developed; from which both storage (elastic) and loss (viscous) moduli were obtained for in vitro and ex vivo porcine liver samples over a broad frequency range. Eliminating the need for a model describing the tissue (or cell) reduces inaccuracies due to rheology approximations and assumptions, thus a more direct mechanical phenotype might be obtained. Despite this success, the method [111] carries its own limiting assumptions that is by basing the analytical solution on Navier's equation, they inherently assume the wave to propagate in isotropic, linear, and homogeneous media. This method is best suited for macroscopic homogenous samples. Within the cell, homogeneity does not hold based on structural and content differences, for example, the isolated nucleus being stiffer and more viscous than the intact cell [142], and cortex mechanics dominating over cytoplasm during mitosis [174]. Nonetheless, the development of model-independent rheology techniques that consider heterogeneous samples is a promising path in elucidating the viscoelastic nature of cells, while minimizing approximations necessary.

Most imaging techniques focus on single cells, however, these methods may be limited in clinical applications where a single sample may contain thousands or millions of cells. For example, a biopsy sample may contain thousands of normal cells and only a small percentage of malignant cells. Processing such a sample one cell at a time is inefficient given that physical properties of cells could be altered in the external environment of buffer or media. High-throughput methods are thus necessary to efficiently translate viscoelasticity as a biomarker to clinical applications. The emerging tools are variant in throughput rates ranging from 1 cell/minute to 20,000 cells/second [175]. Adapting single-cell methods to high-throughput applications would be the natural first step. Optical trapping is one such technique whose utility has been stretched to characterize a large number of cells. Measuring cells in rapid succession, by flushing them through a microfluidic channel through an optical trap, allowed statistically relevant numbers of single-cell analyses, yielding rates of approximately 1 cell/second to 20 cells/second [176,177]. Increasing the throughput rate of the experimentally ubiquitous AFM is limited largely by the time taken for the indentation process. However, a promising approach made use of a feedback system where the information from microscopic images would allow robotic positioning [178]. Automation of the scan through registration of cells relatively increased the speed to 3 seconds/cell; however this method is still riddled by the indentation time, thus it has not been pursued largely. Perhaps, opportunity for machine-learning approaches to increase efficiency of registration may occur.

Impressive throughput rates have been achieved by hydrodynamic and transit through constriction methods. Hydrodynamic approaches rely on intrinsic fluid-dynamic stresses that are regulated by the design of the microfluidic channels, yielding rates of up to 20,000 cells/second [179]. These channels can be divided into two designs, those that support lift of cells induced by the deformability in flow [180,181] and imaging of hydrodynamic-stretched cells [179,182–184]. A major limitation is that these methods highly depend on the cell shape and size, thus altered shapes and sizes cannot be captured within the same measurement. Transit through constriction methods are explained in the naming, where changes in cell size and shape are measured as cells travel through pores [175]. Achievements of this method include rates of 1 cell/second to 100 cells/

second depending on the automation. Automation can occur on readout methods, that is, by implementing high-speed imaging or higher precision sensing cantilever [185–190]. Improved automation may consequent in improved throughput rates. Given that the constriction size has to be calibrated to the size of the cell of interest, this technique is limited to homogenously sized cells. Beyond calibration to the cell of interest, most of high-throughput techniques are promising in efficiently translating viscoelastic measurements to clinical adaptations where processing millions of cells from a sample is necessary. Success in using the label-free mechanical biomarker would lead to more accurate phenotypes based on the natural state of the cell.

## 10.6 Summary

The living cell is a mechanical structure with the ability to detect external forces, internal forces from reorganization of subcellular structures and generate responsive forces that maintain functionality or induce growth and motility. The ability to survive mechanical changes largely depends on the deformability of cells, that is, elasticity, where increased or decreased elasticity can inform on the pathophysiological state. In fact, numerous disease states and developmental abnormalities clinically manifest as reduced elasticity or heightened stiffness in whole organs or tissues. Tissues and organs are fundamentally constituted of a collection of cells, thus by assessing the elasticity of single cells, a label-free biomarker informative of the disease state, functionality, and developmental status can be developed.

Methods to quantify the elasticity of cells are developed and emerging trends attempt to increase spatial and temporal resolutions. Increased spatial resolution makes accessible subcellular structures with techniques increasingly being able to distinguish the mechanical contribution of distinct structures such as the membrane, CSK, and the nucleus, whereas increased temporal resolution captures dynamic changes that rapidly occur. Most rheological techniques rely on a model to verify experimental results through simulations. Choosing a technique and accompanying model largely depends on approximations within reach and their appropriateness to the cell being studied. Variations in cell shape, size, nucleation, adhesive properties, and function dictate the choices in measuring techniques.

Elasticity alone as a biomechanical marker can result in ambiguous characterization, as cells are not entirely solid structures. The presence of a cytoplasm filled with the fluid-like cytosol gives it viscous property. Viscoelasticity is thus necessary to accurately represent the mechanical phenotype of the cell. Of late, frequency-based techniques are most promising in quantifying viscoelasticity with the development of higher frequency methods within which the viscous nature is measurable. Moreover, to reduce the number of approximations, model-independent techniques have been developed of late, aiming to reduce inaccuracies introduced by fitting experimental results to a model. The overall goal of cellular imaging is in determining functional, pathophysiological, and developmental states of living cells. If such techniques are to be translated to clinical applications, increased efficiency in the number of cells processed is necessary as biological samples typically include thousands or millions of cells. High-throughput techniques are a recent topic of interest, whether it includes adapting single cellular techniques or developing new rheological approaches. Recently, the viscoelasticity of cells could be quantified at the impressive rate of 20,000 cells/second through hydrodynamic approaches.

## Acknowledgment

We would like to thank Rodin Chemalat, Thomas Gibaud, and Thomas Dehoux for useful discussions about cell structure, microrheology, and Brillouin scattering, respectively. Partial funding for the cell microelastography work was obtained from a Strategic grant of the Quebec Bioimaging Network.

## References

[1] S.-Y. Tee, A.R. Bausch, P.A. Janmey, The mechanical cell, Curr. Biol. 19 (17) (2009) R745–R748.
[2] M. Khalilian, et al., Alteration in the mechanical properties of human ovum zona pellucida following fertilization: experimental and analytical studies, Exp. Mech. 51 (2) (2011) 175–182.
[3] G.Y. Lee, C.T. Lim, Biomechanics approaches to studying human diseases, Trends Biotechnol. 25 (3) (2007) 111–118.
[4] P. Carl, H. Schillers, Elasticity measurement of living cells with an atomic force microscope: data acquisition and processing, Pflug. Arch. 457 (2) (2008) 551–559.
[5] M.B. Ginzberg, R. Kafri, M. Kirschner, On being the right (cell) size, Science 348 (6236) (2015) 1245075.
[6] R. Changede, M. Sheetz, Integrin and cadherin clusters: a robust way to organize adhesions for cell mechanics, BioEssays 39 (1) (2017) 1–12.

[7] G. Karp, Cell and Molecular Biology: Concepts and Experiments, John Wiley & Sons, 2009.

[8] W. Helfrich, Elastic properties of lipid bilayers: theory and possible experiments, Z. für Naturforschung C. 28 (11-12) (1973) 693–703.

[9] W. Helfrich, Steric interaction of fluid membranes in multilayer systems, Z. für Naturforschung A 33 (3) (1978) 305–315.

[10] E.A. Evans, Bending resistance and chemically induced moments in membrane bilayers, Biophysical J. 14 (12) (1974) 923–931.

[11] C.C. DuFort, M.J. Paszek, V.M. Weaver, Balancing forces: architectural control of mechanotransduction, Nat. Rev. Mol. Cell Biol. 12 (5) (2011) 308–319.

[12] S. Kumar, et al., Viscoelastic retraction of single living stress fibers and its impact on cell shape, cytoskeletal organization, and extracellular matrix mechanics, Biophysical J. 90 (10) (2006) 3762–3773.

[13] T.P. Lele, et al., Mechanical forces alter zyxin unbinding kinetics within focal adhesions of living cells, J. Cell. Physiol. 207 (1) (2006) 187–194.

[14] C.-M. Lo, et al., Cell movement is guided by the rigidity of the substrate, Biophysical J. 79 (1) (2000) 144–152.

[15] B. Alberts, et al., Molecular Biology of the Cell: Reference Edition, Garland Science, 2007, p. 2.

[16] K. Sengupta, A.-S. Smith, Adhesion of biological membranes, in: P. Bassereau, P. Sens (Eds.), Physics of Biological Membranes, Springer International Publishing, Cham, 2018, pp. 499–535.

[17] J.Z. Kechagia, J. Ivaska, P. Roca-Cusachs, Integrins as biomechanical sensors of the microenvironment, Nat. Rev. Mol. Cell Biol. 20 (2019) 457–473.

[18] R.O. Hynes, Integrins: bidirectional, allosteric signaling machines, Cell 110 (6) (2002) 673–687.

[19] M. Kim, C.V. Carman, T.A. Springer, Bidirectional transmembrane signaling by cytoplasmic domain separation in integrins, Science 301 (5640) (2003) 1720–1725.

[20] J.D. Humphries, et al., Signal transduction via integrin adhesion complexes, Curr. Opin. Cell Biol. 56 (2019) 14–21.

[21] Z. Sun, S.S. Guo, R. Fässler, Integrin-mediated mechanotransduction, J. Cell Biol. 215 (4) (2016) 445–456.

[22] E. Sackmann, Advanced concepts and perspectives of membrane physics, in: P. Bassereau, P. Sens (Eds.), Physics of Biological Membranes, Springer International Publishing, Cham, 2018, pp. 45–70.

[23] A. Kusumi, et al., Dynamic organizing principles of the plasma membrane that regulate signal transduction: commemorating the fortieth anniversary of Singer and Nicolson's fluid-mosaic model, Annu. Rev. Cell Dev. Biol. 28 (2012) 215–250.

[24] M. Fritzsche, et al., Actin kinetics shapes cortical network structure and mechanics, Sci. Adv. 2 (4) (2016) e1501337.

[25] P. Sens, J. Plastino, Membrane tension and cytoskeleton organization in cell motility, J. Phys.: Condens. Matter 27 (27) (2015) 273103.

[26] H.M. Nussenzveig, Cell membrane biophysics with optical tweezers, Eur. Biophys. J. 47 (5) (2018) 499–514.

[27] K. Rottner, M. Schaks, Assembling actin filaments for protrusion, Curr. Opin. Cell Biol. 56 (2019) 53–63.

[28] A. Gefen, D. Weihs, Cytoskeleton and plasma-membrane damage resulting from exposure to sustained deformations: a review of the mechanobiology of chronic wounds, Med. Eng. Phys. 38 (9) (2016) 828–833.

[29] S. Boulant, et al., Actin dynamics counteract membrane tension during clathrin-mediated endocytosis, Nat. Cell Biol. 13 (9) (2011) 1124–1131.

[30] P.A. Janmey, C.A. McCulloch, Cell mechanics: integrating cell responses to mechanical stimuli, Annu. Rev. Biomed. Eng. 9 (1) (2007) 1–34.
[31] M. Enrique, M.L. Gardel, Actin mechanics and fragmentation, J. Biol. Chem. 290 (28) (2015) 17137–17144.
[32] G. Salbreux, G. Charras, E. Paluch, Actin cortex mechanics and cellular morphogenesis, Trends Cell Biol. 22 (10) (2012) 536–545.
[33] P. Kollmannsberger, B. Fabry, Linear and nonlinear rheology of living cells, Annu. Rev. Mater. Res. 41 (2011) 75–97.
[34] P. Isermann, J. Lammerding, Nuclear mechanics and mechanotransduction in health and disease, Curr. Biol. 23 (24) (2013) R1113–R1121.
[35] D. Ingber, Mechanobiology and diseases of mechanotransduction, Ann. Med. 35 (8) (2003) 564–577.
[36] M.M. Yallapu, et al., The roles of cellular nanomechanics in cancer, Med. Res. Rev. 35 (1) (2015) 198–223.
[37] C. Rianna, M. Radmacher, Cell mechanics as a marker for diseases: biomedical applications of AFM, AIP Conference Proceedings, AIP Publishing, 2016.
[38] G. Burgstaller, et al., The instructive extracellular matrix of the lung: basic composition and alterations in chronic lung disease, Eur. Respir. J. 50 (1) (2017) 1601805.
[39] J.K. Burgess, et al., The extracellular matrix—the under-recognized element in lung disease? J. Pathol. 240 (4) (2016) 397–409.
[40] J. Mead, I. Lindgren, E. Gaensler, The mechanical properties of the lungs in emphysema, J. Clin. Invest. 34 (7) (1955) 1005–1016.
[41] M. Lekka, et al., Elasticity of normal and cancerous human bladder cells studied by scanning force microscopy, Eur. Biophys. J. 28 (4) (1999) 312–316.
[42] J. Guck, et al., Optical deformability as an inherent cell marker for testing malignant transformation and metastatic competence, Biophys. J. 88 (5) (2005) 3689–3698.
[43] J.J. Fredberg, Frozen objects: small airways, big breaths, and asthma, J. Allergy Clin. Immunol. 106 (4) (2000) 615–624.
[44] J. Li, et al., Cytoskeletal dynamics of human erythrocyte, Proc. Natl Acad. Sci. U. S. A. 104 (12) (2007) 4937–4942.
[45] M. Puig-de-Morales-Marinkovic, et al., Viscoelasticity of the human red blood cell, Am. J. Physiol.-Cell Physiol. 293 (2) (2007) C597–C605.
[46] L. Wang, et al., Tissue and cellular rigidity and mechanosensitive signaling activation in Alexander disease, Nat. Commun. 9 (1) (2018) 1899.
[47] A.J. Booth, et al., Acellular normal and fibrotic human lung matrices as a culture system for in vitro investigation, Am. J. Respir. Crit. Care Med. 186 (9) (2012) 866–876.
[48] H. Parameswaran, A. Majumdar, B. Suki, Linking microscopic spatial patterns of tissue destruction in emphysema to macroscopic decline in stiffness using a 3D computational model, PLoS Comput. Biol. 7 (4) (2011) e1001125.
[49] R. Annoni, et al., Extracellular matrix composition in COPD, Eur. Respir. J. 40 (6) (2012) 1362–1373.
[50] Y.S. Oh, et al., A special report on the NHLBI initiative to study cellular and molecular mechanisms of arterial stiffness and its association with hypertension, Circ. Res. 121 (11) (2017) 1216–1218.
[51] S. Suresh, et al., Connections between single-cell biomechanics and human disease states: gastrointestinal cancer and malaria, Acta Biomater. 1 (1) (2005) 15–30.
[52] N. Schierbaum, J. Rheinlaender, T.E. Schäffer, Viscoelastic properties of normal and cancerous human breast cells are affected differently by contact to adjacent cells, Acta Biomater. 55 (2017) 239–248.
[53] L. Winter, W.H. Goldmann, Biomechanical characterization of myofibrillar myopathies, Cell Biol. Int. 39 (4) (2015) 361–363.

[54] R. Krishnan, et al., Cellular biomechanics in drug screening and evaluation: mechanopharmacology, Trends Pharmacol. Sci. 37 (2) (2016) 87–100.
[55] M. Li, et al., Nanoscale monitoring of drug actions on cell membrane using atomic force microscopy, Acta Pharmacol. Sin. 36 (7) (2015) 769–782.
[56] A.M. Kolodziejczyk, M. Targosz-Korecka, M. Szymonski, Nanomechanical testing of drug activities at the cellular level: case study for endothelium-targeted drugs, Pharmacol. Rep. 69 (6) (2017) 1165–1172.
[57] J.D. Mih, et al., A multiwell platform for studying stiffness-dependent cell biology, PLoS One 6 (5) (2011) e19929.
[58] S. Seyedpour, et al., Effects of an antimitotic drug on mechanical behaviours of the cytoskeleton in distinct grades of colon cancer cells, J. Biomech. 48 (6) (2015) 1172–1178.
[59] O.Y. Selyutina, et al., Influence of glycyrrhizin on permeability and elasticity of cell membrane: perspectives for drugs delivery, Drug Delivery 23 (3) (2016) 848–855.
[60] T. Auth, S. Dasgupta, G. Gompper, Interaction of particles and pathogens with biological membranes, Physics of Biological Membranes, Springer, 2018, pp. 471–498.
[61] M.J. Whitfield, J.P. Luo, W.E. Thomas, Yielding elastic tethers stabilize robust cell adhesion, PLoS Comput. Biol. 10 (12) (2014) e1003971.
[62] A. Persat, et al., The mechanical world of bacteria, Cell 161 (5) (2015) 988–997.
[63] M. Otto, Physical stress and bacterial colonization, FEMS Microbiol. Rev. 38 (6) (2014) 1250–1270.
[64] I. Tardieux, J. Baum, Reassessing the mechanics of parasite motility and host-cell invasion, J. Cell Biol. 214 (5) (2016) 507–515.
[65] A. Elbourne, et al., Bacterial-nanostructure interactions: the role of cell elasticity and adhesion forces, J. Colloid Interface Sci. 546 (2019) 192–210.
[66] R. Ebady, et al., Biomechanics of *Borrelia burgdorferi* vascular interactions, Cell Rep. 16 (10) (2016) 2593–2604.
[67] M.A. Dragovich, et al., Biomechanical characterization of TIM protein–mediated *Ebola* virus–host cell adhesion, Sci. Rep. 9 (1) (2019) 267.
[68] Y. Sun, T.-L. Sun, H.W. Huang, Physical properties of *Escherichia coli* spheroplast membranes, Biophys. J. 107 (9) (2014) 2082–2090.
[69] S. Hillringhaus, et al., Importance of erythrocyte deformability for the alignment of malaria parasite upon invasion, Biophys. J. 117 (2019) 1202–1214.
[70] S. Hasim, et al., β-(1, 3)-Glucan unmasking in some Candida albicans mutants correlates with increases in cell wall surface roughness and decreases in cell wall elasticity, Infect. Immun. 85 (1) (2017) e00601-16.
[71] M. Beckmann, et al., Measuring cell surface elasticity on enteroaggregative Escherichia coli wild type and dispersin mutant by AFM, Ultramicroscopy 106 (8-9) (2006) 695–702.
[72] N.M. Ayad, S. Kaushik, V.M. Weaver, Tissue mechanics, an important regulator of development and disease, Philos. Trans. R. Soc. B 374 (1779) (2019) 20180215.
[73] A. Elosegui-Artola, X. Trepat, P. Roca-Cusachs, Control of mechanotransduction by molecular clutch dynamics, Trends Cell Biol. 28 (5) (2018) 356–367.
[74] L. Chin, et al., Mechanotransduction in cancer, Curr. Opin. Chem. Eng. 11 (2016) 77–84.
[75] A.Y. Malkin, A.I. Isayev, Rheology: Concepts, Methods, and Applications, Elsevier, 2017.
[76] E. Evans, A.J.Bj Yeung, Apparent viscosity and cortical tension of blood granulocytes determined by micropipet aspiration, Biophys. J. 56 (1) (1989) 151–160.
[77] J.P. Brody, Y. Han, R.H. Austin, M. Bitensky, Deformation and flow of red blood cells in a synthetic lattice: evidence for an active cytoskeleton, Biophys. J. 68 (6) (1995) 2224–2232.

[78] W.A. Lam, M.J. Rosenbluth, D.A. Fletcher, Chemotherapy exposure increases leukemia cell stiffness, Blood 109 (8) (2007) 3505–3508.
[79] N. Wang, J.P. Butler, D.E. Ingber, Mechanotransduction across the cell surface and through the cytoskeleton, Science 260 (5111) (1993) 1124–1127.
[80] J. Guck, R. Ananthakrishnan, H. Mahmood, T.J. Moon, C.C. Cunningham, J. Käs, The optical stretcher: a novel laser tool to micromanipulate cells, Biophys. J. 81 (2) (2001) 767–784.
[81] M. Khalilian, M. Navidbakhsh, M.R. Valojerdi, M. Chizari, P.E. Yazdi, Estimating Young's modulus of zona pellucida by micropipette aspiration in combination with theoretical models of ovum, J. R. Soc. Interface 7 (45) (2009) 687–694.
[82] Y. Sun, K.T. Wan, K.P. Roberts, J.C. Bischof, B.J. Nelson, Mechanical property characterization of mouse zona pellucida, IEEE Trans. Nanobioscience 2 (4) (2003) 279–286.
[83] Y. Murayama, et al., Elasticity measurement of zona pellucida using a micro tactile sensor to evaluate embryo quality, J. Mammalian Ova Res. 25 (1) (2008) 8–16.
[84] A. Ashkin, J.M. Dziedzic, Optical trapping and manipulation of viruses and bacteria, Science 235 (4795) (1987) 1517–1520.
[85] A. Ashkin, K. Schütze, J.M. Dziedzic, U. Euteneuer, M. Schliwa, Force generation of organelle transport measured in vivo by an infrared laser trap, Nature 348 (6299) (1990) 346–348.
[86] S.M. Block, D.F. Blair, H.C. Berg, Compliance of bacterial flagella measured with optical tweezers, Nature 338 (6215) (1989) 514–518.
[87] J. Wu, Acoustical tweezers, J. Acoustical Soc. Am. 89 (5) (1991) 2140–2143.
[88] A. Ozcelik, J. Rufo, F. Guo, Y. Gu, P. Li, J. Lata, et al., Acoustic tweezers for the life sciences, Nat. Methods 15 (12) (2018) 1021–1028.
[89] A.R. Bausch, W. Möller, E. Sackmann, Measurement of local viscoelasticity and forces in living cells by magnetic tweezers, Biophys. J. 76 (1) (1999) 573–579.
[90] F.J. Alenghat, B. Fabry, K.Y. Tsai, W.H. Goldmann, D.E. Ingber, Analysis of cell mechanics in single vinculin-deficient cells using a magnetic tweezer, Biochem. Biophys. Res. Commun. 277 (1) (2000) 93–99.
[91] A. Emad, W.F. Heinz, M.D. Antonik, N.P. D'Costa, S. Nageswaran, C.A. Schoenenberger, et al., Relative microelastic mapping of living cells by atomic force microscopy, Biophys. J. 74 (3) (1998) 1564–1578.
[92] C.A. Putman, K.O. Van der Werf, B.G. de Grooth, N.F. van Hulst, J. Greve, Viscoelasticity of living cells allows high resolution imaging by tapping mode atomic force microscopy, Biophys. J. 67 (4) (1994) 1749–1753.
[93] M. Radmacher, R.W. Tillamnn, M. Fritz, H.E. Gaub, From molecules to cells: imaging soft samples with the atomic force microscope, Science 257 (5078) (1992) 1900–1905.
[94] X. Liu, J. Shi, Z. Zong, K.T. Wan, Y. Sun, Elastic and viscoelastic characterization of mouse oocytes using micropipette indentation, Ann. Biomed. Eng. 40 (10) (2010) 2122–2130.
[95] M. Radmacher, M. Fritz, C.M. Kacher, J.P. Cleveland, P.K. Hansma, Measuring the viscoelastic properties of human platelets with the atomic force microscope, Biophys. J. 70 (1) (1996) 556–567.
[96] T.G. Mason, K. Ganesan, J.H. Van Zanten, D. Wirtz, S.C. Kuo, Particle tracking microrheology of complex fluids, Phys. Rev. Lett. 79 (17) (1997) 3282–3285.
[97] J.C. Crocker, M.T. Valentine, E.R. Weeks, T. Gisler, P.D. Kaplan, A.G. Yodh, et al., Two-point microrheology of inhomogeneous soft materials, Phys. Rev. Lett. 85 (4) (2000) 888–891.
[98] B. Fabry, et al., Time scale and other invariants of integrative mechanical behavior in living cells, Phys. Rev. E 68 (4) (2003) 041914.

[99] G.N. Maksym, B. Fabry, J.P. Butler, D. Navajas, D.J. Tschumperlin, J.D. Laporte, et al., Mechanical properties of cultured human airway smooth muscle cells from 0.05 to 0.4 Hz, J. Appl. Physiol. 89 (4) (2000) 1619−1632.
[100] Y. Tseng, T.P. Kole, D. Wirtz, Micromechanical mapping of live cells by multiple-particle-tracking microrheology, Biophys. J. 83 (6) (2002) 3162−3176.
[101] Y. Tseng, J.S. Lee, T.P. Kole, I. Jiang, D. Wirtz, Micro-organization and visco-elasticity of the interphase nucleus revealed by particle nanotracking, J. Cell Sci. 117 (10) (2004) 2159−2167.
[102] G. Scarcelli, W.J. Polacheck, H.T. Nia, K. Patel, A.J. Grodzinsky, R.D. Kamm, et al., Noncontact three-dimensional mapping of intracellular hydromechanical properties by Brillouin microscopy, Nat. Methods 12 (12) (2015) 1132−1134.
[103] G. Scarcelli, S.H. Yun, Confocal Brillouin microscopy for three-dimensional mechanical imaging, Nat. Photonics 2 (1) (2008) 39−43.
[104] P.J. Wu, et al., Water content, not stiffness, dominates Brillouin spectroscopy measurements in hydrated materials, Nat. Methods 15 (8) (2018) 561−562.
[105] T.A. Krouskop, D.R. Dougherty, F.S. Vinson, A pulsed Doppler ultrasonic system for making noninvasive measurements of the mechanical properties of soft tissue, J. Rehabilitation Res. Dev. 24 (2) (1987) 1−8.
[106] S. Catheline, F. Yu, M. Fink, A solution to diffraction biases in sonoelasticity: the acoustic impulse technique, J. Acoustical Soc. Am. 105 (5) (1999) 2941−2950.
[107] R. Muthupillai, P.J. Rossman, D.J. Lomas, J.F. Greenleaf, S.J. Riederer, R.L. Ehman, Magnetic resonance imaging of transverse acoustic strain waves, Magnetic Reson. Med. 36 (2) (1996) 266−274.
[108] L. Sandrin, et al., Transient elastography: a new noninvasive method for assessment of hepatic fibrosis, Ultrasound Med. Biol. 29 (12) (2003) 1705−1713.
[109] J. Bercoff, et al., In vivo breast tumor detection using transient elastography, Ultrasound Med. Biol. 29 (10) (2003) 1387−1396.
[110] F. Sebag, J. Vaillant-Lombard, J. Berbis, V. Griset, J.F. Henry, P. Petit, et al., Shear wave elastography: a new ultrasound imaging mode for the differential diagnosis of benign and malignant thyroid nodules, J. Clin. Endocrinol. Metab. 95 (12) (2010) 5281−5288.
[111] S. Kazemirad, S. Bernard, S. Hybois, A. Tang, G. Cloutier, Ultrasound shear wave viscoelastography: model-independent quantification of the complex shear modulus, IEEE Trans. Ultrasonics, Ferroelectrics, Frequency Control. 63 (9) (2016) 1399−1408.
[112] P. Grasland-Mongrain, A. Zorgani, S. Nakagawa, S. Bernard, L. Gomes Paim, G. Fitzharris, S. Catheline, G. Cloutier, Ultrafast imaging of cell elasticity with optical microelastography, Proc. Natl Acad. Sci. U. S. A. 115 (5) (2018) 861−866.
[113] M.M. Brandao, A. Fontes, M.L. Barjas-Castro, L.C. Barbosa, F.F. Costa, C.L. Cesar, et al., Optical tweezers for measuring red blood cell elasticity: application to the study of drug response in sickle cell disease, Eur. J. Haematol. 70 (4) (2003) 207−211.
[114] L. Tetard, A. Passian, R.M. Lynch, B.H. Voy, G. Shekhawat, V. Dravid, et al., Elastic phase response of silica nanoparticles buried in soft matter, Appl. Phys. Lett. 93 (13) (2008) 133113.
[115] C.T. Lim, E.H. Zhou, S.T. Quek, Mechanical models for living cells—a review, J. Biomech. 39 (2) (2006) 195−216.
[116] J. Chen, Nanobiomechanics of living cells: a review, Interface Focus. 4 (2) (2014) 20130055.
[117] B. Fuller, Tensegrity. Portfolio and Art News Annual. 1961.
[118] F.R. Buckminster, Tensile-integrity structures. 1962, Google Patents.

[119] R.E. Skelton, et al., An introduction to the mechanics of tensegrity structures. in Proceedings of the 40th IEEE Conference on Decision and Control (Cat. No. 01CH37228). 2001. IEEE.
[120] D.E. Ingber, Tensegrity: the architectural basis of cellular mechanotransduction, Annu. Rev. Physiol. 59 (1) (1997) 575–599.
[121] D.E. Ingber, Cellular tensegrity: defining new rules of biological design that govern the cytoskeleton, J. Cell Sci. 104 (3) (1993) 613–627.
[122] L. Wang, W. Chen, Modelling cell origami via a tensegrity model of the cytoskeleton in adherent cells, Appl. Bionics Biomech. (2019) 8541303.
[123] K.Y. Volokh, On tensegrity in cell mechanics, Mol. Cell. Biomech. 8 (2011) 195–214.
[124] D.E. Ingber, N. Wang, D. Stamenović, Tensegrity, cellular biophysics, and the mechanics of living systems, Rep. Prog. Phys. 77 (4) (2014) 046603.
[125] D. Stamenović, M.F. Coughlin, The role of prestress and architecture of the cytoskeleton and deformability of cytoskeletal filaments in mechanics of adherent cells: a quantitative analysis, J. Theor. Biol. 201 (1) (1999) 63–74.
[126] M.F. Coughlin, D. Stamenovic, A quantitative model of cellular elasticity based on tensegrity, J. Biomech. Eng. 122 (1) (2000) 39–43.
[127] P. Canadas, et al., A cellular tensegrity model to analyse the structural viscoelasticity of the cytoskeleton, J. Theor. Biol. 218 (2) (2002) 155–173.
[128] G.-K. Xu, et al., A tensegrity model of cell reorientation on cyclically stretched substrates, Biophys. J. 111 (7) (2016) 1478–1486.
[129] N. Bonakdar, et al., Mechanical plasticity of cells, Nat. Mater. 15 (10) (2016) 1090–1094.
[130] L. Zhang, X. Feng, S. Li, Review and perspective on soft matter modeling in cellular mechanobiology: cell contact, adhesion, mechanosensing, and motility, Acta Mech. 228 (12) (2017) 4095–4122.
[131] M. Fraldi, et al., Buckling soft tensegrities: Fickle elasticity and configurational switching in living cells, J. Mech. Phys. Solids 124 (2019) 299–324.
[132] Y.D. Bansod, et al., A finite element bendo-tensegrity model of eukaryotic cell, J. Biomech. Eng. 140 (10) (2018) 101001.
[133] G. Forgacs, On the possible role of cytoskeletal filamentous networks in intracellular signaling: an approach based on percolation, J. Cell Sci. 108 (6) (1995) 2131–2143.
[134] M.J. Saxton, The membrane skeleton of erythrocytes. A percolation model, Biophys. J. 57 (6) (1990) 1167–1177.
[135] D. Stamenović, D.E. Ingber, Models of cytoskeletal mechanics of adherent cells, Biomech. Modeling Mechanobiol. 1 (1) (2002) 95–108.
[136] T.G. Fai, et al., Image-based model of the spectrin cytoskeleton for red blood cell simulation, PLoS Comput. Biol. 13 (10) (2017) e1005790.
[137] V. Privman, et al., Lattice percolation approach to numerical modelling of tissue aging, Int. J. Parallel, Emergent Distrib. Syst. 31 (1) (2016) 1–19.
[138] E. Evans, B. Kukan, Passive material behavior of granulocytes based on large deformation and recovery after deformation tests, Blood 64 (5) (1984) 1028–1035.
[139] G. Schmid-Schonbein, Y.Y. Shih, S.J.B. Chien, Morphometry of human leukocytes, Blood 56 (5) (1980) 866–875.
[140] D. Needham, R.M. Hochmuth, A sensitive measure of surface stress in the resting neutrophil, Biophys. J. 61 (6) (1992) 1664–1670.
[141] C. Dong, R. Skalak, K.L. Sung, Cytoplasmic rheology of passive neutrophils, Biorheology 28 (6) (1991) 557–567.
[142] F. Guilak, et al., Viscoelastic properties of the cell nucleus, Biochem. Biophys. Res. Commun. 269 (3) (2000) 781–786.

[143] N. Caille, et al., Contribution of the nucleus to the mechanical properties of endothelial cells, J. Biomech. 35 (2) (2002) 177–187.
[144] H.-C. Kan, et al., Hydrodynamics of a compound drop with application to leukocyte modeling, Phys. Fluids 10 (4) (1998) 760–774.
[145] R. Tran-Son-Tay, et al., Time-dependent recovery of passive neutrophils after large deformation, Biophys. J. 60 (4) (1991) 856–866.
[146] R. Hochmuth, et al., Viscosity of passive human neutrophils undergoing small deformations, Biophys. J. 64 (5) (1993) 1596–1601.
[147] R. Tran-Son-Tay, et al., Rheological modelling of leukocytes, Med. Biol. Eng. Comput. 36 (2) (1998) 246–250.
[148] M.A. Tsai, R.S. Frank, R.E.J.Bj Waugh, Passive mechanical behavior of human neutrophils: power-law fluid, Biophys. J. 65 (5) (1993) 2078–2088.
[149] J.L. Drury, M.J.Bj Dembo, Hydrodynamics of micropipette aspiration, Biophys. J. 76 (1) (1999) 110–128.
[150] J.L. Drury, M. Dembo, Aspiration of human neutrophils: effects of shear thinning and cortical dissipation, Biophys. J. 81 (6) (2001) 3166–3177.
[151] C. Dong, et al., Passive deformation analysis of human leukocytes, J. Biomech. Eng. 110 (1) (1988) 27–36.
[152] C. Dong, R. Skalak, Leukocyte deformability: finite element modeling of large viscoelastic deformation, J. Theor. Biol. 158 (2) (1992) 173–193.
[153] J.D. Ferry, Viscoelastic Properties of Polymers, John Wiley & Sons, 1980.
[154] D.P. Theret, M.J. Levesque, M. Sato, R.M. Nerem, L.T. Wheeler, The application of a homogeneous half-space model in the analysis of endothelial cell micropipette measurements, J. Biochem. Eng. 110 (3) (1988) 190–199.
[155] W.R. Jones, et al., Alterations in the Young's modulus and volumetric properties of chondrocytes isolated from normal and osteoarthritic human cartilage, J. Biomech. 32 (2) (1999) 119–127.
[156] G.G. Bilodeau, Regular pyramid punch problem, J. Mech. Phys. Solids 59 (3) (1992) 519–523.
[157] S.M. Mijailovich, et al., A finite element model of cell deformation during magnetic bead twisting, J. Appl. Physiol. 93 (4) (2002) 1429–1436.
[158] M. Sato, D.P. Theret, L.T. Wheeler, N. Ohshima, R.M. Nerem, Application of the micropipette technique to the measurement of cultured porcine aortic endothelial cell viscoelastic properties, J. Biochem. Eng. 112 (3) (1990) 263–268.
[159] E.J. Koay, A.C. Shieh, K.A. Athanasiou, Creep indentation of single cells, J. Biomech. Eng. 125 (3) (2003) 334–341.
[160] M.A. Haider, F. Guilak, An axisymmetric boundary integral model for incompressible linear viscoelasticity: application to the micropipette aspiration contact problem, J. Biomech. Eng. 122 (3) (2000) 236–244.
[161] P.I. Milner, R.J. Wilkins, J.S. Gibson, Cellular physiology of articular cartilage in health and disease, Principles of Osteoarthritis-Its Definition, Character, Derivation and Modality-Related Recognition, IntechOpen, 2012.
[162] J. Alcaraz, et al., Microrheology of human lung epithelial cells measured by atomic force microscopy, Biophys. J. 84 (3) (2003) 2071–2079.
[163] B. Fabry, et al., Scaling the microrheology of living cells, Phys. Rev. Lett. 87 (14) (2001) 148102.
[164] V.D. Djordjevic, et al., Fractional derivatives embody essential features of cell rheological behavior, Ann. Biomed. Eng. 31 (6) (2003) 692–699.
[165] P. Cai, et al., Quantifying cell-to-cell variation in power-law rheology, Biophys. J. 105 (5) (2013) 1093–1102.
[166] P. Cai, et al., Temporal variation in single-cell power-law rheology spans the ensemble variation of cell population, Biophys. J. 113 (3) (2017) 671–678.

[167] V.C. Mow, S.C. Kuei, W.M. Lai, C.G. Armstrong, Biphasic creep and stress relaxation of articular cartilage in compression: theory and experiments, J. Biomech. Eng. 102 (1) (1980) 73–84.
[168] J.A. Wahlquist, et al., Indentation mapping revealed poroelastic, but not viscoelastic, properties spanning native zonal articular cartilage, Acta Biomater. 64 (2017) 41–49.
[169] J.Z. Wu, W. Herzog, M. Epstein, Modelling of location- and time-dependent deformation of chondrocytes during cartilage loading, J. Biomech. 32 (6) (1999) 563–572.
[170] F. Guilak, V.C. Mow, The mechanical environment of the chondrocyte: a biphasic finite element model of cell-matrix interactions in articular cartilage, J. Biomech. 33 (12) (2000) 1663–1673.
[171] A. Rigato, et al., High-frequency microrheology reveals cytoskeleton dynamics in living cells, Nat. Phys. 13 (8) (2017) 771–775.
[172] L. Deng, et al., Fast and slow dynamics of the cytoskeleton, Nat. Mater. 5 (8) (2006) 636–640.
[173] S. Bernard, S. Kazemirad, G. Cloutier, A frequency-shift method to measure shear-wave attenuation in soft tissues, IEEE Trans. Ultrasonics, Ferroelectrics, Frequency Control. 64 (3) (2017) 514–524.
[174] E. Fischer-Friedrich, et al., Rheology of the active cell cortex in mitosis, Biophys. J. 111 (3) (2016) 589–600.
[175] E.M. Darling, D. Di Carlo, High-throughput assessment of cellular mechanical properties, Annu. Rev. Biomed. Eng. 17 (2015) 35–62.
[176] I. Sraj, et al., Cell deformation cytometry using diode-bar optical stretchers, J. Biomed. Opt. 15 (4) (2010) 047010.
[177] T. Sawetzki, et al., Viscoelasticity as a biomarker for high-throughput flow cytometry, Biophys. J. 105 (10) (2013) 2281–2288.
[178] Z. Wang, et al., A fully automated system for measuring cellular mechanical properties, J. Laboratory Autom. 17 (6) (2012) 443–448.
[179] J.S. Dudani, et al., Pinched-flow hydrodynamic stretching of single-cells, Lab. a Chip 13 (18) (2013) 3728–3734.
[180] H. Amini, W. Lee, D. Di Carlo, Inertial microfluidic physics, Lab. a Chip 14 (15) (2014) 2739–2761.
[181] S.C. Hur, et al., Deformability-based cell classification and enrichment using inertial microfluidics, Lab. a Chip 11 (5) (2011) 912–920.
[182] D. Di Carlo, et al., Continuous inertial focusing, ordering, and separation of particles in microchannels, Proc. Natl Acad. Sci. U. S. A. 104 (48) (2007) 18892–18897.
[183] S. Cha, et al., Cell stretching measurement utilizing viscoelastic particle focusing, Anal. Chem. 84 (23) (2012) 10471–10477.
[184] Y. Zheng, et al., Characterization of red blood cell deformability change during blood storage, Lab. a Chip 14 (3) (2013) 577–583.
[185] A. Adamo, et al., Microfluidics-based assessment of cell deformability, Anal. Chem. 84 (15) (2012) 6438–6443.
[186] S. Byun, et al., Characterizing deformability and surface friction of cancer cells, Proc. Natl Acad. Sci. U. S. A. 110 (19) (2013) 7580–7585.
[187] J. Chen, et al., Classification of cell types using a microfluidic device for mechanical and electrical measurement on single cells, Lab. a Chip 11 (18) (2011) 3174–3181.
[188] H. Bow, et al., A microfabricated deformability-based flow cytometer with application to malaria, Lab. a Chip 11 (6) (2011) 1065–1073.

[189] A.C. Rowat, et al., Nuclear envelope composition determines the ability of neutrophil-type cells to passage through micron-scale constrictions, J. Biol. Chem. 288 (12) (2013) 8610–8618.
[190] O. Otto, et al., Real-time deformability cytometry: on-the-fly cell mechanical phenotyping, Nat. Methods 12 (3) (2015) 199–202.

# Index

*Note*: Page numbers followed by "*f*" and "*t*" refer to figures and tables, respectively.

## N

Kolibri 360 GmbH
Ferdinand-Jühlke-Straße 7
99095 Erfurt, Deutschland
produktsicherheit@kolibri360.de